ALASKA

THE GREAT COUNTRY

BY

ELLA HIGGINSON

This edition published by Read Books Ltd.

British Library Cataloguing-in-Publication Data
A catalogue record for this book is available
from the British Library

CONTENTS

Photo by E. W. Merrill, Sitka
Courtesy of G. Kostrometinoff
Alexander Baranoff

TO
MR. AND MRS. HENRY ELLIOTT HOLMES

FOREWORD

When the Russians first came to the island of Unalaska, they were told that a vast country lay to the eastward and that its name was Al-ay-ek-sa. Their own island the Aleuts called Nagun-Alayeksa, meaning "the land lying near Alayeksa."

The Russians in time came to call the country itself Alashka; the peninsula, Aliaska; and the island, Unalashka. Alaska is an English corruption of the original name.

A great Russian moved under inspiration when he sent Vitus Behring out to discover and explore the continent lying to the eastward; two great Americans—Seward and Sumner—were inspired when, nearly a century and a half later, they saved for us, in the face of the bitterest opposition, scorn, and ridicule, the country that Behring discovered and which is now coming to be recognized as the most glorious possession of any people; but, first of all, were the gentle, dark-eyed Aleuts inspired when they bestowed upon this same country—with the simplicity and dignified repression for which their character is noted—the beautiful and poetic name which means "the great country."

ALASKA

THE GREAT COUNTRY

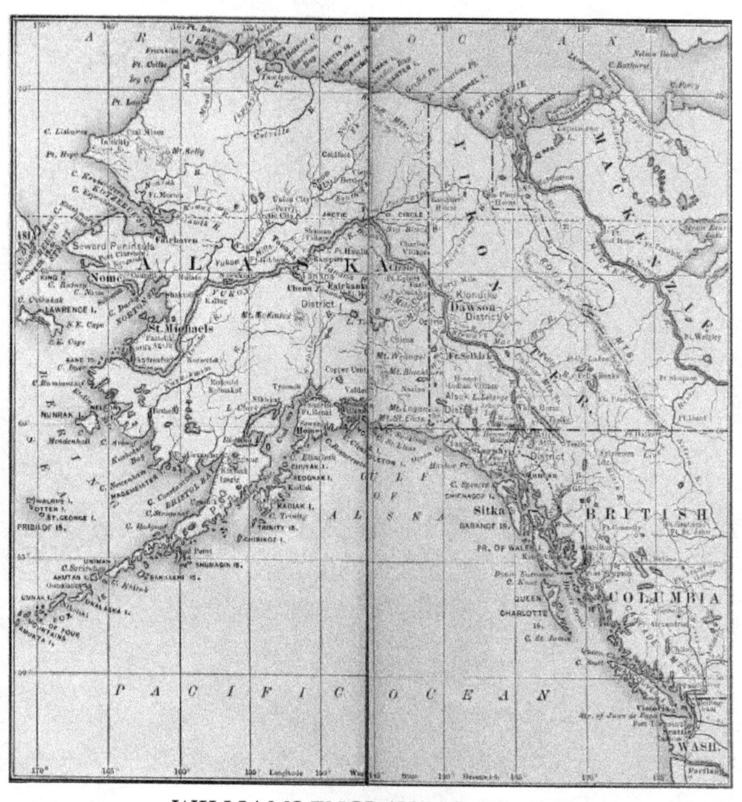

WILLIAMS ENGRAVING CO., N.Y.
Alaska

CHAPTER I

Every year, from June to September, thousands of people "go to Alaska." This means that they take passage at Seattle on the most luxurious steamers that run up the famed "inside passage" to Juneau, Sitka, Wrangell, and Skaguay. Formerly this voyage included a visit to Muir Glacier; but because of the ruin wrought by a recent earthquake, this once beautiful and marvellous thing is no longer included in the tourist trip.

This ten-day voyage is unquestionably a delightful one; every imaginable comfort is provided, and the excursion rate is reasonable. However, the person who contents himself with this will know as little about Alaska as a foreigner who landed in New York, went straight to Niagara Falls and returned at once to his own country, would know about America.

Enchanting though this brief cruise may be when the weather is favorable, the real splendor, the marvellous beauty, the poetic and haunting charm of Alaska, lie west of Sitka. "To Westward" is called this dream-voyage past a thousand miles of snow-mountains rising straight from the purple sea and wrapped in coloring that makes it seem as though all the roses, lilies, and violets of heaven had been pounded to a fine dust and sifted over them; past green islands and safe harbors; past the Malaspina and the Columbia glaciers; past Yakutat, Kyak, Cordova, Valdez, Seward, and Cook Inlet; and then, still on "to Westward"—past Kodiak Island, where the Russians made their first permanent settlement in America in 1784 and whose sylvan and idyllic charm won the heart of the great naturalist, John Burroughs; past the Aliaska Peninsula, with its smoking Mount Pavloff; past Unimak Island, one of whose active volcanoes, Shishaldin, is the most perfect and symmetrical cone on the Pacific Coast, not even excepting Hood—and on and in among the divinely pale green Aleutian Islands to Unalaska, where enchantment broods in a mist of rose and lavender and where one may scarcely step without crushing violets and bluebells.

The spell of Alaska falls upon every lover of beauty who has voyaged along those far northern snow-pearled shores with the violet waves of the North Pacific Ocean breaking splendidly upon them; or who has drifted down the mighty rivers of the interior which flow, bell-toned and lonely, to the sea.

I know not how the spell is wrought; nor have I ever met one who could put the miracle of its working into words. No writer has ever described Alaska; no one writer ever will; but each must do his share, according to the spell that the country casts upon him.

Some parts of Alaska lull the senses drowsily by their languorous charm; under their influence one sinks to a passive delight and drifts unresistingly on through a maze of tender loveliness. Nothing irritates. All is soft, velvety, soothing. Wordless lullabies are played by different shades of blue, rose, amber, and green; by the curl of the satin waves and the musical kiss of their cool and faltering lips; by the mists, light as thistle-down and delicately tinted as wild-rose petals, into which the steamer pushes leisurely; by the dreamy poise of sea-birds on white or lavender wings high in the golden atmosphere; by the undulating flight of purple Shadow, tiptoe, through the dim fiords; by the lap of waves on shingle, the song of birds along the wooded shore, the pressure of soft winds on the temples and hair, the sparkle of the sea weighing the eyelids down. The magic of it all gets into the blood.

Copper Smelter in Southeastern Alaska

The steamer slides through green and echoing reaches; past groups of totems standing like ghosts of the past among the dark spruce or cedar trees; through stone-walled canyons where the waters move dark and still; into open, sunlit seas.

But it is not until one sails on "to Westward" that the spell of Alaska falls upon one; sails out into the wild and splendid North Pacific Ocean. Here are the majesty, the sublimity, that enthrall; here are the noble spaces, the Titanic forces, the untrodden heights, that thrill and inspire.

The marvels here are not the marvels of men. They are wrought of fire and stone and snow by the tireless hand that has worked through centuries unnumbered and unknown.

He that would fall under the spell of Alaska, will sail on "to Westward," on to Unalaska; or he will go Northward and drift down the Yukon—that splendid, lonely river that has its birth within a few miles of the sea, yet flows twenty-three hundred miles to find it.

Alaskan steamers usually sail between eight o'clock in the

evening and midnight, and throngs of people congregate upon the piers of Seattle to watch their departure. The rosy purples and violets of sunset mix with the mists and settle upon the city, climbing white over its hills; as hours go by, its lights sparkle brilliantly through them, yet still the crowds sway upon the piers and wait for the first still motion of the ship as it slides into the night and heads for the far, enchanted land—the land whose sweet, insistent calling never ceases for the one who has once heard it.

Passengers who stay on deck late will be rewarded by the witchery of night on Puget Sound—the soft fragrance of the air, the scarlet, blue, and green lights wavering across the water, the glistening wake of the ship, the city glimmering faintly as it is left behind, the dim shores of islands, and the dark shadows of bays.

One by one the lighthouses at West Point on the starboard side, and at Point-No-Point, Marrowstone, and Point Wilson, on the port, flash their golden messages through the dusk. One by one rise, linger, and fade the dark outlines of Magnolia Bluff, Skagit Head, Double Bluff, and Liplip Point. If the sailing be early in the evening, midnight is saluted by the lights of Port Townsend, than which no city on the Pacific Coast has a bolder or more beautiful situation.

The splendid water avenue—the burning "Opal-Way"—that leads the ocean into these inland seas was named in 1788 by John Meares, a retired lieutenant of the British navy, for Juan de Fuca (whose real name was Apostolos Valerianos), a Greek pilot who, in 1592, was sent out in a small "caravela" by the Viceroy of Mexico in search of the fabled "Strait of Anian," or "Northwest Passage"—supposed to lead from the Pacific to the Atlantic north of forty degrees of latitude.

As early as the year 1500 this strait was supposed to have been discovered by a Portuguese navigator named Cortereal, and to have been named by him for one of his brothers who accompanied him.

The names of certain other early navigators are mentioned in connection with the "Strait of Anian." Cabot is reported vaguely as having located it "neere the 318 meridian, between 61 and 64 degrees in the eleuation, continuing the same bredth about 10 degrees West, where it openeth Southerly more and more, until it come under the tropicke of Cancer, and so runneth into Mar del Zur, at least 18 degrees more in bredth there than where it began;" Frobisher; Urdaneta, "a Fryer of Mexico, who came out of Mar del Zur this way into Germanie;" and several others whose stories of having sailed the dream-strait that was then supposed to lead from ocean to ocean are not now considered seriously until we come to Juan de Fuca, who claimed that in his "caravela" he followed the coast "vntill hee came to the latitude of fortie seuen degrees, and that there finding that the land trended North and Northeast, with a broad Inlet of Sea between 47 and 48 degrees of Latitude, hee entered thereinto, sayling therein more than twenty days, and found that land trending still sometime Northwest and Northeast and North, and also East and Southeastward, and very much broader sea then was at said entrance, and that hee passed by diuers Ilands in that sayling. And that at the entrance of this said Strait, there is on the Northwest coast thereof, a great Hedland or Iland, with an exceeding high pinacle or spired Rocke, like a pillar, thereupon."

He landed and saw people clothed in the skins of beasts; and he reported the land fruitful, and rich in gold, silver, and pearl.

Bancroft and some other historians consider the story of Juan de Fuca's entrance to Puget Sound the purest fiction, claiming that his descriptions are inaccurate and that no pinnacled or spired rock is to be found in the vicinity mentioned.

Meares, however, and many people of intelligence gave it credence; and when we consider the differences in the descriptions of other places by early navigators, it is not difficult to believe that Juan de Fuca really sailed into the strait that now bears his name. Schwatka speaks of him as, "An explorer—if such he may be called—who never entered this beautiful sheet of water, and who owes his immortality to an audacious guess, which came so near the truth as to deceive the scientific world for many a century."

The Strait of Juan de Fuca is more than eighty miles long and from ten to twelve wide, with a depth of about six hundred feet. At the eastern end it widens into an open sea or sound where beauty blooms like a rose, and from which forest-bordered water-ways wind slenderly in every direction.

From this vicinity, on clear days, may be seen the Olympic Mountains floating in the west; Mount Rainier, in the south; the lower peaks of the Crown Mountains in the north; and Mount Baker—or Kulshan, as the Indians named it—in the east.

The Island of San Juan, lying east of the southern end of Vancouver Island, is perhaps the most famous, and certainly the most historic, on the Pacific Coast. It is the island that barely escaped causing a declaration of war between Great Britain and the United States, over the international boundary, in the late fifties. For so small an island,—it is not more than fifteen miles long, by from six to eight wide,—it has figured importantly in large affairs.

The earliest trouble over the boundary between Vancouver Island and Washington arose in 1854. Both countries claimed ownership of San Juan and other islands near by, the Oregon Treaty of 1846 having failed to make it clear whether the boundary was through the Canal de Haro or the Strait of Rosario.

I. N. Ebey, American Collector of Customs, learning that several thousand head of sheep, cattle, and hogs had been shipped to San Juan without compliance with customs regulations, visited the island and was promptly insulted by a British justice of the peace. The *Otter* made her appearance in the harbor, bearing James Douglas, governor of Vancouver Island and vice-admiral of the British navy; but nothing daunted, Mr. Ebey stationed Inspector Webber upon the island, declaring that he would continue to discharge his official duties. The final trouble arose, however, in 1859, when an American resident shot a British pig; and serious trouble was precipitated as swiftly as when a United States warship was blown up in Havana Harbor. General Harney hastily established military quarters on one end of the island, known as the American Camp, Captain Pickett transferring his company from Fort Bellingham for this purpose. English Camp was established on the northern end. Warships kept guard in the harbors. Joint occupation was agreed upon, and until 1871 the two camps were maintained, the friendliest social relations existing between them. In that year the Emperor of Germany was chosen as arbitrator, and decided in favor of the United States, the British withdrawing the following year.

Until 1895 the British captain's house still stood upon its beautiful bluff, a thousand feet above the winding blue bay, the shore descending in steep, splendid terraces to the water, stairwayed in stone, and grown with old and noble trees. Macadam roads led several miles across the island; the old block-house of pioneer days remained at the water's edge; and clustered around the old parade ground—now, alas! a meadow of hay—were the quarters of the officers, overgrown with English ivy. The captain's house, which has now been destroyed by fire, was a low, eight-roomed house with an immense fireplace in each room; the old claret- and ivory-striped wall-paper—which had been brought "around the Horn" at immense cost—was still on the walls. Gay were the scenes and royal the hospitalities of this house in the good days of the sixties. Its site, commanding the straits, is one of the most effective on the Pacific Coast; and at the present writing it is extremely probable that a captain's house may again rise among the old trees on the terraced bluff—but not for the occupancy of a British captain.

Every land may occasionally have a beautiful sunset, and many lands have gorgeous and brilliant ones; but nowhere have they such softly burning, milky-rose, opaline effects as on this inland sea.

Their enchanting beauty is doubtless due to the many wooded islands which lift dark green forestated hills around open sweeps of water, whereon settle delicate mists. When the fires of sunrise or of sunset sink through these mists, the splendor of coloring is marvellous and not equalled anywhere. It is as though the whole sound were one great opal, which had broken apart and flung its escaping fires of rose, amethyst, amber, and green up through the maze of trembling pearl above it. The unusual beauty of its sunsets long ago gave Puget Sound the poetic name of Opal-Sea or Sea of Opal.

Kasa-an

CHAPTER II

After passing the lighthouse on the eastern end of Vancouver Island, Alaskan steamers continue on a northerly course and enter the Gulf of Georgia through Active Pass, between Mayne and Galiana islands. This pass is guarded by a light on Mayne Island, to the steamer's starboard, going north.

The Gulf of Georgia is a bold and sweeping body of water. It is usually of a deep violet or a warm purplish gray in tone. At its widest, it is fully sixty miles—although its average width is from twenty to thirty miles—and it rolls between the mainland and Vancouver Island for more than one hundred miles.

The real sea lover will find an indescribable charm in this gulf, and will not miss an hour of it. It has the boldness and the sweep of the ocean, but the setting, the coloring, and the fragrance of the forest-bordered, snow-peaked sea. A few miles above the boundary, the Fraser River pours its turbulent waters into the gulf, upon whose dark surface they wind and float for many miles, at sunrise and at sunset resembling broad ribbons of palest old rose crinkled over waves of silvery amber silk. At times these narrow streaks widen into still pools of color that seem to float suspended over the heavier waters of the gulf. Other times they draw lines of different color everywhere, or drift solid banks of smoky pink out to meet others of clear blue, with only the faintest thread of pearl to separate them. These islands of color constitute one of the charms of this part of the voyage to Alaska; along with the velvety pressure of the winds; the picturesque shores, high and wooded in places, and in others sloping down into the cool shadowy bays where the shingle is splashed by spent waves; and the snow-peaks linked above the clouds on either side of the steamer.

Splendid phosphorescent displays are sometimes witnessed in

the gulf, but are more likely to occur farther north, in Grenville, or one of the other narrow channels, where their brilliancy is remarkable.

Tourists to whom a whale is a novelty will be gratified, without fail, in this vicinity. They are always seen sporting about the ships,—sometimes in deadly conflict with one another,—and now and then uncomfortably near.

In December, 1907, an exciting battle between a whale and a large buck was witnessed by the passengers and crew of the steamer *Cassiar*, in one of the bays north of Vancouver, on the vessel's regular run from that city to northern ports.

When the *Cassiar* appeared upon the scene, the whale was making furious and frequent attacks upon the buck. Racing through the water, which was lashed into foam on all sides by its efforts, it would approach close to its steadily swimming prey and then disappear, only to come to the surface almost under the deer. This was repeated a number of times, strangely enough without apparent injury to the deer. Again, the whale would make its appearance at the side of the deer and repeatedly endeavor to strike it with its enormous tail; but the deer was sufficiently wise to keep so close to the whale that this could not be accomplished, notwithstanding the crushing blows dealt by the monster.

The humane passengers entreated the captain to go to the rescue of the exhausted buck and save it from inevitable death. The captain ordered full speed ahead, and at the approach of the steamer the whale curved up out of the water and dived gracefully into the sea, as though making a farewell, apologetic bow on its final disappearance.

Whereupon the humane passengers shot the helpless and worn-out buck at the side of the steamer, and he was hauled aboard.

It may not be out of place to devote a few pages to the average tourist. To the one who loves Alaska and the divinely blue, wooded, and snow-pearled ways that lead to its final and sublime beauty, it is an enduring mystery why certain persons—usually women—should make this voyage. Their minds and their desires never rise above a whale or an Indian basket; and unless the one is to be seen and the other to be priced, they spend their time in the cabin, reading, playing cards, or telling one another what they have at home.

"Do you know," said one of these women, yawning into the full glory of a sunset, "we have sailed this whole day past Vancouver Island. Not a thing to be seen but it and this water you call the Gulf of Georgia! I even missed the whales, because I went to sleep, and I'd rather have seen them than anything. If they don't hurry up some towns and totem-poles, I'll be wishing I'd stayed at home. Do you play five hundred?"

The full length of the *Jefferson* was not enough to put between this woman and the woman who had enjoyed every one of those purple water-miles; every pearly cloud that had drifted across the pale blue sky; every bay and fiord indenting the shore of the largest island on the Pacific Coast; every humming-bird that had throbbed about us, seeking a rose at sea; every thrilling scent that had blown down the northern water-ways, bearing the far, sweet call of Alaska to senses awake and trembling to receive it; who had felt her pulses beating full to the throb of the steamer that was bearing her on to the land of her dreams—to the land of Far Delight.

If only the players of bridge and the drinkers of pink tea would stay at home, and leave this enchanted voyage for those who understand! There be enough of the elect in the world who possess the usual five senses, as well as that sixth sense which is of the soul, to fill every steamer that sails for Alaska.

Or, the steamship companies might divide their excursions into classes—some for those who love beauty, and some for those who love bridge.

For the sea lover, it is enough only to stand in the bow of a steamer headed for Alaska and hear the kiss and the rippling murmur of the waves as they break apart when the sharp cut-water pierces them, and then their long, musical rush along the steamer's sides, ere they reunite in one broad wake of bowing silver that leads across the purple toward home.

The mere vibration of a ship in these still inland seas is a physical pleasure by day and a sensuous lullaby at night; while, in summer, the winds are so soft that their touches seem like caresses.

The inlets and fiords extending for many miles into the mainland in this vicinity are of great beauty and grandeur, many winding for forty or fifty miles through walls of forestation and snow that rise sheer to a height of eight or ten thousand feet. These inlets are very narrow, sometimes mere clefts, through which the waters slip, clear, still, and of deepest green. They are of unknown depth; the mountains are covered with forests, over which rise peaks of snow. Cascades are numerous, and their musical fall is increased in these narrow fastnesses to a roar that may be heard for miles.

Passing Burrard Inlet, on which the city of Vancouver is situated, the more important inlets are Howe, Jervis, from which Sechelt Arm leads southward and is distinguished by the wild thunder of its rapids; Homery Channel, Price Channel, which, with Lewis Channel on the west, forms Redonda Island; Bute Inlet, which is the most beautiful and the most important; Knight, Seymour, Kingcome, and Belize inlets.

The wild and picturesque beauty of these inlets has been praised by tourists for many years. The Marquis of Lorne was charmed by the scenery along Bute Inlet, which he extolled. It is about fifty miles in length and narrows in places to a width of a half-mile. The shores rise in sheer mountain walls, heavily forestated, to a height of seven and eight thousand feet, their snowy crests overhanging the clear, green-black waters of the narrow fiord. Many glaciers stream down from these peaks.

The Gulf of Georgia continues for a distance of one hundred miles in a northwesterly direction between the mainland and Vancouver Island. Texada, Redonda, and Valdes are the more important islands in the gulf. Texada appears on the starboard, opposite Comox; the narrow strait separating it from the mainland is named Malaspina, for the Italian explorer. The largest glacier in the world, streaming into the sea from Mount St. Elias, more than a thousand miles to the northwestward from this strait, bears the same name.

Texada Island is twenty-eight miles long, with an average width of three miles. It is wooded and mountainous, the leading peak—Mount Shepard—rising to a height of three thousand feet. The lighthouse on its shore is known as "Three Sisters Light."

Along the shores of Vancouver Island and the mainland are many ranches owned and occupied by "remittance men." In these beautiful, lonely solitudes they dwell with all the comforts of "old England," forming new ties, but holding fast to old memories.

It is said that the woman who should have one day been the Queen of England, lived near the city of Vancouver a few years ago. Before the death of his elder brother, the present Prince of Wales passionately loved the young and beautiful daughter of Admiral Seymour. His infatuation was returned, and so desperately did the young couple plead with the present King and the Admiral, that at last the prince was permitted to contract a morganatic marriage.

The understanding and agreement were that, should the prince ever become the heir to the throne of England, neither he nor his wife would oppose the annulment of the marriage.

There was only one brief year of happiness, when the elder brother of the prince died, and the latter's marriage to the Princess May was demanded.

No murmur of complaint was ever heard from the unhappy morganatic wife, nor from the royal husband; and when the latter's marriage was solemnized, it was boldly announced that no bar to the union existed.

Here, in the western solitude, lived for several years—the veriest remittance woman—the girl who should now, by the right of love and honor, be the Princess of Wales; and whose infant daughter should have been the heir to the throne.

To Vancouver, a few years ago, came, with his princess, the Prince of Wales. The city was gay with flags and flowers, throbbing with music, and filled with joyous and welcoming people. Somewhere, hidden among those swaying throngs, did a pale young woman holding a child by the hand, gaze for the last time upon the man she loved and upon the woman who had taken her place? And did her long-tortured heart in that hour finally break? It is said that she died within a twelvemonth.

Passing Cape Mudge lighthouse, Discovery Passage, sometimes called Valdes Narrows, is entered. It is a narrow pass, twenty-four miles long, between Vancouver and Valdes islands. Halfway through it is Seymour Narrows, one of the most famous features of the "inside route," or passage, to Alaska. Passengers are awakened, if they desire, that they may be on deck while passing through these difficult narrows.

The Indian name of this pass is Yaculta.

"Yaculta is a wicked spirit," said the pilot, pacing the bridge at four o'clock of a primrose dawn. "She lives down in the clear depths of these waters and is supposed to entice guileless sailors to their doom. Yaculta sleeps only at slack-tide, and then boats, or ships, may slip through in safety, provided they do not make sufficient noise to awaken her. If they try to go through at any other stage of the tide, Yaculta stirs the whole pass into action, trying to get hold of them. Many's the time I've had to back out and wait for Yaculta to quiet down."

If the steamer attempts the pass at an unfavorable hour, fearful seas are found racing through at a fourteen-knot speed; the steamer is flung from side to side of the rocky pass or sucked down into the boiling whirlpools by Yaculta. The brown, shining strands of kelp floating upon Ripple Reef, which carries a sharp edge down the centre of the pass, are the wild locks of Yaculta's luxuriant hair.

Pilots figure, upon leaving Seattle, to reach the narrows during the quarter-hour before or after slack-tide, when the water is found as still and smooth as satin stretched from shore to shore, and not even Yaculta's breathing disturbs her liquid coverlet.

Many vessels were wrecked here before the dangers of the narrows had become fully known: the steamer *Saranac*, in 1875, without loss of life; the *Wachusett*, in 1875; the *Grappler*, in 1883, which burned in the narrows with a very large loss of life, including that of the captain; and several less appalling disasters have occurred in these deceptive waters.

Three miles below Cape Mudge the tides from Juan de Fuca meet those from Queen Charlotte Sound, and force a fourteen-knot current through the narrows. The most powerful steamers are frequently overcome and carried back by this current.

Discovery Passage merges at Chatham Point into Johnstone Strait. Here the first Indian village, Alert Bay, is seen to starboard on the southern side of Cormorant Island. These are the Kwakiutl Indians, who did not at first respond to the advances of civilization so readily as most northern tribes. They came from their original village at the mouth of the Nimpkish River, to work in the canneries on the bay, but did not take kindly to the ways of the white man. A white child, said to have been stolen from Vancouver, was taken from these Indians a few years ago.

Some fine totem-poles have been erected here, and the graveyard has houses built over the graves. From the steamer the little village presents an attractive appearance, situated on a curving beach, with wooded slopes rising behind it.

Gorgeous potlatches are held here; and until the spring of 1908 these orgies were rendered more repulsive by the sale of young girls.

Howkan

Dr. Franz Boas, in his "Kwakiutl Texts," describes a game formerly played with stone disks by the Kwakiutls. They also had a myth that a game was played with these disks between the birds of the upper world and the myth-people, that is, "all the animals and all the birds." The four disks were called the "mist-covered gambling stone," the "rainbow gambling stone," the "cloud-covered gambling stone," and the "carrier of the world." The woodpecker and the other myth-birds played on one side; the Thunder-bird and the birds of the upper air on the other. The contestants were ranged in two rows; the gambling stones were thrown along the middle between them, and they speared them with their beaks. The Thunder-bird and the birds of the upper air were beaten. This myth is given as an explanation of the reason for playing the game with the gambling stones, which are called lælæ.

The Kwakiutls still play many of their ancient and picturesque gambling games at their potlatches.

Johnstone Strait is fifty-five miles long, and is continued by Broughton Strait, fifteen miles long, which enters Queen Charlotte Sound.

Here is a second, and smaller, Galiana Island, and on its western end is a spired rock which, some historians assert, may be "the great headland or island with an exceeding high pinnacle or spired rock thereon," which Juan de Fuca claimed to discover, and which won for him the charge of being an "audacious guesser" and an "unscrupulous liar." His believers, however, affirm that, having sailed for twenty days in the inland sea, he discovered this pinnacle at the entrance to what he supposed to be the Atlantic Ocean; and so sailed back the course he had come, believing himself to have been successful in discovering the famed strait of Anian. Why Vancouver's mistakes, failures, and faults should all be condoned, and Juan de Fuca's most uncompromisingly condemned, is difficult to understand.

Fort Rupert, on the northern end of Vancouver Island, beyond Broughton Strait, is an old Hudson's Bay post, situated on Beaver Harbor. The fort was built in 1849, and was strongly defended, troubles frequently arising from the attacks of Kwakiutl and Haidah Indians. Great potlatches were held there, and the chief's lodge was as notable as was the "Old-Man House" of Chief Seattle. It was one hundred feet long and eighty feet wide, and rested on carved corner posts. There was an immense wooden potlatch dish that held food for one hundred people.

Queen Charlotte Sound is a splendid sweep of purple water; but tourists do not, usually, spend much time enjoying its beauty. Their berths possess charms that endure until shelter of the islands is once more assured, after the forty miles of open exposure to the swell of the ocean which is not always mild, notwithstanding its name. Those who miss it, miss one of the most beautiful features of the inland voyage. The warm breath of the Kuro Siwo, penetrating all these inland seas and passages, is converted by the great white peaks of the horizon into pearl-like mist that drifts in clouds and fragments upon the blue waters.

Nowhere are these mists more frequent, nor more elusive, than in Queen Charlotte Sound. They roll upon the sparkling surface like thistle-down along a country lane—here one instant, vanished the next. At sunrise they take on the delicate tones of the primrose or the pinkish star-flower; at sunset, all the royal rose and purple blendings; all the warm flushes of amber, orange, and gold. Through a maze of pale yellow, whose fine cool needles sting one's face and set one's hair with seed-pearls, one passes into a little open water-world where a blue sky sparkles above a bluer sea, and the air is like clear, washed gold. But a mile ahead a solid wall of amethyst closes in this brilliant sea; and presently the steamer glides into it, shattering it into particles that set the hair with amethysts, instead of pearls. Sometimes these clear spaces resemble rooms walled in different colors, but ceiled and floored in blue. Other times, the whole sound is clear, blue, shining; while exquisite gossamers of changeful tints wrap and cling about the islands, wind scarfs around the green hills, or set upon the brows of majestic snow-monarchs crowns as jewelled and as evanescent as those worn by the real kings of the earth. Now and then a lofty fir or cedar may be seen draped with slender mist-veils as a maiden might wind a scarf of cobwebby lace about her form and head and arms—so lightly and so gracefully, and with such art, do the delicate folds trail in and out among the emerald-green branches of the tree.

It is this warm and excessive moisture—this daily mist-shower—that bequeaths to British Columbia and Alaska their marvellous and luxuriant growth of vegetation, their spiced sweetness of atmosphere, their fairness and freshness of complexion—blending and constituting that indescribable charm which inspires one, standing on the deck of a steamer at early dawn, to give thanks to God that he is alive and sailing the blue water-ways of this sublime country.

"I don't know what it is that keeps pulling me back to this country," said a man in the garb of a laborer, one day. He stood down in the bow of the steamer, his hands were in his pockets, his throat was bared to the wind; his blue eyes—sunken, but burning with that fire which never dies in the eyes of one who loves nature—were gazing up the pale-green narrow avenue named Grenville Channel. "It's something that you can't exactly put into words. You don't know that it's got hold of you while you're up here, but before you've been 'outside' a month, all at once you find it pulling at you—and after it begins, it never lets up. You try to think what it is up here that you want so; what it is keeps begging at you to come back. Maybe there ain't a darn soul up here you care particular about! Maybe you ain't got an interest in a claim worth hens' teeth! Maybe you're broke and know you'll have to work like a go-devil when you get here! It don't make any difference. It's just Alaska. It calls you and calls you and calls you. Maybe you can't come, so you keep pretending you don't hear—but Lord, you do hear! Maybe somebody shakes hands as if he liked you—and there's Alaska up and calling right through you, till you feel your heart shake! Maybe a phonograph sets up a tune they used to deal out at Magnuson's roadhouse on the trail—and you hear that blame lonesome waterfall up in Keystone Canyon calling you as plain as you hear the phonograph! Maybe you smell something like the sun shining on snow, all mixed up with tundra and salt air—and there's double quick action on your eyes and a lump in your throat that won't be swallowed down! Maybe you see a white mountain, or a green valley, or a big river, or a blue strait, or a waterfall—and like a flash your heart opens, and shuts in an ache for Alaska that stays!... No, I don't know *what* it is, but I do know *how* it is; and so does every other poor devil that ever heard that something calling him that's just Alaska. It wakes you up in the middle of the night, just as plain as if somebody had said your name out loud, and you just lay there the rest of the night aching to go. I tell you what, if ever a country had a spirit, it's Alaska; and when it once gets hold of you and gets to calling you to come, you might just as well get up and start, for it calls you and follows you, and haunts you till you do."

It is the pleading of the mountains and the pleading of the

sea woven into one call and sent floating down laden with the sweetness of the splendid spaces. No mountaineer can say why he goes back to the mountains; no sailor why he cannot leave the sea. No one has yet seen the spirit that dwells in the waterfall, but all have heard it calling and have known its spell.

Distant View of Davidson Glacier

"If you love the sea, you've got to follow it," said a sea-rover, "and that's all there is to it. A man can get along without the woman he loves best on earth if he has to, but he can't get along without the sea if he once gets to loving it. It gets so it seems like a thing alive to him, and it makes up for everything else that he don't have. And it's just like that with Alaska. When a man has made two-three trips to Alaska, you can't get him off on a southern run again, as long as he can help himself."

It is an unimaginative person who can wind through these intricate and difficult sounds, channels, and passes without a strange, quickened feeling, as of the presence of those dauntless navigators who discovered and charted these waters centuries ago. From Juan de Fuca northward they seem to be sailing with us, those grim, brave spectres of the past—Perez, Meares, Cuadra, Valdes, Malaspina, Duncan, Vancouver, Whidbey—and all the others who came and went through these beautiful ways, leaving their names, or the names of their monarchs, friends, or sweethearts, to endure in blue stretches of water or glistening domes of snow.

We sail in safety, ease, luxury, over courses along which they felt their perilous way, never knowing whether Life or Death waited at the turn of the prow. Nearly a century and a quarter ago Vancouver, working his way cautiously into Queen Charlotte Sound, soon came to disaster, both the *Discovery* and her consort, the *Chatham*, striking upon the rocks that border the entrance. Fortunately the return of the tide in a few hours released them from their perilous positions, before they had sustained any serious damage.

But what days of mingled indecision, hope, and despair—what nights of anxious watching and waiting—must have been spent in these places through which we glide so easily now; and the silent spirits of the grim-peopled past take hold of our heedless hands and lead us on. Does a pilot sail these seas who has never on wild nights felt beside him on the bridge the presence of those early ones who, staring ever ahead under stern brows, drove their vessels on, not knowing what perils lay beyond? Who, asked, "What shall we do when hope be gone?" made answer, "Why, sail on, and on, and on."

From Queen Charlotte Sound the steamer passes into Fitzhugh Sound around Cape Calvert, on Calvert Island. Off the southern point of this island are two dangerous clusters of rocks, to which, in 1776, by Mr. James Hanna, were given the interesting names of "Virgin" and "Pearl." In this poetic vicinage, and nearer the island than either, is another cluster of rocks, upon which some bold and sacrilegious navigator has bestowed the name of "Devil."

"It don't sound so pretty and ladylike," said the pilot who pointed them out, "but it's a whole lot more appropriate. Rocks *are* devils—and that's no joke; and what anybody should go and name them 'virgins' and 'pearls' for, is more than a man can see, when he's standing at a wheel, hell-bent on putting as many leagues between him and them as he can. It does seem as if some men didn't have any sense at all about naming things. Now, if I were going to name anything 'virgin'"—his blue eyes narrowed as they stared into the distance ahead—"it would be a mountain that's always white; or a bay that gets the first sunshine in the morning; or one of those little islands down in Puget Sound that's just*covered* with flowers."

Just inside Fitzhugh Sound, on the island, is Safety Cove, or Oatsoalis, which was named by Mr. Duncan in 1788, and which has ever since been known as a safe anchorage and refuge for ships in storm. Vancouver, anchoring there in 1792, found the shores to be bold and steep, the water from twenty-three to thirty fathoms, with a soft, muddy bottom. Their ships were steadied with hawsers to the trees. They found a small beach, near which was a stream of excellent water and an abundance of wood. Vessels lie here at anchor when storms or fogs render the passage across Queen Charlotte Sound too perilous to be undertaken.

Fitzhugh Sound is but a slender, serene water-way running directly northward thirty miles. On its west, lying parallel with the mainland, are the islands of Calvert, Hecate, Nalau, and Hunter, separated by the passages of Kwakshua, Hakai, and Nalau, which connect Fitzhugh with the wide sweep of Hecate Strait.

Burke Channel, the second link in the exquisite water chain that winds and loops in a northwesterly course between the islands of the Columbian and the Alexander archipelagoes and the mainland of British Columbia and Alaska, is scarcely entered by the Alaskan steamer ere it turns again into Fisher Channel, and from this, westward, into the short, very narrow, but most beautiful Lama Pass.

From Burke Channel several ribbonlike passages form King Island.

Lama Pass is more luxuriantly wooded than many of the others, and is so still and narrow that the reflections of the trees, growing to the water's edge, are especially attractive. Very effective is the graveyard of the Bella Bella Indians, in its dark forest setting, many totems and curious architectures of the dead showing plainly from the steamer when an obliging captain passes under slow bell. Near by, on Campbell Island, is the village of the Bella Bellas, who, with the Tsimpsians and the Alert Bay Indians, were formerly regarded as the most treacherous and murderous Indians of the Northwest Coast. Now, however, they are gathered into a model village, whose houses, church, school, and stores shine white and peaceful against a dark background.

Lama Pass is one of the most poetic of Alaskan water-ways.

Seaforth Channel is the dangerous reach leading into Millbank Sound. It is broken by rocks and reefs, on one of which, Rejetta Reef, the *Willapa* was stranded ten years ago. Running off Seaforth and Millbank are some of the finest fiords of the inland passage—Spiller, Johnston, Dean, Ellerslie, and Portlock channels, Cousins and Cascades inlets, and many others. Dean and Cascades channels are noted for many waterfalls of wonderful beauty. The former is ten miles long and half a mile wide. Cascades Inlet extends for the same distance in a northeasterly direction, opening into Dean. Innumerable cataracts fall sheer and foaming down their great precipices; the narrow canyons are filled with their musical, liquid thunder, and the prevailing color seems to be palest green, reflected from the color of the water underneath the beaded foam. Vancouver visited these canals and named them in 1793, and although, seemingly, but seldom moved by beauty, was deeply impressed by it here. He considered the cascades "extremely grand, and by much the largest and most tremendous we had ever beheld, their impetuosity sending currents of air across the canal."

These fiords are walled to a great height, and are of magnificent beauty. Some are so narrow and so deep that the sunlight penetrates only for a few hours each day, and eternal mist and twilight fill the spaces. In others, not disturbed by cascades, the waters are as clear and smooth as glass, and the stillness is so profound that one can hear a cone fall upon the water at a distance of many yards. Covered with constant moisture, the vegetation is of almost tropic luxuriance. In the shade, the huge leaves of the devil's-club seem to float, suspended, upon the air, drooping slightly at the edges when touched by the sun. Raspberries and salmon-berries grow to enormous size, but are so fragile and evanescent that they are gone at a breath, and the most delicate care must be exercised in securing them. They tremble for an instant between the tongue and the palate, and are gone, leaving a sensation as of dewdrops flavored with wine; a memory as haunting and elusive as an exquisite desire known once and never known again.

In Dean Canal, Vancouver found the water almost fresh at low tide, on account of the streams and cascades pouring into it.

There he found, also, a remarkable Indian habitation; a square, large platform built in a clearing, thirty feet above the ground. It was supported by several uprights and had no covering, but a fire was burning upon one end of it.

In Cascade Canal he visited an Indian village, and found the construction of the houses there very curious. They apparently backed straight into a high, perpendicular rock cliff, which supported their rears; while the fronts and sides were sustained by slender poles about eighteen feet in height.

Vancouver leaves the method of reaching the entrances to these houses to the reader's imagination.

It was in this vicinity that Vancouver first encountered "split-lipped" ladies. Although he had grown accustomed to distortions and mutilations among the various tribes he had visited, he was quite unprepared for the repulsive style which now confronted him.

A horizontal incision was made about three-tenths of an inch below the upper part of the lower lip, extending from one corner of the mouth to the other, entirely through the flesh; this orifice was then by degrees stretched sufficiently to admit an ornament made of wood, which was confined close to the gums of the lower jaws, and whose external surface projected horizontally.

These wooden ornaments were oval, and resembled a small platter, or dish, made concave on both sides; they were of various lengths, the smallest about two inches and a half; the largest more than three inches long, and an inch and a half broad.

They were about one-fifth of an inch thick, and had a groove along the middle of the outside edge to receive the lip.

These hideous things were made of fir, and were highly polished. Ladies of the greatest distinction wore the largest labrets. The size also increased with age. They have been described by Vancouver, Cook, Lisiansky, La Pérouse, Dall, Schwatka, Emmans, and too many others to name here; but no description can quite picture them to the liveliest imagination. When the "wooden trough" was removed, the incision gave the appearance of two mouths.

All chroniclers unite as to the hideousness and repulsiveness of the practice.

Of the Indians in the vicinity of Fisher Channel, Vancouver remarks, without a glimmer of humor himself, that the vivacity of their countenance indicated a lively genius; and that, from their frequent bursts of laughter, it would appear that they were great humorists, for their mirth was not confined to their own people, but was frequently at the expense of his party. They seemed a happy, cheerful people. This is an inimitable English touch; a thing that no American would have written, save with a laugh at himself.

Poison Cove in Mussel Canal, or Portlock Canal, was so named by Vancouver, whose men ate roasted mussels there. Several were soon seized with numbness of the faces and extremities. In spite of all that was done to relieve their sufferings, one—John Carter—died and was buried in a quiet bay which was named for him.

Millbank Sound, named by Mr. Duncan before Vancouver's arrival, is open to the ocean, but there is only an hour's run before the shelter of the islands is regained; so that, even when the weather is rough, but slight discomfort is experienced by the most susceptible passengers. The finest scenery on the regular steamer route, until the great snow fields and glaciers are reached, is considered by many well acquainted with the route, to lie from Millbank on to Dixon Entrance. The days are not long enough now for all the beauty that weighs upon the senses like caresses. At evening, the sunset, blooming like a rose upon these splendid reaches, seems to drop perfumed petals of color, until the still air is pink with them, and the steamer pushes them aside as it glides through with faint throbbings that one feels rather than hears.

Through Finlayson Channel, Heikish Narrows, Graham, Fraser, and McKay reaches, Grenville Channel,—through all these enchanting water avenues one drifts for two hundred miles, passing from one reach to another without suspecting the change, unless familiar with the route, and so close to the wooded shores that one is tormented with the desire to reach out one's hand and strip the cool green spruce and cedar needles from the drooping branches.

Each water-way has its own distinctive features. In Finlayson Channel the forestation is a solid mountain of green on each side, growing down to the water and extending over it in feathery, flat sprays. Here the reflections are so brilliant and so true on clear days, that the dividing line is not perceptible to the vision. The mountains rise sheer from the water to a great height, with snow upon their crests and occasional cataracts foaming musically down their fissures. Helmet Mountain stands on the port side of the channel, at the entrance.

There's something about "Sarah" Island! I don't know what it is, and none of the mariners with whom I discussed this famous island seems to know; but the fact remains that they are all attached to "Sarah."

Down in Lama Pass, or possibly in Fitzhugh Sound, one hears casual mention of "Sarah" in the pilot-house or chart-room. Questioned, they do not seem to be able to name any particular feature that sets her apart from the other islands of this run.

"Well, there she is!" exclaimed the captain, at last. "Now, you'll see for yourself what there is about Sarah."

It is a long, narrow island, lying in the northern end of Finlayson Channel. Tolmie Channel lies between it and Princess Royal Island; Heikish Narrows—a quarter of a mile wide—between it and Roderick Island. Through Heikish the steamer passes into the increasing beauty of Graham Reach.

"Now, there!" said the captain. "If you can tell me what there is about that island, you can do more than any skipper *I* know can do; but just the same, there isn't one of us that doesn't look forward to passing Sarah, that doesn't give her particular attention while we are passing, and look back at her after we're in Graham Reach. She isn't so little ... nor so big.... The Lord knows she isn't so pretty!" He was silent for a moment. Then he burst out suddenly: "I'm blamed if *I* know what it is! But it's just so with some women. There's something about a woman, now and then, and a man can't tell, to save his soul, what it is; only, he doesn't forget her. You see, a captain meets hundreds of women; and he has to be nice to every one. If he is smart, he can make every woman think she is just running the ship—but Lord! he wouldn't know one of them if he met her next week on the street ... only now and then ... in years and years ... *one!* And that one he can't forget. He doesn't know what there is about her, any more than he knows what there is about 'Sarah.' Maybe he doesn't know the color of her eyes nor the color of her hair. Maybe she's married, and maybe she's single—for that isn't it. He isn't in love with her—at least I guess he isn't. It's just that she has a way of coming back to him. Say he sees the Northern Lights along about midnight—and that woman comes like a flash and stands there with him. After a while it gets to be a habit with him when he gets into a port, to kind of look over the crowds for some one.

For a minute or two he feels almost as if he *expected* some one to meet him; then he knows he's disappointed about somebody not being there. He asks himself right out who it is. And all at once he remembers. Then he calls himself an ass. If she was the kind of woman that runs to docks to see boats come in, he'd laugh and gas with her—but he wouldn't be thinking of her till she pushed herself on him again."

The captain sighed unconsciously, and taking down a chart from the ceiling, spread it out upon a shelf and bent over it. I looked at Sarah, with her two lacy cascades falling like veils from her crown of snow. Already she was fading in the distance—yet how distinguished was she! How set apart from all others!

Then I fell to thinking of the women. What kind are they—*the ones that stay!* The one that comes at midnight and stands silent beside a man when he sees the Northern Lights, even though he is not in love with her—what kind of woman is she?

"Captain," I said, a little later, "I want to add something to Sarah's name."

"What is it?" said he, scowling over the chart.

"I want to name her '*Sarah, the Remembered*.'"

He smiled.

"All right," said he, promptly. "I'll write that on the chart."

And what an epitaph that would be for a woman—"The Remembered!" If one only knew upon whose bit of marble to grave it.

Fraser and McKay reaches follow Graham, and then is entered Wright Sound, a body of water of great, and practically unknown, depth. This small sound feeds six channels leading in different directions, one of which—Verney Pass—leads through Boxer Reach into the famed magnificence and splendor of Gardner Canal, whose waters push for fifty miles through dark and towering walls. An immense, glaciered mountain extends across the end of the canal.

Gardner Canal—named by Vancouver for Admiral Sir Alan Gardner, to whose friendship and recommendation he was indebted for the command of the expedition to Nootka and the Northwest Coast—is doubtless the grandest of British Columbian inlets or fiords. At last, the favorite two adjectives of the Vancouver expedition—"tremendous" and "stupendous"—seem to have been most appropriately applied. Lieutenant Whidbey, exploring it in the summer of 1793, found that it "presented to the eye one rude mass of almost naked rocks, rising into rugged mountains, more lofty than he had before seen, whose towering summits, seeming to overhang their bases, gave them a *tremendous*appearance. The whole was covered with perpetual ice and snow that reached, in the gullies formed between the mountains, close down to the high-water mark; and many waterfalls of various dimensions were seen to descend in every direction."

This description is quoted in full because it is an excellent example of the descriptions given out by Vancouver and his associates, who, if they ever felt a quickening of the pulses in contemplation of these majestic scenes, were certainly successful in concealing such human emotions from the world. True, they did occasionally chronicle a "pleasant" breeze, a "pleasing" landscape which "reminded them of England;" and even, in the vicinity of Port Townsend, they were moved to enthusiasm over a "landscape almost as enchantingly beautiful as the most elegantly finished pleasure-grounds in Europe," which called to their remembrance "certain delightful and beloved situations in Old England."

But apparently, having been familiar only with pleasing pastoral scenes, they were not able to rise to an appreciation of the sublime in nature. "Elegant" is the mincing and amusing adjective applied frequently to snow mountains by Vancouver; he mentions, also, "spacious meadows, elegantly adorned with trees;" but when they arrive at the noble beauty which arouses in most beholders a feeling of exaltation and an appreciation of the marvellous handiwork of God, Vancouver and his associates, having never seen anything of the kind in England, find it only "tremendous," or "stupendous," or a "rude mass." They would have probably described the chaste, exquisite cone of Shishaldin on Unimak Island—as peerless and apart in its delicate beauty among mountains as Venice is among cities—as "a mountain covered with snow to the very sea and having a most elegant point."

There are many mountains more than twice the height of Shishaldin, but there is nowhere one so beautiful.

Great though our veneration must be for those brave mariners of early years, their apparent lack of appreciation of the scenery of Alaska is to be deplored. It has fastened upon the land an undeserved reputation for being "rugged" and "gloomy"—two more of their adjectives; of being "ice-locked, ice-bound, and ice-bounded." We may pardon them much, but scarcely the adjective "grotesque," as applied to snow mountains.

Grenville Channel is a narrow, lovely reach, extending in a northwestward direction from Wright Sound for forty-five miles, when it merges into Arthur Passage. In its slender course it curves neither to the right nor to the left.

In this reach, at one o'clock one June day, the thrilling cry of "man overboard" ran over the decks of the *Santa Ana*. There were more than two hundred passengers aboard, and instantly an excited and dangerous stampede to starboard and stern occurred; but the captain, cool and stern on the bridge, was equal to the perilous situation. A life-boat was ordered lowered, and the steerage passengers were quietly forced to their quarters forward. Life-buoys, life-preservers, chairs, ropes, and other articles were flung overboard, until the water resembled a junk-shop. Through them all, the man's dark, closely shaven head could be seen, his face turned from the steamer, as he swam fiercely toward the shore against a strong current. The channel was too narrow for the steamer to turn, but a boat was soon in hot pursuit of the man who was struggling fearfully for the shore, and who was supposed to be too bewildered to realize that he was headed in the wrong direction. What was our amazement, when the boat finally reached him, to discover, by the aid of glasses, that he was resisting his rescuers. There was a long struggle in the water before he was overcome and dragged into the boat.

He was a pitiable sight when the boat came level with the hurricane deck; wild-eyed, gray-faced, shuddering like a dog; his shirt torn open at the throat and exposing its tragic emaciation; his glance flashing wildly from one face to another, as though in search of one to be trusted—he was an object to command the pity of the coldest heart. In his hand was still gripped his soft hat which he had taken from his head before jumping overboard.

"What is it, my man?" asked the captain, kindly, approaching him.

The man's wild gaze steadied upon the captain and seemed to recognize him as one in authority.

"They've been trying to kill me, sir, all the way up."

"Who?"

The poor fellow shuddered hard.

"They," he said. "They're on the boat. I had to watch them night and day. I didn't dast go to sleep. It got too much; I couldn't stand it. I had to get ashore. I'd been waiting for this channel because it was so narrow. I thought the current 'u'd help me get away. I'm a good swimmer."

"A better one never breasted a wave! Take him below. Give him dry clothes and some whiskey, and set a watch over him."

The poor wretch was led away; the crowd drifted after him. Pale and quiet, the captain went back to the chart-room and resumed his slow pacing forth and back.

"I wish tragedies of body and soul would not occur in such beautiful lengths of water," he said at last. "I can never sail through Grenville Channel again without seeing that poor fellow's haggard face and wild, appealing eyes. And after Gardner Canal, there is not another on the route more beautiful than this!"

Two inlets open into Grenville Channel on the starboard going north, Lowe and Klewnuggit,—both affording safe anchorage to vessels in trouble. Pitt Island forms almost the entire western shore—a beautifully wooded one—of the channel. There is a salmon cannery in Lowe Inlet, beside a clear stream which leaps down from a lake in the mountains. The waters and shores of Grenville have a clear, washed green, which is springlike. In many of the other narrow ways the waters are blue, or purple, or a pale blue-gray; but here they suddenly lead you along the palest of green, shimmering avenues, while mountains of many-shaded green rise steeply on both sides, glimmering away into drifts of snow, which drop threads of silver down the sheer heights.

This shaded green of the mountains is a feature of Alaskan landscapes. Great landslides and windfalls cleave their way from summit to sea, mowing down the forests in their path. In time the new growth springs up and streaks the mountain side with lighter green.

Probably one-half of the trees in southeastern Alaska are the Menzies spruce, or Sitka pine. Their needles are sharp and of a bluish green.

The Menzies spruce was named for the Scotch botanist who accompanied Vancouver.

The Alaska cedar is yellowish and lacy in appearance, with a graceful droop to the branches. It grows to an average height of one hundred and fifty feet. Its wood is very valuable.

Arbor-vitæ grows about the glaciers and in cool, dim fiords. Birch, alder, maple, cottonwood, broom, and hemlock-spruce are plentiful, but are of small value, save in the cause of beauty.

The Menzies spruce attains its largest growth in the Alexander Archipelago, but ranges as far south as California. The Douglas fir is not so abundant as it is farther south, nor does it grow to such great size.

The Alaska cedar is the most prized of all the cedars. It is in great demand for ship-building, interior finishing, cabinet-making, and other fine work, because of its close texture, durable quality, and aromatic odor, which somewhat resembles that of sandalwood. In early years it was shipped to Japan, where it was made into fancy boxes and fans, which were sold under guise of that scented Oriental wood. Its lasting qualities are remarkable—

sills having been found in perfect preservation after sixty years' use in a wet climate. Its pleasant odor is as enduring as the wood. The long, slender, pendulous fruits which hang from the branches in season, give the tree a peculiarly graceful and appealing appearance.

The western white pine is used for interior work. It is a magnificent tree, as seen in the forest, having bluish green fronds and cones a foot long.

The giant arbor-vitæ attains its greatest size close to the coast. The wood splits easily and makes durable shingles. It takes a brilliant polish and is popular for interior finishing. Its beauty of growth is well known.

Wherever there is sufficient rainfall, the fine-fronded hemlock may be found tracing its lacelike outlines upon the atmosphere. There is no evergreen so delicately lovely as the hemlock. It stands apart, with a little air of its own, as a fastidious small maid might draw her skirts about her when common ones pass by.

The spruces, firs, and cedars grow so closely together that at a distance they appear as a solid wall of shaded green, varying from the lightest beryl tints, on through bluish grays to the most vivid and dazzling emerald tones. At a distance canyons and vast gulches are filled so softly and so solidly that they can scarcely be detected, the trees on the crests of the nearer hills blending into those above, and concealing the deep spaces that sink between.

These forests have no tap-roots. Their roots spread widely upon a thin layer of soil covering solid stone in many cases, and more likely than not this soil is created in the first place by the accumulation of parent needles. Trees spring up in crevices of stone where a bit of sand has sifted, grow, fruit, and shed their needles, and thrive upon them. The undergrowth is so solid that one must cut one's way through it, and the progress of surveyors or prospectors is necessarily slow and difficult.

These forests are constantly drenched in the warm mists precipitated by the Kuro Siwo striking upon the snow, and in this quickening moisture they reach a brilliancy of coloring that is remarkable. At sunset, threading these narrow channels, one may see mountain upon mountain climbing up to crests of snow, their lower wooded slopes covered with mists in palest blue and old rose tones, through which the tips of the trees, crowded close together, shine out in brilliant, many-shaded greens.

After Arthur Passage is that of Malacca, which is dotted by several islands. "Lawyer's," to starboard, bears a red light; "Lucy," to port, farther north, a fixed white light. Directly opposite "Lucy"—who does not rival "Sarah," or who in the pilot's words "has nothing about her"—is old Metlakahtla.

Davidson Glacier

CHAPTER III

The famous ukase of 1821 was issued by the Russian Emperor on the expiration of the twenty-year charter of the Russian-American Company. It prohibited "to all foreign vessels not only to land on the coasts and islands belonging to Russia, as stated above" (including the whole of the northwest coast of America, beginning from Behring Strait to the fifty-first degree of northern latitude, also from the Aleutian Islands to the eastern coast of Siberia, as well as along the Kurile Islands from Behring Strait to the south cape of the Island of Urup) "but also to approach them within less than one hundred miles."

After the Nootka Convention in 1790, the Northwest Coast was open to free settlement and trade by the people of any country. It was claimed by the Russians to the Columbia, afterward to the northern end of Vancouver Island; by the British, from the Columbia to the fifty-fifth degree; and by the United States, from the Rocky Mountains to the Pacific, between Forty-two and Fifty-four, Forty. By the treaty of 1819, by which Florida was ceded to us by Spain, the United States acquired all of Spanish rights and claims on the coast north of the forty-second degree. By its trading posts and regular trading vessels, the United States was actually in possession.

By treaty with the United States in 1824, and with Great Britain in 1825, Russia, realizing her mistake in issuing the ukase of 1821, agreed to Fifty-four, Forty as the limit of her possessions to southward. Of the interior regions, Russia claimed the Yukon region; England, that of the Mackenzie and the country between Hudson Bay and the Rocky Mountains; the United States, all west of the Rockies, north of Forty-two.

The year previous to the one in which the United States acquired Florida and all Spanish rights on the Pacific Coast north of Forty-two, the United States and England had agreed to a joint occupation of the region. In 1828 this was indefinitely extended, but with the emigration to Oregon in the early forties, this country demanded a settlement of the boundary question.

President Tyler, in his message to Congress in 1843, declared that "the United States rights appertain to all between forty-two degrees and fifty-four degrees and forty minutes."

The leading Democrats of the South were at that time advocating the annexation of Texas. Mr. Calhoun was an ardent champion of the cause, and was endeavoring to effect a settlement with the British minister, offering the forty-ninth parallel as a compromise on the boundary dispute, in his eagerness to acquire Texas without danger of interference.

The compromise was declined by the British minister.

In 1844 slave interests defeated Mr. Van Buren in his aspirations to the presidency. Mr. Clay was nominated instead. The latter opposed the annexation of Texas and advised caution and compromise in the Oregon question; but the Democrats nominated Polk and under the war-cry of "Fifty-four, Forty, or Fight," bore him on to victory. The convention which nominated him advocated the reannexation of Texas and the reoccupation of Oregon; the two significant words being used to make it clear that Texas had belonged to us before, through the Louisiana purchase; and Oregon, before the treaty of joint occupation with Great Britain.

President Polk, in his message, declared that, "beyond all question, the protection of our laws and our jurisdiction, civil and criminal, ought to be immediately extended over our citizens in Oregon."

He quoted from the convention which had nominated him that "our title to the country of Oregon as far as Fifty-four, Forty, is clear and unquestionable;" and he boldly declared "for all of Oregon or none."

John Quincy Adams eloquently supported our title to the country to the line of Fifty-four, Forty in a powerful speech in the House of Representatives.

Yet it soon became apparent that both the Texas policy and the Oregon question could not be successfully carried out during the administration. "Fifty-four, Forty, or Fight" as a watchword in a presidential campaign was one thing, but as a challenge to fight flung in the face of Great Britain, it was quite another.

In February, 1846, the House declared in favor of giving notice to Great Britain that the joint occupancy of the Oregon country must cease. The Senate, realizing that this resolution was practically a declaration of war, declined to adopt it, after a very bitter and fiery controversy.

Those who retreated from their first position on the question were hotly denounced by Senator Hannegan, the Democratic senator from Indiana. He boldly attacked the motives which led to their retreat, and angrily exclaimed:—

"If Oregon were good for the production of sugar and cotton, it would not have encountered this opposition."

The resolution was almost unanimously opposed by the Whig senators. Mr. Webster, while avoiding the point of our actual rights in the matter, urged that a settlement on the line of the forty-ninth parallel be recommended, as permitting both countries to compromise with dignity and honor. The resolution that was finally passed by the Senate and afterward by the House, authorized the president to give notice at his discretion to Great Britain that the treaty should be terminated, "in order that the attention of the governments of both countries may be the more earnestly directed to the adoption of all proper measures for a speedy and amicable adjustment of the differences and disputes in regard to said territory."

Forever to their honor be it remembered that a few of the Southern Democrats refused to retreat from their first position—among them, Stephen A. Douglas. Senator Hannegan reproached his party for breaking the pledges on which it had marched to victory.

The passage of the milk-and-water resolution restored to the timid of the country a feeling of relief and security; but to the others, and to the generations to come after them, helpless anger and undying shame.

The country yielded was ours. We gave it up solely because to retain it we must fight, and we were not in a position at that time to fight Great Britain.

When the Oregon Treaty, as it was called, was concluded by Secretary Buchanan and Minister Pakenham, we lost the splendid country now known as British Columbia, which, after our purchase of Alaska from Russia, would have given us an unbroken frontage on the Pacific Ocean from Southern California to Behring Strait, and almost to the mouth of the Mackenzie River on the Frozen Ocean.

Many reasons have been assigned by historians for the retreat of the Southern Democrats from their former bold and flaunting position; but in the end the simple truth will be admitted—that they might brag, but were not in a position to fight. They were like Lieutenant Whidbey, whom Vancouver sent out to explore Lynn Canal in a small boat. Mr. Whidbey was ever ready and eager, when he deemed it necessary, to fire upon a small party of Indians; but when they met him, full front, in formidable numbers and with couched spears, he instantly fell into a panic and deemed it more "humane" to avoid a conflict with those poor, ignorant people.

A Phantom Ship

The Southern Democrats who betrayed their country in 1846 were the Whidbeys of the United States. For no better reason than that of "humanity," they gave nearly four hundred thousand square miles of magnificent country to Great Britain.

Another problem in this famous boundary settlement

question has interested American historians for sixty years: Why England yielded so much valuable territory to the United States, after protecting what she claimed as her rights so boldly and so unflinchingly for so many years.

Professor Schafer, the head of the Department of American History at the University of Oregon, claims to have recently found indisputable proof in the records of the British Foreign Office and those of the old Hudson's Bay Company, in London, that the abandonment of the British claim was influenced by the presence of American pioneers who had pushed across the continent and settled in the disputed territory, bringing their families and founding homes in the wilderness.

England knew, in her heart, that the whole disputed territory was ours; and as our claims were strengthened by settlement, she was sufficiently far-sighted to be glad to compromise at that time. If the Oregon Treaty had been delayed for a few years, British Columbia would now be ours. Proofs which strengthen our claim were found in the winter of 1907-1908 in the archives of Sitka.

There would be more justice in our laying claim to British Columbia now, than there was in the claims of Great Britain in the famous *lisière* matter which was settled in 1903.

By the treaties of 1824, between Russia and the United States, and of 1825, between Russia and Great Britain, the limits of Russian possessions are thus defined, and upon our purchase of Alaska from Russia, were repeated in the Treaty of Washington in 1867:—

"Commencing from the southernmost point of the island called Prince of Wales Island, which point lies in the parallel of fifty-four degrees and forty minutes north latitude, and between the one hundred and thirty-first and the one hundred and thirty-third degree of west longitude (meridian of Greenwich), the said line shall ascend to the North along the channel called Portland Channel, as far as the point of the continent where it strikes the fifty-sixth degree of north latitude; from this last mentioned point, the line of demarcation shall follow the summit of the mountains situated parallel to the coast as far as the point of intersection of the one hundred and forty-first degree of west longitude (of the same meridian); and finally, from the said point of intersection, the said meridian line of the one hundred and forty-first degree, in its prolongation as far as the Frozen Ocean, shall form the limit between the Russian and British possessions on the Continent of America to the northwest.

"With reference to the line of demarcation laid down in the preceding article, it is understood:—

"First, That the island called Prince of Wales Island shall belong wholly to Russia.

"Second, That whenever the summit of the mountains which extend parallel to the coast from the fifty-sixth degree of north latitude to the point of intersection of the one hundred and forty-first degree of west longitude shall prove to be at the distance of more than ten marine leagues from the ocean, the limit between the British possessions and the line of coast which is to belong to Russia as above mentioned shall be formed by a line parallel to the windings of the coast, and which shall never exceed the distance of ten marine leagues therefrom.

"The western limit within which the territories and dominion conveyed are contained, passes through a point in Behring Strait on the parallel of sixty-five degrees, thirty minutes, north latitude, at its intersection by the meridian which passes midway between the islands of Krusenstern, or Ignalook, and the island of Ratmanoff, or Noonarbook, and proceeds due north, without limitation, into the same Frozen Ocean. The same western limit, beginning at the same initial point, proceeds thence in a course nearly southwest, through Behring Strait and Behring Sea, so as to pass midway between the northwest point of the island of St. Lawrence and the southeast point of Cape Choukotski, to the meridian of one hundred and seventy-two west longitude; thence, from the intersection of that meridian in a southwesterly direction, so as to pass midway between the island of Attou and the Copper Island of the Kormandorski couplet or group in the North Pacific Ocean, to the meridian of one hundred and ninety-three degrees west longitude, so as to include in the territory conveyed the whole of the Aleutian Islands east of that meridian."

In the cession was included the right of property in all public lots and squares, vacant lands, and all public buildings, fortifications, barracks, and other edifices, which were not private individual property. It was, however, understood and agreed that the churches which had been built in the ceded territory by the Russian government should remain the property of such members of the Greek Oriental Church resident in the territory as might choose to worship therein. All government archives, papers, and documents relative to the territory and dominion aforesaid which were existing there at the time of transfer were left in possession of the agent of the United States; with the understanding that the Russian government or any Russian subject may at any time secure an authenticated copy thereof.

The inhabitants of the territory were given their choice of returning to Russia within three years, or remaining in the territory and being admitted to the enjoyment of all rights, advantages, and immunities of citizens of the United States, protected in the free enjoyment of their liberty, property, and religion.

It must be confessed with chagrin that very few Russians availed themselves of this opportunity to free themselves from the supposed oppression of their government, to unite with the vaunted glories of ours.

Before 1825, Great Britain, Spain, Portugal, and the United States had no rights of occupation and assertion on the Northwest Coast. Different nations had "planted bottles" and "taken possession" wherever their explorers had chanced to land, frequently ignoring the same ceremony on the part of previous explorers; but these formalities did not weigh against the rights of discovery and actual occupation by Russia—else Spain's rights would have been prior to Great Britain's.

Between the years of 1542 and 1774 Spanish explorers had examined and traced the western coast of America as far north as fifty-four degrees and forty minutes, Perez having reached that latitude in 1774, discovering Queen Charlotte Islands on the 16th of June, and Nootka Sound on the 9th of August.

Although he did not land, he had friendly relations with the natives, who surrounded his ship, singing and scattering white feathers as a beautiful token of peace. They traded dried fish, furs, and ornaments of their own making for knives and old iron; and two, at least, boarded the ship.

Perez named the northernmost point of Queen Charlotte Islands Point Santa Margarita.

Proceeding south, he made a landfall and anchored in a roadstead in forty-nine degrees and thirty minutes, which he called San Lorenzo—afterward the famous Nootka of Vancouver Island. He also discovered the beautiful white mountain which dignifies the entrance to Puget Sound, and named it Santa Rosalia. It was renamed Mount Olympus fourteen years later by John Meares.

This was the first discovery of the Northwest Coast, and when Cook and Vancouver came, it was to find that the Spanish had preceded them.

Not content with occupying the splendid possessions of the United States through the not famous, but infamous, Oregon Treaty, Canada, upon the discovery of gold in the Cassiar district of British Columbia, brought up the question of the *lisière*, or thirty-mile strip. This was the strip of land, "not exceeding ten marine leagues in width," which bordered the coast from the southern limit of Russian territory at Portland Canal (now the southern boundary of Alaska) to the vicinity of Mount St. Elias. The purpose of this strip was stated by the Russian negotiations to be "the establishment of a barrier at which would be stopped, once for all, to the North as to the West of the coast allotted to our American Company, the encroachments of the English agents of the Amalgamated Hudson Bay and Northwest English Company."

In 1824, upon the proposal of Sir Charles Bagot to assign to Russia a strip with the uniform width of ten marine leagues from the shore, limited on the south by a line between thirty and forty miles north from the northern end of the Portland Canal, the Russian Plenipotentiaries replied:—

"The motive which caused the adoption of the principle of mutual expediency to be proposed, and the most important advantage of this principle, is to prevent the respective establishments on the Northwest Coast from injuring each other and entering into collision.

"The English establishments of the Hudson Bay and Northwest companies have a tendency to advance westward along the fifty-third and fifty-fourth degrees of north latitude.

"The Russian establishments of the American Company have a tendency to descend southward toward the fifty-fifth parallel and beyond; for it should be noted that, if the American Company has not yet made permanent establishments on the mathematical line of the fifty-fifth degree, it is nevertheless true that by virtue of its privilege of 1799, against which privilege no power has ever protested, it is exploiting the hunting and the fishing in these regions, and that it regularly occupies the islands and the neighboring coasts during the season, which allows it to send its hunters and fishermen there.

"It was, then, to the mutual advantage of the two Empires to assign just limits to this advance on both sides, which, in time, could not fail to cause most unfortunate complications.

"It was also to their mutual advantage to fix their limits according to natural partitions, which always constitute the most distinct and certain frontiers.

"For these reasons the Plenipotentiaries of Russia have proposed as limits upon the coast of the continent, to the South, Portland Channel, the head of which lies about (par) the fifty-sixth degree of north latitude, and to the East, the chain of mountains which follows at a very short distance the sinuosities of the coast."

Sir Charles Bagot urged the line proposed by himself and offered, on the part of Great Britain, to include the Prince of Wales Island within the Russian line.

Russia, however, insisted upon having her *lisière* run to the Portland Canal, declaring that the possession of Wales Island, without a slice (portion) of territory upon the coast situated in front of that island, could be of no utility whatever to Russia; that any establishment formed upon said island, or upon the surrounding islands, would find itself, as it were, flanked by the English establishments on the mainland, and completely at the mercy of these latter.

England finally yielded to the Russian demand that the *lisière* should extend to the Portland Canal.

The claim that the Canadian government put forth, after the discovery of gold had made it important that Canada should secure a short line of traffic between the northern interior and the ocean, was that the wording of certain parts of the treaty of 1825 had been wrongly interpreted. The Canadians insisted that it was not the meaning nor the intention of the Convention of 1825 that there should remain in the exclusive possession of Russia a continuous fringe, or strip—the *lisière*—of coast, separating the British possessions from the bays, ports, inlets, havens, and waters of the ocean.

Or, if it should be decided that this was the meaning of the treaty, they maintained that the width of the *lisière* was to be measured from the line of the general direction of the mainland coast, and not from the heads of the many inlets.

They claimed, also, that the broad and beautiful "Portland's Canal" of Vancouver and the "Portland Channel" of the Convention of 1825, were the Pearse Channel or Inlet of more recent times. This contention, if sustained, would give them our Wales and Pearse islands.

It was early suspected, however, that this claim was only made that they might have something to yield when, as they hoped, their later claim to Pyramid Harbor and the valley of the Chilkaht

River should be made and upheld. This would give them a clear route into the Klondike territory.

In 1898 a Joint High Commission was appointed for the consideration of Pelagic Fur Sealing, Commercial Reciprocity, and the Alaska Boundary. The Commission met in Quebec. The discussion upon the boundary continued for several months, the members being unable to agree upon the meaning of the wording of the treaty of 1825.

The British and Canadian members, thereupon, unblushingly proposed that the United States should cede to Canada Pyramid Harbor and a strip of land through the entire width of the *lisière.*

To Americans who know that part of our country, this proposal came as a shock. Pyramid Harbor is the best harbor in that vicinity; and its cession, accompanied by a highway through the *lisière* to British possessions, would have given Canada the most desirable route at that time to the Yukon and the Klondike—the rivers upon which the eyes of all nations were at that time set. Many routes into that rich and picturesque region had been tested, but no other had proved so satisfactory.

It has since developed that the Skaguay route is the real prize. Had Canada foreseen this, she would not have hesitated to demand it.

From the disagreement of the Joint High Commission of 1898 arose the modus vivendi of the following year. There has been a very general opinion that the temporary boundary points around the heads of the inlets at the northern end of Lynn Canal, laid down in that year, were fixed for all time—although it seems impossible that this opinion could be held by any one knowing the definition of the term "modus vivendi."

By the modus vivendi Canada was given temporary possession of valuable Chilkaht territory, and her new maps were made accordingly.

Road through Cut-off Canyon

In 1903 a tribunal composed of three American members and three representing Great Britain, two of whom were Canadians, met in Great Britain, to settle certain questions relating to the *lisière*.

The seven large volumes covering the arguments and decisions of this tribunal, as published by the United States government, make intensely interesting and valuable reading to one who cares for Alaska.

The majority of the tribunal, that is to say, Lord Alverstone and the three members from the United States, decided that the Canadians have no rights to the waters of any of the inlets, and that it was the meaning of the Convention of 1825 that the *lisière* should for all time separate the British possessions from the bays, ports, inlets, and waters of the ocean north of British Columbia; and that, furthermore, the width of the *lisière* was not to be measured from the line of the general direction of the mainland coast, leaping the bays and inlets, but from a line running around the heads of such indentations.

The tribunal, however, awarded Pearse and Wales islands, which belonged to us, to Canada; it also narrowed the *lisière* in several important points, notably on the Stikine and Taku rivers.

The fifth question, however, was the vital one; and it was answered in our favor, the two Canadian members dissenting. The boundary lines have now been changed on both United States and Canadian maps, in conformity with the decisions of the tribunal.

Blaine, Bancroft, and Davidson have made the clearest statements of the boundary troubles.

CHAPTER IV

The first landing made by United States boats after leaving Seattle is at Ketchikan. This is a comparatively new town. It is seven hundred miles from Seattle, and is reached early on the third morning out. It is the first town in Alaska, and glistens white and new on its gentle hills soon after crossing the boundary line in Dixon Entrance—which is always saluted by the lifting of hats and the waving of handkerchiefs on the part of patriotic Americans.

Ketchikan has a population of fifteen hundred people. It is the distributing point for the mines and fisheries of this section of southeastern Alaska. It is the present port of entry, and the Customs Office adds to the dignity of the town. There is a good court-house, a saw-mill with a capacity of twenty-five thousand feet daily, a shingle mill, salmon canneries, machine shops, a good water system, a cold storage plant, two excellent hotels, good schools and churches, a progressive newspaper, several large wharves, modern and well-stocked stores and shops, and a sufficient number of saloons. The town is lighted by electricity and many of the buildings are heated by steam. A creditable chamber of commerce is maintained.

There are seven salmon canneries in operation which are tributary to Ketchikan. The most important one "mild-cures" fish for the German market.

Among the "shipping" mines, which are within a radius of fifty miles, and which receive mails and supplies from Ketchikan, are the Mount Andrews, the Stevenston, the Mamies, the Russian Brown, the Hydah, the Niblack, and the Sulzer. From fifteen to twenty prospects are under development.

There are smelters in operation at Hadley and Copper Mountain, on Prince of Wales Island. From Ketchikan to all points in the mining and fishing districts safe and commodious steamers are regularly operated. The chief mining industries are silver, copper, and gold.

The residences are for the most part small, but, climbing by green terraces over the hill and surrounded by flowers and neat lawns, they impart an air of picturesqueness to the town. There are several totem-poles; the handsomest was erected to the memory of Chief "Captain John," by his nephew, at the entrance to the house now occupied by the latter. The nephew asserts that he paid $2060 for the carving and making of the totem. Owing to its freshly painted and gaudy appearance, it is as lacking in interest as the one which stands in Pioneer Square, Seattle, and which was raped from a northern Indian village.

Four times had I landed at Ketchikan on my way to far beautiful places; with many people had I talked concerning the place; folders of steamship companies and pamphlets of boards of trade had I read; yet never from any person nor from any printed page had I received the faintest glimmer that this busy, commercially described northwestern town held, almost in its heart, one of the enduring and priceless jewels of Alaska. To the beauty-loving, Norwegian captain of the steamship *Jefferson* was I at last indebted for one of the real delights of my life.

It was near the middle of a July night, and raining heavily, when the captain said to us:—

"Be ready on the stroke of seven in the morning, and I'll show you one of the beautiful things of Alaska."

"But—at Ketchikan, captain!"

"Yes, at Ketchikan."

I thought of all the vaunted attractions of Ketchikan which had ever been brought to my observation; and I felt that at seven o'clock in the morning, in a pouring rain, I could live without every one of them. Then—the charm of a warm berth in a gray hour, the cup of hot coffee, the last dream to the drowsy throb of the steamer—

"It will be raining, captain," one said, feebly.

The look of disgust that went across his expressive face!

"What if it is! You won't know it's raining as soon as you get your eyes filled with what I want to show you. But if you're one of *that* kind—"

He made a gesture of dismissal with his hands, palms outward, and turned away.

"Captain, I shall be ready at seven. I'm not one of that kind," we all cried together.

"All right; but I won't wait five minutes. There'll be two hundred passengers waiting to go."

Scene on the White Pass

"You know that letter that Thomas Bailey Aldrich wrote to Professor Morse," spoke up a lady from Boston, who had overheard. "You know Professor Morse wrote a hand that couldn't be deciphered, and among other things, Mr. Aldrich wrote: 'There's a singular and perpetual charm in a letter of yours; it never grows old; it never loses its novelty. One can say to one's self every day: "There's that letter of Morse's. I have not read it yet. I think I shall take another shy at it." Other letters are read and thrown away and forgotten; but yours are kept forever—unread!' Now, that letter, somehow, in the vaguest kind of way, suggests itself when one considers this getting up anywhere from three to six in the morning to see things in Alaska. There's *always* something to be seen during these unearthly hours. Every night we are convinced that we will be on deck early, to see something, and we leave an order to be wakened; but when the dreaded knocking comes upon the door, and a hoarse voice announces 'Wrangell Narrows,' or 'Lama Pass,' our berths suddenly take on curves and attractions they possess at no other time. The side-rails into which we have been bumping seem to be cushioned with down, the space between berths to grow wider, the air in the room sweeter and more drowsily delicious. We say, 'Oh, we'll get up to-morrow morning and see something,' and we pull the berth-curtain down past our faces and go to sleep. After a while, it grows to be one of the perpetual charms of a trip to Alaska—this always going to get up in the morning and this never getting up. It never grows old; it never loses its novelty. One can say to one's self every morning: 'There's that little matter to decide now about getting up. Shall I, or shall I not?' I have been to Alaska three times, but I've never seen Ketchikan. Other places are seen and admired and forgotten; but it remains forever—unseen.... Now, I'll go and give an order to be called at half-past six, to see this wonderful thing at Ketchikan!"

I looked around for her as I went down the slushy deck the next morning on the stroke of seven; but she was not in sight. It was raining heavily and steadily—a cold, thick rain; the wind was so strong and so changeful that an umbrella could scarcely be held.

Alas for the captain! Out of his boasted two hundred passengers, there came forth, dripping and suspicious-eyed, openly scenting a joke, only four women and one man. But the captain was undaunted. He would listen to no remonstrances.

"Come on, now," he cried, cheerfully, leading the way. "You told me you came to Alaska to see things, and as long as you travel with me, you are going to see all that is worth seeing. Let the others sleep. Anybody can sleep. You can sleep at home; but you can't see what I am going to show you now anywhere but in Alaska. Do you suppose I would get up at this hour and waste my time on you, if I didn't know you'd thank me for it all the rest of your life?"

So on and on we went; up one street and down another; around sharp corners; past totem-poles, saloons, stylish shops, windows piled with Indian baskets and carvings; up steps and down terraces; along gravelled roads; and at last, across a little bridge, around a wooded curve,—and then—

Something met us face to face. I shall always believe that it was the very spirit of the woods that went past us, laughing and saluting, suddenly startled from her morning bath in the clear, amber-brown stream that came foaming musically down over smooth stones from the mountains.

It was so sudden, so unexpected. One moment, we were in the little northern fishing- and mining-town, which sits by the sea, trumpeting its commercial glories to the world; the next, we were in the forest, and under the spell of this wild, sweet thing that fled past us, returned, and lured us on.

For three miles we followed the mocking call of the spirit of the brown stream. Her breath was as sweet as the breath of wild roses covered with dew. Never in the woods have I been so impressed, so startled, with the feeling that a living thing was calling me.

We could find no words to express our delight as we climbed the path beside the brown stream, whose waters came laughingly down through a deep, dim gorge. They fell sheer in sparkling cataracts; they widened into thin, singing shallows of palest amber, clinking against the stones; narrow and foaming, they wound in and out among the trees; they disappeared completely under wide sprays of ferns and the flat, spreading branches of trees, only to "make a sudden sally" farther down.

At first we were level with them, walked beside them, and paused to watch the golden gleams in their clear depths; but gradually we climbed, until we were hundreds of feet above them.

Down in those purple shadows they went romping on to the sea; sometimes only a flash told us where they curved; other times, they pushed out into open spaces, and made pause in deep pools, where they whirled and eddied for a moment before drawing together and hurrying on. But always and everywhere the music of their wild, sweet, childish laughter floated up to us.

In the dim light of early morning the fine mist of the rain sinking through the gorge took on tones of lavender and purple. The tall trees climbing through it seemed even more beautiful than they really were, by the touch of mystery lent by the rain.

I wish that Max Nonnenbruch, who painted the adorable, compelling "Bride of the Wind," might paint the elfish sprite that dwells in the gorge at Ketchikan. He, and he alone, could paint her so that one could hear her impish laughter, and her mocking, fluting call.

The name of the stream I shall never tell. Only an unimaginative modern Vancouver or Cook could have bestowed upon it the name that burdens it to-day. Let it be the "brown stream" at Ketchikan.

If the people of the town be wise, they will gather this gorge to themselves while they may; treasure it, cherish it, and keep it "unspotted from the world"—yet *for* the world.

Metlakahtla means "the channel open at both ends." It was here that Mr. William Duncan came in 1857, from England, as a lay worker for the Church Mission Society. It had been represented that existing conditions among the natives sorely demanded high-minded missionary work. The savages at Fort Simpson were considered the worst on the coast at that time, and he was urged not to locate there. Undaunted, however, Mr. Duncan, who was then a very young man, filled with the fire and zeal of one who has not known failure, chose this very spot in which to begin his work—among Indians so low in the scale of human intelligence that they had even been accused of cannibalism.

Port Simpson was then an important trading-post of the Hudson Bay Company. It had been established in the early thirties about forty miles up Nass River, but a few years later was removed to a point on the Tsimpsian Peninsula. In 1841 Sir George Simpson found about fourteen thousand Indians, of various tribes, living there. He found them "peculiarly comely, strong, and well-grown ... remarkably clever and ingenious."

They carved neatly in stone, wood, and ivory. Sir George Simpson relates with horror that the savages frequently ate the dead bodies of their relatives, some of whom had died of smallpox, even after they had become putrid. They were horribly diseased in other ways; and many had lost their eyes through the ravages of smallpox or other disease. They fought fiercely and turbulently with other tribes.

Such were the Indians among whom Mr. Duncan chose to work. He was peculiarly fitted for this work, being possessed of certain unusual qualities and attributes of character which make for success.

The unselfishness and integrity of his nature made themselves visible in his handsome face, and particularly in the direct gaze of his large and intensely earnest blue eyes; his manners were simple, and his air was one of quiet command; he had unfailing cheerfulness, faith, and that quality which struggles on under the heaviest discouragement with no thought of giving up.

His word was as good as his bond; his energy and enthusiasm were untiring, and he never attempted to work his Indians harder than he himself worked. The entire absence of that trait which seeks self-praise or self-glory,—in fact, his absolute self-effacement, his devotion of self and self-interest to others, and to hard and humble work for others,—all these high and noble parts of an unusual and lovable character, added to a most winning and attractive personality, gradually won for young William Duncan the almost Utopian success which many others in various parts of the world have so far worked for in vain.

The Indians grew to trust his word, to believe in his sincerity and single-heartedness, to accept his teachings, to love him, and finally, and most reluctantly of all, to work for him.

At first only fifty of the Tsimsheans, or Tsimpsians, accompanied him to the site of his first community settlement. Here the land was cleared and cultivated; neat two-story cottages, a church, a schoolhouse, stores on the coöperative plan, a saw-mill, and a cannery, were erected by Mr. Duncan and the Indians. At first a corps of able assistants worked with Mr. Duncan, instructing the Indians in various industries and arts, until the young men were themselves able to carry along the different branches of work,—such as carpentry, shoemaking, cabinet building, tanning, rope-making, and boat building. The village band was instructed by a German, until one among them was qualified to become their band-master. The women were taught to cook, to sew, to keep house, to weave, and to care for the sick.

Here was a model village, an Utopian community, an ideal life,—founded and carried on by the genius of one young, simple-hearted, high-minded, earnest, and self-devoted English gentleman.

But William Duncan's way, although strewn with the full sweet roses of success, was not without its bitter, stinging thorns. Mr. Duncan was not an ordained minister, and in 1881 it was decided by the Church of England authorities who had sent Mr. Duncan out, that his field should be formed into a separate diocese, and as this decision necessitated the residence of a bishop, Bishop Ridley was sent to the field—a man whose name will ever stand as a dark blot upon the otherwise clean page whereon is written the story which all men honor and all men praise—the story of the exalted life-work of William Duncan.

Mr. Duncan, being a layman, had conducted services of

the simplest nature, and had not considered it advisable to hold communion services which would be embarrassing of explanation to people so recently won from the customs of cannibalism. Bigoted and opinionated, and failing utterly to understand the Indians, to win their confidence, or to exercise patience with them, Bishop Ridley declined to be under the direction of a man who was not ordained, and criticised the form of service held by Mr. Duncan. The latter, having been in sole charge of his work for more than thirty years, and being conscious of its full and unusual results, chafed under the Bishop's supervision and superintendence.

In the meantime, seven other missions had been established at various stations in southeastern Alaska. The Bishop undertook to inaugurate communion services. This was strongly opposed by Mr. Duncan, and he was supported by the Indians, who were sincerely attached to him, the Society in England sympathizing with the Bishop. Friction between the two was ceaseless and bitter, and continued until 1887. This has been given out as the cause of the withdrawal of Mr. Duncan to New Metlakahtla; but his own people—graduates of Eastern universities—claim that it is not the true reason. He and his Indians had for some time desired to be under the laws of the United States, and in 1887 Mr. Duncan went to Washington City to negotiate with the United States for Annette Island. The Bishop established himself in residence, but failed ignominiously to win the respect of the Indians. He quarrelled with them in the commonest way, struck them, went among them armed, and finally appealed to a man-of-war for protection from people whom he considered bloodthirsty savages.

Mr. Duncan, having been successful in his mission to Washington, his faithful followers, during his absence, removed to Annette Island, and here he found on his return all but one hundred out of the original eight hundred which had composed his village on the Bishop's arrival—the few having been persuaded to remain with the latter at Old Metlakahtla. Those who went to the new location on Annette were allowed by the Canadian government to take nothing but their personal property; all their houses, public buildings, and community interests being sacrificed to their devotion to William Duncan—and this is, perhaps, the highest, even though a wordless, tribute that this great man will, living or dead, ever receive.

This story, brief and incomplete, of which we gather up the threads as best we may—for William Duncan dwells in this world to work, and not to talk about his work—is one of the most pathetic in history. When one considers the low degree of savagery from which they had struggled up in thirty years of hardest, and at times most discouraging, labor, to a degree of civilization which, in one respect, at least, is reached by few white people in centuries, if ever; when one considers how they had grown to a new faith and to a new form of religious services, to confidence in the possession of homes and other community property, and to believe their title to them to be enduring; when one considers the tenacity of an Indian's attachment to his home and belongings, and his sorrowful and heart-breaking reluctance to part with them—this shadowy, silent migration through northern waters to a new home on an uncleared island, taking almost nothing with them but their religion and their love for Mr. Duncan, becomes one of the sublime tragedies of the century.

On Annette Island, then, twenty years ago, Mr. Duncan's work was taken up anew. Homes were built; a saw-mill, schools, wharf, cannery, store, town hall, a neat cottage for Mr. Duncan, and finally, in 1895, the large and handsome church, rose in rapid succession out of the wilderness. Roads were built, and sidewalks. A trading schooner soon plied the near-by waters. All was the work of the Indians under the direct supervision of Mr. Duncan, who, in 1870, had journeyed to England for the purpose of learning several simple trades which he might, in turn, teach to the Indians whom he fondly calls his "people." Thus personally equipped, and with such implements and machinery as were required, he had returned to his work.

To-day, at the end of twenty years, the voyager approaching Annette Island, beholds rising before his reverent eyes the new Metlakahtla—the old having sunken to ruin, where it lies, a vanishing stain on the fair fame of the Church of England of the past; for the church of to-day is too broad and too enlightened to approve of the action of its Mission Society in regard to its most earnest and successful worker, William Duncan.

The new town shines white against a dark hill. The steamer lands at a good wharf, which is largely occupied by salmon canneries. Sidewalks and neat gravelled paths lead to all parts of the village. The buildings are attractive in their originality, for Mr. Duncan has his own ideas of architecture. The church, adorned with two large square towers, has a commanding situation, and is a modern, steam-heated building, large enough to seat a thousand people, or the entire village. It is of handsome interior finish in natural woods. Above the altar are the following passages: *The angel said unto them: Fear not, for behold, I bring you good tidings of great joy which shall be to all people.... Thou shalt call his name Jesus, for he shall save his people from their sins.*

The cottages are one and two stories in height, and are surrounded by vegetable and flower gardens, of which the women seem to be specially proud. They and the smiling children stand at their gates and on corners and offer for sale baskets and other articles of their own making. These baskets are, without exception, crudely and inartistically made; yet they have a value to collectors by having been woven at Metlakahtla by Mr. Duncan's Indian women, and no tourist fails to purchase at least one, while many return to the steamer laden with them.

There is a girls' school and a boys' school; a hotel, a town hall, several stores, a saw-mill, a system of water-works, a cannery capable of packing twenty thousand cases of salmon in a season, a wharf, and good warehouses and steam-vessels.

The community is governed by a council of thirty members,

having a president. There is a police force of twenty members. Taxes are levied for public improvements, and for the maintenance of public institutions. The land belongs to the community, from which it may be obtained by individuals for the purpose of building homes. The cannery and the saw-mill, which is operated by water, belong to companies in which stock is held by Indians who receive dividends. The employees receive regular wages.

The people seem happy and contented. They are deeply attached to Mr. Duncan, and very proud of their model town. They have an excellent band of twenty-one pieces, at the mere mention of which their dark faces take on an expression of pride and pleasure, and their black eyes shine into their questioner's eyes with intense interest; in fact, if one desires to steady the gaze and hold the attention of a Metlakahtla Indian, he can most readily accomplish his purpose by introducing the subject of the village band.

It is a surprise that these Indians do not, generally, speak English more fluently; but this is coming with the younger generations. Some of these young men and young women have been graduated from Eastern colleges, and have returned to take up missionary work in various parts of Alaska. Meeting one of these young men on a steamer, I asked him if he knew Mr. Duncan. The smile of affection and pride that went across his face! "*I am one of his boys*," he replied, simply. This was the Reverend Edward Marsden, who, returning from an Eastern college in 1898, began missionary work at Saxman, near Juneau, where he has been very successful.

Mr. Duncan is exceedingly modest and unassuming in manner and bearing, seeming to shrink from personal attention, and to desire that his work shall speak for itself. He is frequently called "Father," which is exceedingly distasteful to him. Visitors seeking information are welcome to spend a week or two at the guest-house and learn by observation and by conversation with the people what has been accomplished in this ideal community;

but, save on rare occasions, he cannot be persuaded to dwell upon his own work, and after he has given his reasons for this attitude, only a person lost to all sense of decency and delicacy would urge him to break his rule of silence.

"I am here to work, and not to talk or write about my work," he says, kindly and cordially. "If I took the time to answer one-tenth of the questions I am asked, verbally and by letter, I would have no time left for my work, and my time for work is growing short. I am an old man,"—his beautiful, intensely blue eyes smiled as he said this, and he at once shook his white-crowned head,—"that is what they are saying of me, but it is not true. I am young, I *feel* young, and have many more years of work ahead of me. Still, I must confess that I do not work so easily, and my cares are multiplying. Some to whom I make this explanation will not respect my wishes or understand my silence. They press me by letter, or personally, to answer only this question or only that. They are inconsiderate and hamper me in my work."

Possibly this is the key-note to Mr. Duncan's success. "Here is my work; let it speak for itself." He has devoted his whole life to his work, with no thought for the fame it may bring him. For the latter, he cares nothing.

This is the reason that pilgrims voyage to Metlakahtla as reverently as to a shrine. It is the noble and unselfish life-work of a man who has not only accomplished a great purpose, but who is great in himself. When he passes on, let him be buried simply among the Indians he has loved and to whom he has given his whole life, and write upon his headstone: "Let his work speak."

The settlement on Annette Island was provided for in the act of Congress, 1891, as follows:—

"That, until otherwise provided for by law, the body of lands known as Annette Islands, situated in Alexander Archipelago in southeastern Alaska, on the north side of Dixon Entrance, be, and the same is hereby, set apart as a reservation for the Metlakahtla Indians, and those people known as Metlakahtlans, who have recently emigrated from British Columbia to Alaska, and such other Alaskan natives as may join them, to be held and used by them in common, under such rules and regulations, and subject to such restrictions, as may be prescribed from time to time by the Secretary of the Interior."

The Indians of the Community are required to sign, and to fulfil the terms of, the following Declaration:—

"We, the people of Metlakahtla, Alaska, in order to secure to ourselves and our posterity the blessings of a Christian home, do severally subscribe to the following rules for the regulation of our conduct and town affairs:—

"To reverence the Sabbath and to refrain from all unnecessary secular work on that day; to attend divine worship; to take the Bible for our rule of faith; to regard all true Christians as our brethren; and to be truthful, honest, and industrious.

"To be faithful and loyal to the Government and laws of the United States.

"To render our votes when called upon for the election of the Town Council, and to promptly obey the by-laws and orders imposed by the said Council.

"To attend to the education of our children and keep them at school as regularly as possible.

"To totally abstain from all intoxicants and gambling, and never attend heathen festivities or countenance heathenish customs in surrounding villages.

"To strictly carry out all sanitary regulations necessary for the health of the town.

"To identify ourselves with the progress of the settlement, and to utilize the land we hold.

"Never to alienate, give away, or sell our land, or any portion thereof, to any person or persons who have not subscribed to these rules."

CHAPTER V

Dixon Entrance belongs to British Columbia, but the boundary crosses its northern waters about three miles above Whitby Point on Dundas Island, and the steamer approaches Revilla-Gigedo Island. It is twenty-five by fifty miles, and was named by Vancouver in honor of the Viceroy of New Spain, who sent out several of the most successful expeditions. It is pooled by many bits of turquoise water which can scarcely be dignified by the name of lakes.

Carroll Inlet cleaves it half in twain. The exquisite gorges and mountains of this island are coming to their own very slowly, as compared with its attractions from a commercial point of view.

The island is in the centre of a rich salmon district, and during the "running" season the clear blue waters flash underneath with the glistening silver of the struggling fish. In some of the fresh-water streams where the hump-backed salmon spawn, the fortunate tourist may literally make true the frequent Western assertion that at certain times "one can walk across on the solid silver bridge made by the salmon"—so tightly are they wedged together in their desperate and pathetic struggles to reach the spawning-ground.

Vancouver found these "hunch-backs," as he called them, not to his liking,—probably on account of finding them at the spawning season.

Leaving Ketchikan, Revilla and Point Higgins are passed to starboard—Higgins being another of Vancouver's choice namings for the president of Chile.

"Did you ever see such a cluttering up of a landscape with odds and ends of names?" said the pilot one day. "And all the ugliest by Vancouver. Give *me* an Indian name every time. It always means something. Take this Revilly-Gig Island; the

Indians called it 'Na-a,' meaning 'the far lakes,' for all the little lakes scattered around. I don't know as we're doing much better in our own day, though," he added, staring ahead with a twinkle in his eyes. "They've just named a couple of mountains *Mount Thomas Whitten* and *Mount Shoup*! Now those names are all right for men—even congressmen—but they're not worth shucks for mountains. Why, the Russians could do better! Take Mount St. Elias—named by Behring because he discovered it on St. Elias' day. I actually tremble every time I pass that mountain, for fear I'll look up and see a sign tacked on it, stating that the name has been changed to Baker or Bacon or Mudge, so that Vancouver's bones will rest more easily in the grave. Now look at that point! It's pretty enough in itself; but—*Higgins!*"

The next feature of interest, however, proved to be blessed with a name sweet enough to take away the bitterness of many others—Clover Pass. It was not named for this most fragrant and dear of all flowers, but for Lieutenant, now Rear-Admiral, Clover, of the United States Navy.

Beyond Clover Pass, at the entrance to Naha Bay, is Loring, a large and important cannery settlement of the Alaska Packers' Association. There is only one salmon-canning establishment in Alaska, or even on the Northwest Coast, more picturesquely situated than this, and it is nearly two thousand miles "to Westward," at the mouth of the famed Karluk River, where the same company maintains large canneries and successful hatcheries. It will be described in another chapter.

A trail leads from Loring through the woods to Dorr Waterfall, in a lovely glen. In Naha Bay thousands of fish are taken at every dip of the seine in the narrowest cove, which is connected with a chain of small lakes linked by the tiniest of streams. In summer these waters seem to be of living silver, so thickly are they swarmed with darting and curving salmon.

Not far from Naha Bay is Traitor's Cove, where Vancouver and his men were attacked in boats by savages in the masks of animals, headed by an old hag who commanded and urged them to bloodthirsty deeds.

This vixen seemed to be a personage of prestige and influence, judging both by the immense size of her lip ornament and her air of command. She seized the lead line from Vancouver's boat and made it fast to her own canoe, while another stole a musket.

Vancouver, advancing to parley with the chief, made the mistake of carrying his musket; whereupon about fifty savages leaped at him, armed with spears and daggers.

The chief gave him to understand by signs that they would lay down their arms if he would set the example; but the terrible old woman, scenting peace and scorning it, violently and turbulently harangued the tribe and urged it to attack.

The brandishing of spears and the flourishing of daggers became so uncomfortably close and insistent, that Vancouver finally overcame his "humanity," and fired into the canoes.

The effect was electrical. The Indians in the small canoes instantly leaped into the water and swam for the shore; those in the larger ones tipped the canoes to one side, so that the higher side shielded them while they made the best of their way to the shore.

There they ascended the rocky cliffs and stoned the boats. Several of Vancouver's men were severely wounded, one having been speared completely through the thigh.

The point at the northern entrance to Naha Bay, where they landed to dress wounds and take account of stock not stolen, was named Escape Point; a name which it still retains.

Kasa-an Bay is an inlet pushing fifteen miles into the eastern coast of Prince of Wales Island, which is two hundred miles in length and averages forty in width. Cholmondeley Sound penetrates almost as far, and Moira Sound, Niblack Anchorage on North Arm, Twelve Mile Arm, and Skowl Arm, are all storied and lovely inlets. Skowl was an old chief of the Eagle Clan, whose sway was questioned by none. He was the greatest chief of his time, and ruled his people as autocratically as the lordly, but blustering, Baranoff ruled his at Sitka. Skowl repulsed the advances of missionaries and scorned all attempts at Christianizing himself and his tribe. His was a powerful personality which is still mentioned with a respect not unmixed with awe. To say that a chief is as fearless as Skowl is a fine compliment, indeed, and one not often bestowed.

Although not on the regular run of steamers, Howkan, now a Presbyterian missionary village on Cordova Bay, on the southwestern part of Prince of Wales Island, must not be entirely neglected. In early days the village was a forest of totems, and the graves were almost as interesting as the totems. Both are rapidly vanishing and losing their most picturesque features before the march of civilization and Christianity; but Howkan is still one of the show-places of Alaska. The tourist who is able to make this side trip on one of the small steamers that run past there, is the envy of the unfortunate ones who are compelled to forego that pleasure.

Steel Cantilever Bridge, near Summit of White Pass

Totemism is the poetry of the Indian—or would be if it possessed any religious significance.

I once asked an educated Tsimpsian Indian what the Metlakahtla people believed,—meaning the belief that Mr. Duncan had taught them. He put the tips of his fingers together, and with an expression of great earnestness, replied:—

"They believed in a great Spirit, to whom they prayed and whom they worshipped everywhere, believing that this beautiful Spirit was everywhere and could hear. They worshipped it in the forest, in the trees, in the flowers, in the sun and wind, in the blades of grass,—alone and far from every one,—in the running water and the still lakes."

"Oh, how beautiful!" I said, in all sincerity. "It must be the same as my own belief; only I never heard it put into words before. And that is what Mr. Duncan has taught them?"

He turned and looked at me squarely and steadily. It was a look of weariness, of disgust.

"Oh, no," he replied, coldly; "that was what they believed before they knew better; before they were taught the truth; before Christianity was explained to them. That is what they believed *while they were savages*!"

We were in the library of the *Jefferson*. The room is always warm, and at that moment it was warmer than I had ever known it to be. Under the steady gaze of those shining dark eyes it presently became too warm to be endured. With my curiosity quite satisfied, I withdrew to the hurricane deck, where there is always air.

Of the Indians in the territory of Alaska there are two stocks—the Thlinkits, or Coast Indians, and the Tinneh, or those inhabiting the vast regions of the interior. The Thlinkits comprise the Tsimpsians, or Chimsyans, the Kygáni, or Haidahs, the true Thlinkits, or Koloshes, and the Yakutats.

The Kygáni, or Haidah, Indians inhabit the Queen Charlotte Archipelago, which, although belonging to British Columbia, must be taken into consideration in any description of the

Indians of Alaska. They were formerly a warlike, powerful, and treacherous race, making frequent attacks upon neighboring tribes, even as far south as Puget Sound. They are noted, not only for these savage qualities, but also for the grace and beauty of their canoes and for their delicate and artistic carvings. Their small totems, pipes, and other articles carved out of a dark gray, highly polished slate stone obtained on their own islands, sometimes inlaid with particles of shell, are well known and command fancy prices. Haidah basketry and hats are of unusual beauty and workmanship. The peculiar ornamentation is painted upon the hats and not woven in. The designs which are most frequently seen are the head, wings, tail, and feet of a duck,—certain details somewhat resembling a large oyster-shell, or a human ear,—painted in black and rich reds. The hats are usually in the plain twined weaving, and of such fine, even workmanship that they are entirely waterproof. The Haidahs formerly wore the nose- and ear-rings, or other ornaments, and the labret in the lower lip.

The Thlinkits,—or Koloshians, as the Russians and Aleuts called them, from their habit of wearing the labret,—are divided into two tribes, the Stikines and the Sitkans; the former inhabiting the mainland in the vicinity of the Stikine River, straggling north and south for some distance along the coast.

The Sitkans dwell in the neighborhood of Sitka and on the near-by islands. They are among the tribes of Indians who gave Baranoff much trouble. They formerly painted with vermilion or lamp-black mixed with oil, traced on their faces in startling patterns. At the present time they dress almost like white people, except for the everlasting blanket on the older ones. Some of the younger women are very handsome—clean, light-brown of skin, red-cheeked, of good figure, and having large, dark eyes, at once soft and bright. They also have good, white teeth, and are decidedly attractive in their coquettish and saucy airs and graces. The young Indian women at Sitka, Yakutat, and Dundas are the prettiest and the most attractive in Alaska; nor have I seen any in the Klondike, or along the Yukon, to equal them in appearance. Also, one can barter with them for their fascinating wares without praying to heaven to be deprived of the sense of smell for a sufficient number of hours.

Among the Thlinkits, as well as among many of the Innuit, or Eskimo tribes, the strange and cruel custom prevails of isolating young girls approaching puberty in a hut set aside for this purpose. The period of isolation varies from a month to a year, during which they are considered unclean and are allowed only liquid food, which soon reduces them to a state of painful emaciation. No one is permitted to minister to their needs but a mother or a female slave, and they cannot hold conversation with any one.

When a maiden finally emerges from her confinement there is great rejoicing, if she be of good family, and feasting. A charm of peculiar design is hung around her neck, called a "Virgin Charm," or "Virtue Charm," which silently announces that she is "clean" and of marriageable age. Formerly, according to Dall and other authorities, the lower lip was pierced and a silver pin shaped like a nail inserted. This made the same announcement.

The chief diet of the Thlinkit is fish, fresh or smoked. Unlike the Aleutians, they do not eat whale blubber, as the whale figures in their totems, but are fond of the porpoise and seal. The women are fond of dress, and a voyager who will take a gay last year's useless hat along in her steamer trunk, will be sure to "swap" it for a handsome Indian basket. In many places they still employ their early methods of fishing—raking herring and salmon out of the streams, during a run, with long poles into which nails are driven, like a rake.

They are fond of game of all kinds. They weave blankets out of the wool of the mountain sheep. Large spoons, whose handles are carved in the form and designs of totems, are made out of the horns of sheep and goats.

The Thlinkits are divided into four totems—the whale, the eagle, the raven, and the wolf. The raven, which by the Tinnehs is considered an evil bird, is held in the highest respect by the Thlinkits, who believe it to be a good spirit.

Totemism is defined as the system of dividing a tribe into clans according to their totems. It comprises a class of objects which the savage holds in superstitious awe and respect, believing that it holds some relation to, and protection over, himself. There is the clan totem, common to a whole clan; the sex totem, common to the males or females of a clan; and the individual totem, belonging solely to one person and not descending to any member of the next generation. It is generally believed that the totem has some special religious significance; but this is not true, if we are to believe that the younger and educated Indians of to-day know what totemism means. Some totems are veritable family trees. The clan totem is reverenced by a whole clan, the members of which are known by the name of their totem, and believe themselves to be descended from a common animal ancestor, and bound together by ties closer and more sacred than those of blood.

Old Russian Building, Sitka

The system of totemism is old; but the word itself, according to J. G. Frazer, first appeared in literature in the nineteenth century, being introduced from an Ojibway word by J. Long, an interpreter. The same authority claims that it had a religious aspect; but this is denied, so far, at least, as the Thlinkits are concerned.

The Eagle clan believe themselves to be descended from an eagle, which they, accordingly, reverence and protect from harm or death, believing that it is a beneficent spirit that watches over them.

Persons of the same totem may neither marry nor have sexual intercourse with each other. In Australia the usual penalty for the breaking of this law was death. With the Thlinkits, a man might marry a woman of any save his own totem clan. The raven represented woman, and the wolf, man. A young man selected his individual totem from the animal which appeared most frequently and significantly in his dreams during his lonely fast and vigil in the heart of the forest for some time before reaching the state of puberty. The animals representing a man's different totems—clan, family, sex, and individual—were carved and painted on his tall totem-pole, his house, his paddles, and other objects; they were also woven into hats, basketry, and blankets, and embroidered upon moccasins with beads. Some of the Haidah canoes have most beautifully carven and painted prows, with the totem design appearing. These canoes are far superior to those of Puget Sound. The very sweep of the prow, strong and graceful, as it cleaves the golden air above the water, proclaims its northern home. Their well-known outlines, the erect, rigid figures of the warriors kneeling in them, and the strong, swift, sure dip of the paddles, sent dread to the hearts of the Puget Sound Indians and the few white settlers in the early part of the last century. The cry of "Northern Indians!" never failed to create a panic. They made many marauding expeditions to the south in their large and splendid canoes. The inferior tribes of the sound held them in the greatest fear and awe.

A child usually adopts the mother's totem, and at birth receives a name significant of her family. Later on he receives one from his father's family, and this event is always attended with much solemnity and ceremony.

A man takes wives in proportion to his wealth. If he be the possessor of many blankets, he takes trouble unto himself by the dozen. There are no spring bonnets, however, to buy. They do not indulge themselves with so many wives as formerly; nor do they place such implicit faith in the totem, now that they are becoming "Christianized."

Dall gives the following interesting description of a Thlinkit wedding ceremony thirty years ago: A lover sends to his mistress's relations, asking for her as a wife. If he receives a favorable reply, he sends as many presents as he can get together to her father. On the appointed day he goes to the house where she lives, and sits down with his back to the door.

The father has invited all the relations, who now raise a song, to allure the coy bride out of the corner where she has been sitting. When the song is done, furs or pieces of new calico are laid on the floor, and she walks over them and sits down by the side of the groom. All this time she must keep her head bowed down. Then all the guests dance and sing, diversifying the entertainment, when tired, by eating. The pair do not join in any of the ceremonies. That their future life may be happy, they fast for two days more. Four weeks afterward they come together, and are then recognized as husband and wife.

The bridegroom is free to live with his father-in-law, or return to his own home. If he chooses the latter the bride receives a *trousseau* equal in value to the gifts received by her parents from her husband. If the husband becomes dissatisfied with his wife, he can send her back with her dowry, but loses his own gifts. If a wife is unfaithful he may send her back with nothing, and demand his own again. They may separate by mutual consent without returning any property. When the marriage festival is over, the silver pin is removed from the lower lip of the bride and replaced by a plug, shaped like a spool, but not over three-quarters of an inch long, and this plug is afterward replaced by a larger one of wood, bone, or stone, so that an old woman may have an ornament of this kind two inches in diameter. These large ones are of an oval shape, but scooped out above, below, and around the edge, like a pulley-wheel. When very large, a mere strip of flesh goes around the *kalúshka*, or "little trough." From the name which the Aleuts gave the appendage when they first visited Sitka, the nickname "Kolosh" has arisen, and has been applied to this and allied tribes.

Many years ago, when a man died, his brother or his sister's son was compelled to marry the widow.

That seems worth while. Naturally, the man would not desire the woman, and the woman would not desire the man; therefore, the result of the forced union might prove full of delightful surprises. If such a law could have been passed in England, there would have been no occasion for the prolonged agitation over the "Deceased wife's sister" bill, which dragged its weary way through the courts and the papers. Nobody would desire to marry his deceased wife's sister; or, if he did, she would decline the honor.

An ancient Thlinkit superstition is, that once a man—a Thlinkit, of course—had a young wife whom he so idolized that he would not permit her to work. This is certainly the most convincing proof that an Indian could give of his devotion. From morning to night she dwelt in sweet idleness, guarded by eight little redbirds, that flew about her when she walked, or hovered over her when she reclined upon her furs or preciously woven blankets.

These little birds were good spirits, of course, but alas! they resembled somewhat women who are so good that out of their very goodness evil is wrought. In the town in which I dwell there is a good woman, a member of a church, devout, and scorning sin, who keeps "roomers." On two or three occasions this good woman has found letters which belonged to her roomers, and she has done what an honorable woman would not do. She has read letters that she had no right to read, and she has found therein secrets that would wreck families and bow down heads in sorrow to their graves; and yet, out of her goodness, she has felt it to be her duty "to tell," and she has told.

Since knowing the story of the eight little Thlinkit redbirds, I have never seen this woman without a red mist seeming to float round her; her mouth becomes a twittering beak, her feet are claws that carry her noiselessly into secret places, her eyes are little black beads that flash from side to side in search of other people's sins, and her shoulders are folded wings. For what did the little good redbirds do but go and tell the Thlinkit man that his young and pretty and idolized wife had spoken to another man. He took her out into the forest and shut her up in a box. Then he killed all his sister's children because they knew his secret. His sister went in lamentations to the beach, where she was seen by her totem whale, who, when her cause of grief was made known to him, bade her be of good cheer.

"Swallow a small stone," said the whale, "which you must pick up from the beach, drinking some sea-water at the same time."

The woman did as the whale directed. In a few months she gave birth to a son, whom she was compelled to hide from her brother. This child was Yehl (the raven), the beneficent spirit of the Thlinkits, maker of forests, mountains, rivers, and seas; the one who guides the sun, moon, and stars, and controls the winds and floods. His abiding-place is at the head waters of the Nass River, whence the Thlinkits came to their present home. When he grew up he became so expert in the use of the bow and arrow that it is told of his mother that she went clad in the rose, green, and lavender glory of the breasts of humming-birds which he had killed in such numbers that she was able to fashion her entire raiment of their most exquisite parts,—as befitted the mother of the good spirit of men.

Yehl performed many noble and miraculous deeds, the most dazzling of which was the giving of light to the world. He had heard that a rich old chief kept the sun, moon, and stars in boxes, carefully locked and guarded. This chief had an only daughter whom he worshipped. He would allow no one to make love to her, so Yehl, perceiving that only a descendant of the old man could secure access to the boxes, and knowing that the chief examined all his daughter's food before she ate it, and that it would therefore avail him nothing to turn himself into ordinary food, conceived the idea of converting himself into a fragrant grass and by springing up persistently in the maiden's path, he was one day eaten and swallowed. A grandson was then born to the old chief, who wrought upon his affections—as grandsons have a way of doing—to such an extent that he could deny him nothing.

One day the young Yehl, who seems to have been appropriately named, set up a lamentation for the boxes he desired and continued it until one was in his possession. He took it outdoors and opened it. Millions of little milk-white, opaline birds instantly flew up and settled in the sky. They were followed by a large, silvery bird, which was so heavy and uncertain in her flight to the sky that, although she finally reached it, she never appeared twice the same thereafter, and on some nights could not be seen at all. The old chief was very angry, and it was not until Yehl had wept and fasted himself to death's very door that he obtained the sun; whereupon, he changed himself back into a raven, and flying away from the reach of his stunned and temporary grandfather, who had commanded him not to open the box, he straightway lifted the lid—and the world was flooded with light.

One of the most interesting of the Thlinkit myths is the one of the spirits that guard and obey the shamans. The most important are those dwelling in the North. They were warriors; hence, an unusual display of the northern lights was considered an omen of approaching war. The other spirits are of people who died a commonplace death; and the greatest care must be exercised by relatives in mourning for these, or they will have difficulty in reaching their new abode. Too many tears are as bad as none at all; the former mistake mires and gutters the path, the latter leaves it too deep in dust. A decent and comfortable quantity makes it hard and even and pleasant.

Their deluge myth is startling in its resemblance to ours. When their flood came upon them, a few were saved in a great canoe which was made of cedar. This wood splits rather easily, parallel to its grain, under stress of storm, and the one in which the people embarked split after much buffeting. The Thlinkits clung to one part, and all other peoples to the other part, creating a difference in language. Chet'l, the eagle, was separated from his sister, to whom he said, "You may never see me again, but you shall hear my voice forever." He changed himself into a bird of tremendous size and flew away southward. The sister climbed Mount Edgecumbe, which opened and swallowed her, leaving a hole that has remained ever since. Earthquakes are caused by her struggles with bad spirits which seek to drive her away, and by her invariable triumph over them she sustains the poise of the world.

Chet'l returned to Mount Edgecumbe, where he still lives. When he comes forth, which is but seldom, the flapping of his great wings produces the sound which is called thunder. He is, therefore, known everywhere as the Thunder-bird. The glance of his brilliant eyes is the lightning.

Concerning the totem-pole which was taken from an Indian village on Tongas Island, near Ketchikan, by members of the *Post-Intelligencer* business men's excursion to Alaska in 1899—and for which the city of Seattle was legally compelled to pay handsomely afterward—the following letter from a member of the family originally owning the totem is of quaint interest:—

"I have received your letter, and I am going to tell you the story of the totem-pole. Now, the top one is a crow himself, and the next one from the pole top is a man. That crow have told him a story. Crow have told him a good-looking woman want to married some man. So he did marry her. She was a frog. And the fourth one is a mink. One time, the story says, that one time it was a high tide for some time, and so crow got marry to mink, so crow he eats any kind of fishes from the water. After some time crow got tired of mink, and he leave her, and he get married to that whale-killer, and then crow he have all he want to eat. That last one on the totem-pole is the father of the crow. The story says that one time it got dark for a long while. The darkness was all over the world, and only crow's father was the only one can give light to the world. He simply got a key. He keeps the sun and moon in a chest, that one time crow have ask his father if he play with the sun and moon in the house but, was not allowed, so he start crying for many days until he was sick. So his father let him play with it and he have it for many days. And one day he let the moon in the sky by mistake, but he keep the sun, and he which take time before he could get his chances to go outside of the house. As soon as he was out he let sun back to the sky again, and it was light all over the world again. (End of story.)

"Yours respectfully,

"David E. Kinninnook.

"P.S. The Indians have a long story, and one of the chiefs of a village or of a tribe only a chief can put up so many carvings on our totem-pole, and he have to fully know the story of what totem he is made. I may give you the whole story of it sometimes. Crow on top have a quart moon in his mouth, because he have ask his father for a light.

"D. E. K.

"If you can put this story on the *Post-Intelligencer*, of Seattle, Wash., and I think the people will be glad to know some of it."

The Thlinkits burned their dead, with the exception of the shamans, but carefully preserved the ashes and all charred bones from the funeral pyre. These were carefully folded in new blankets and buried in the backs of totems. One totem, when taken down to send to the Lewis and Clark Exposition, was found to contain the remains of a child in the butt-end of the pole which was in the ground; the portion containing the child being sawed off and reinterred.

Greek-Russian Church at Sitka

A totem-pole donated to the exposition by Yannate, a very old Thlinkit, was made by his own hands in honor of his mother. His mother belonged to the Raven Clan, and a large raven is at the crest of the pole; under it is the brown bear—the totem of the Kokwonton Tribe, to which the woman's husband belonged; underneath the bear is an Indian with a cane, representing the woman's brother, who was a noted shaman or sorcerer many years ago; at the bottom are two faces, or masks, representing the shaman's favorite slaves.

The Haidahs did not burn their dead, but buried them, usually in the butts of great cedars. Frequently, however, they were buried at the base of totem-poles, and when in recent years poles have been removed, remains have been found and reinterred.

On the backs of some of the old totem-poles at Wrangell and other places, may be seen the openings that were made to receive the ashes of the dead, the portion that had been sawed out being afterward replaced.

The wealth of a Thlinkit is estimated according to his number of blankets; his honor and importance by the number of potlatches he has given. Every member of his totem is called upon to contribute to the potlatch of the chief, working to that end, and "skimping" himself in his own indulgences for that object, for many years, if necessary. The potlatch is given at the full of the moon; the chief's clan and totem decline all gifts; it is not in good form for any member thereof to accept the slightest gift. Guests are seated and treated according to their rights, and the resentment of a slight is not postponed until the banquet is over and the blood has cooled. An immediate fight to the bitter end is the result; so that the greatest care is exercised in this nice matter—which has proven a pitfall to many a white hostess in the most civilized lands; so seldom does a guest have the right and the honor to feel that where he sits is the head of the table. At these potlatches a "frenzied" hospitâlity prevails; everything is bestowed with a lavish and reckless hand upon the visitors, from food and drink to the host's most precious possession, blankets. His wives are given freely, and without the pang which must go with every blanket. Visitors come and remain for days, or until the host is absolutely beggared and has nothing more to give.

But since every one accepting his potlatch is not only expected, but actually bound by tribal laws as fixed as the stars, to return it, the beggared chief gradually "stocks up" again; and in a few years is able to launch forth brilliantly once more. This is the same system of give and take that prevails in polite society in the matter of party-giving. With neither, may the custom be considered as real hospitality, but simply a giving with the expectation of a sure return. Chiefs have frequently, however, given away fortunes of many thousands of dollars within a few days. These were chiefs who aspired to rise high above their contemporaries in glory; and, therefore, would be disappointed to have their generosity equally returned.

A shaman is a medicine-man who is popularly supposed to be possessed of supernatural powers. A certain mystery, or mysticism, is connected with him. He spends much time in the solitudes of the mountains, working himself into a highly emotional mental state. The shaman has his special masks, carved ivory diagnosis-sticks, and other paraphernalia. The hair of the shaman was never cut; at his death, his body was not burned, but was invariably placed in a box on four high posts. It first reposed for one whole night in each of the four corners of the house in which he died. On the fifth day it was laid to rest by the sea-shore; and every time a Thlinkit passed it, he tossed a small offering into the water, tosecure the favor of the dead shaman, who, even in death, was believed to exercise an influence over the living, for good or ill.

Slavery was common, as—until the coming of the Russians—was cannibalism. The slaves were captives from other tribes. They were forced to perform the most disagreeable duties, and were subjected to cruel treatment, punished for trivial faults, and frequently tortured, or offered in sacrifice. A few very old slaves are said to be in existence at the present time; but they are now treated kindly, and have almost forgotten that their condition is inferior to that of the remainder of the tribe.

The most famous slaves on the Northwest Coast were John Jewitt and John Thompson, sole survivors of the crew of the *Boston*, which was captured in 1802 by the Indians of Nootka Sound, on the western coast of Vancouver Island. The officers and all the other men were most foully murdered, and the ship was burned.

Jewitt and Thompson were spared because one was an armorer and the other a sailmaker. They were held as slaves for nearly three years, when they made their escape.

Jewitt published a book, in which he simply and effectively described many of the curious, cruel, and amusing customs of the people. The two men finally made their escape upon a boat which had appeared unexpectedly in the harbor.

The Yakutats belong to the Thlinkit stock, but have never worn the "little trough," the distinguishing mark of the true Thlinkit. They inhabit the country between Mount Fairweather and Mount St. Elias, and were the cause of much trouble and disaster to Baranoff, Lisiansky, and other early Russians. They have never adopted the totem; and may, therefore, eat the flesh and blubber of the whale, which the Thlinkits respect, because it figures on their totems. The graveyards of the Yakutats are very picturesque and interesting.

The tribes of the Tinneh, or interior Indians, will be considered in another chapter.

Behm Canal is narrow, abruptly shored, and offers many charming vistas that unfold unexpectedly before the tourist's eyes. Alaskan steamers do not enter it and, therefore, New Eddystone Rock is missed by many. This is a rocky pillar that rises straight from the water, with a circumference of about one hundred feet at the base and a height of from two to three hundred feet. It is draped gracefully with mosses, ferns, and vines. Vancouver breakfasted here, and named it for the famous Eddystone Light of England. Unuk River empties its foaming, glacial waters into Behm Canal.

CHAPTER VI

Leaving Ketchikan, Clarence Strait is entered. This was named by Vancouver for the Duke of Clarence, and extends in a northwesterly direction for a hundred miles. The celebrated Stikine River empties into it. On Wrangell Island, near the mouth of the Stikine, is Fort Wrangell, where the steamer makes a stop of several hours.

Fort Wrangell was the first settlement made in southeastern Alaska, after Sitka. It was established in 1834, by Lieutenant Zarembo, who acted under the orders of Baron Wrangell, Governor of the Colonies at that time.

A grave situation had arisen over a dispute between the Russian American Company and the equally powerful Hudson Bay Company, the latter having pressed its operations over the Northwest and seriously undermined the trade of the former. In 1825, the Hudson Bay Company had taken advantage of the clause in the Anglo-Russian treaty of that year,—which provided for the free navigation of streams crossing Russian territory in their course from the British possessions to the sea,—and had pushed its trading operations to the upper waters of the Stikine, and in 1833 had outfitted the brig *Dryad* with colonists, cattle, and arms for the establishing of trading posts on the Stikine.

Lieutenant Zarembo, with two armed vessels, the *Chichagoff* and the *Chilkaht*, established a fort on a small peninsula, on the site of an Indian village, and named it Redoubt St. Dionysius. All unaware of these significant movements, the *Dryad*, approaching the mouth of the Stikine, was received by shots from the shore, as well as from a vessel in the harbor. She at once put back until out of range, and anchored. Lieutenant Zarembo went out in a boat, and, in the name of the Governor and the Emperor, forbade the entrance of a British vessel into the river. Representations from the agents of the Hudson Bay Company were unavailing; they were warned to at once remove themselves and their vessel from the vicinity—which they accordingly did.

This affair was the cause of serious trouble between the two nations, which was not settled until 1839, when a commission met in London and solved the difficulties by deciding that Russia should pay an indemnity of twenty thousand pounds, and lease to the Hudson Bay Company the now celebrated *lisière*, or thirty-mile strip from Dixon Entrance to Yakutat.

In 1840 the Hudson Bay Company raised the British flag and changed the name from Redoubt St. Dionysius to Fort Stikine. Sir George Simpson's men are said to have passed several years of most exciting and adventurous life there, owing to the attacks and besiegements of the neighboring Indians. An attempt to scale the stockade resulted in failure and defeat. The following year the fort's supply of water was cut off and the fort was besieged; but the Britishers saved themselves by luckily seizing a chief as hostage.

A year later occurred another attack, in which the fort would have fallen had it not been for the happy arrival of two armed vessels in charge of Sir George Simpson, who tells the story in this brief and simple fashion:—

"By daybreak on Monday, the 25th of April (1842), we were in Wrangell's Straits, and toward evening, as we approached Stikine, my apprehensions were awakened byobserving the two national flags, the Russian and the English, hoisted half-mast high, while, on landing about seven, my worst fears were realized by hearing of the tragical end of Mr. John McLoughlin, Jr., the gentleman recently in charge. On the night of the twentieth a dispute had arisen in the fort, while some of the men, as I was grieved to hear, were in a state of intoxication; and several shots were fired, by one of which Mr. McLoughlin fell. My arrival at this critical juncture was most opportune, for otherwise the fort might have fallen a sacrifice to the savages, who were assembled round to the number of two thousand, justly thinking that the place could make but a feeble resistance, deprived as it was of its head, and garrisoned by men in a state of complete insubordination."

In 1867 a United States military post was established on a new site. A large stockade was erected and garrisoned by two companies of the Twenty-first Infantry. This post was abandoned in 1870, the buildings being sold for six hundred dollars.

In the early eighties Lieutenant Schwatka found Wrangell "the most tumble-down-looking company of cabins I ever saw." He found its "Chinatown" housed in an old Stikine River steamboat on the beach, which had descended to its low estate as gradually and almost as imperceptibly as Becky Sharpe descended to the "soiled white petticoat" condition of life. As Queen of the Stikine, the old steamer had earned several fortunes for her owners in that river's heyday times; then she was beached and used as a store; then, as a hotel; and, last of all, as a Chinese mess- and lodging-house.

In 1838 another attempt had been made by the Hudson Bay Company to establish a trading post at Dease Lake, about sixty miles from Stikine River and a hundred and fifty from the sea. This attempt also was a failure. The tortures of fear and starvation were vividly described by Mr. Robert Campbell, who had charge of the party making the attempt, which consisted of four men.

"We passed a winter of constant dread from the savage Russian Indians, and of much suffering from starvation. We were dependent for subsistence on what animals we could catch, and, failing that, on *tripe de roche* (moss). We were at one time reduced to such dire straits that we were obliged to eat our parchment windows, and our last meal before abandoning Dease Lake, on the eighth of May, 1839, consisted of the lacings of our snow-shoes."

Had it not been for the kindness and the hospitality of the female chief of the Nahany tribe of Indians, who inhabited the region, the party would have perished.

The Indians of the coast in early days made long trading

excursions into the interior, to obtain furs.

The discovery of the Cassiar mines, at the head of the Stikine, was responsible for the revival of excitement and lawlessness in Fort Wrangell, as it had been named at the time of its first military occupation, and a company of the Fourth Artillery was placed in charge until 1877, the date of the removal of troops from all posts in Alaska.

The first post and the ground upon which it stood were sold to W. K. Lear. The next company occupied it at a very small rental, contrary to the wishes of the owner. In 1884 the Treasury Department took possession, claiming that the first sale was illegal. A deputy collector was placed in charge. The case was taken into the courts, but it was not until 1890 that a decision was rendered in the Sitka court that, as the first sale was unconstitutional, Mr. Lear was entitled to his six hundred dollars with interest compounding for twenty years.

Wrangell gradually fell into a storied and picturesque decay. The burnished halo of early romance has always clung to her. At the time of the gold excitement and the rush to the Klondike, the town revived suddenly with the reopening of navigation on the Stikine. This was, at first, a favorite route to the Klondike. At White Horse may to-day be seen steamers which were built on the Stikine in 1898, floated by piecemeal up that river and across Lake Teslin, and down the Hootalinqua River to the Yukon, having been packed by horses the many intervening miles between rivers and lakes, at fifty cents a pound. Reaching their destination at White Horse, they were put together, and started on the Dawson run.

Looking at these historic steamers, now lying idle at White Horse, the passenger and freight rates do not seem so exorbitant as they do before one comes to understand the tremendous difficulties of securing any transportation at all in these unknown and largely unexplored regions in so short a time. Even a person who owns no stock in steamship or railway corporations, if he be sensible and reasonable, must be able to see the point of view of the men who dauntlessly face such hardships and perils to furnish transportation in these wild and inaccessible places. They take such desperate chances neither for their health nor for sweet charity's sake.

Three years ago Wrangell was largely destroyed by fire. It is partially rebuilt, but the visitor to-day is doomed to disappointment at first sight of the modern frontier buildings. Ruins of the old fort, however, remain, and several ancient totems are in the direction of the old burial ground. One, standing in front of a modern cottage which has been erected on the site of the old lodge, is all sprouted out in green. Mosses, grasses, and ferns spring in April freshness out of the eyes of children, the beaks of eagles, and the open mouths of frogs; while the very crest of the totem is crowned a foot or more high with a green growth. The effect is at once ludicrous and pathetic,—marking, as it does, the vanishing of a picturesque and interesting race, its customs and its superstitions.

The famous chief of the Stikine region was Shakes, a fierce, fighting, bloodthirsty old autocrat, dreaded by all other tribes, and insulted with impunity by none. He was at the height of his power in the forties, but lived for many years afterward, resisting the advances of missionaries and scorning their religion to the day of his death. In many respects he was like the equally famous Skowl of Kasa-an, who went to the trouble and the expense of erecting a totem-pole for the sole purpose of perpetuating his scorn and derision of Christian advances to his people. The totem is said to have been covered with the images of priests, angels, and books.

Shakes was given one of the most brilliant funerals ever held in Alaska; but whether as an expression of irreconcilable grief or of uncontrollable joy in the escape of his people from his tyrannic and overbearing sway, is not known. He belonged to the bear totem, and a stuffed bear figured in the pageant and was left to guard his grave.

The climate of Wrangell is charming, owing to the high

mountains on the islands to the westward which shelter the town from the severity of the ocean storms. The growing of vegetables and berries is a profitable investment, both reaching enormous size, the latter being of specially delicate flavor. Flowers bloom luxuriantly.

The Wrangell shops at present contain some very fine specimens of basketry, and the prices were very reasonable, although most of the tourists from our steamer were speechless when they heard them. Some real Attu and Atka baskets were found here at prices ranging from one hundred dollars up. At Wrangell, therefore, the tourist begins to part with his money, and does not cease until he has reached Skaguay to the northward, or Sitka and Yakutat to the westward; and if he should journey out into the Aleutian Isles, he may borrow money to get home. The weave displayed is mostly twined, but some fine specimens of coiled and coiled imbricated were offered us in the dull, fascinating colors used by the Thompson River Indians of British Columbia, having probably been obtained in trade. These latter are treasures, and always worth buying, especially as Indian baskets are increasing in value with every year that passes. Baskets that I purchased easily for three dollars or three and a half in 1905 were held stubbornly at seven and a half or eight in 1907; while the difference in prices of the more expensive ones was even greater.

Squaws sit picturesquely about the streets, clad in gay colors, with their wares spread out on the sidewalk in front of them. They invariably sit with their backs against buildings or fences, seeming to have an aversion to permitting any one to stand or pass behind them. They have grown very clever at bargaining; and the little trick, which has been practised by tourists for years, of waiting until the gangway is being hauled in and then making an offer for a coveted basket, has apparently been worn threadbare, and is received with jeers and derision,—which is rather discomfiting to the person making the offer if he chances to be upon a crowded steamer. The squaws point their fingers at him, to shame him, and chuckle and tee-hee among themselves, with many guttural cluckings and side-glances so good-naturedly contemptuous and derisive as to be embarrassing beyond words,—particularly as some greatly desired basket disappears into a filthy bag and is borne proudly away on a scornful dark shoulder.

Baskets are growing scarcer and more valuable, and the tourist who sees one that he desires, will be wise to pay the price demanded for it, as the conditions of trading with the Alaskan Indians are rapidly changing. The younger Indians frequently speak and understand English perfectly; while the older ones are adepts in reading a human face; making a combination not easily imposed upon. Even the officers of the ship, who, being acquainted with "Mollie" or "Sallie," "Mrs. Sam" or "Pete's Wife," volunteer to buy a basket at a reduction for some enthusiastic but thin-pursed passenger, do not at present meet with any exhilarating success.

"S'pose she pay my price," "Mrs. Sam" replies, with smiling but stubborn indifference, as she sets the basket away.

CHAPTER VII

Indian basketry is poetry, music, art, and life itself woven exquisitely together out of dreams, and sent out into a thoughtless world in appealing messages which will one day be farewells, when the poor lonely dark women who wove them are no more.

At its best, the basketry of the islands of Atka and Attu in the Aleutian chain is the most beautiful in the world. Most of the basketry now sold as Attu is woven by the women of Atka, we were told at Unalaska, which is the nearest market for these baskets. Only one old woman remains on Attu who understands this delicate and priceless work; and she is so poorly paid that she was recently reported to be in a starving condition, although the velvety creations of her old hands and brain bring fabulous prices to some one. The saying that an Attu basket increases a dollar for every mile as it travels toward civilization, is not such an exaggeration as it seems. I saw a trader from the little steamer *Dora*—the only one regularly plying those far waters—buy a small basket, no larger than a pint bowl, for five dollars in Unalaska; and a month later, on another steamer, between Valdez and Seattle, an enthusiastic young man from New York brought the same basket out of his stateroom and proudly displayed it.

"I got this one at a great bargain," he bragged, with shining eyes. "I bought it in Valdez for twenty-five dollars, just what it cost at Unalaska. The man needed the money worse than the basket. I don't know how it is, but I'm always stumbling on bargains like that!" he concluded, beginning to strut.

Then I was heartless enough to laugh, and to keep on laughing. I had greatly desired that basket myself!

He had the satisfaction of knowing, however, that his little twined bowl, with the coloring of a Behring Sea sunset woven into it, would be worth fifty dollars by the time he reached Seattle, and at least a hundred in New York; and it was so soft and flexible that he could fold it up meantime and carry it in his pocket, if he chose,—to say nothing of the fact that Elizabeth Propokoffono, the young and famed dark-eyed weaver of Atka, may have woven it herself. Like the renowned "Sally-bags," made by Sally, a Wasco squaw, the baskets woven by Elizabeth have a special and sentimental value. If she would weave her initials into them, she might ask, and receive, any price she fancied. Sally, of the Wascos, on the other hand, is very old; no one weaves her special bag, and they are becoming rare and valuable. They are of plain, twined weaving, and are very coarse. A small one in the writer's possession is adorned with twelve fishes, six eagles, three dogs, and two and a half men. Sally is apparently a woman-suffragist of the old school, and did not consider that men counted for much in the scheme of Indian baskets; yet, being a philosopher, as well as a suffragist, concluded that half a man was better than none at all.

At Yakutat "Mrs. Pete" is the best-known basket weaver. Young, handsome, dark-eyed, and clean, with a chubby baby in her arms, she willingly, and with great gravity, posed against the pilot-house of the old *Santa Ana* for her picture. Asked for an address to which I might send one of the pictures, she proudly replied, "Just Mrs. Pete, Yakutat." Her courtesy was in marked contrast to the exceeding rudeness with which the Sitkan women treat even the most considerate and deferential photographers; glaring at them, turning their backs, covering their heads, hissing, and even spitting at them.

However, the Yakutats do not often see tourists, who, heaven knows, are not one of the novelties of the Sitkans' lives.

According to Lieutenant G. T. Emmons, who is the highest authority on Thlinkit Indians, not only so far as their basketry is concerned, but their history, habits, and customs, as well, nine-tenths of all their basketwork is of the open, cylindrical type which throws the chief wear and strain upon the borders. These are, therefore, of greater variety than those of any other Indians, except possibly the Haidahs.

As I have elsewhere stated, nearly all Thlinkit baskets are of the twined weave, which is clearly described by Otis Tufton Mason in his precious and exquisite work, "Aboriginal American Basketry"; a work which every student of basketry should own. If anything could be as fascinating as the basketry itself, it would be this charmingly written and charmingly illustrated book.

Basketry is either hand-woven or sewed. Hand-woven work is divided into checker work, twilled work, wicker work, wrapped work, and twined work. Sewed work is called coiled basketry.

Twined work is found on the Pacific Coast from Attu to Chile, and is the most delicate and difficult of all woven work. It has a set of warp rods, and the weft elements are worked in by two-strand or three-strand methods. Passing from warp to warp, these weft elements are twisted in half-turns on each other, so as to form a two-strand or three-strand twine or braid, and usually with a deftness that keeps the glossy side of the weft outward.

"The Thlinkit, weaving," says Lieutenant Emmons, "sits with knees updrawn to the chin, feet close to the body, bent-shouldered, with the arms around the knees, the work held in front. Sometimes the knees fall slightly apart, the work held between them, the weft frequently held in the mouth, the feet easily crossed. The basket is held bottom down. In all kinds of weave, the strands are constantly dampened by dipping the fingers in water." The finest work of Attu and Atka is woven entirely under water. A rude awl, a bear's claw or tooth, are the only implements used. The Attu weaver has her basket inverted and suspended by a string, working from the bottom down toward the top.

Almost every part of plants is used—roots, stems, bark, leaves, fruit, and seeds. The following are the plants chiefly used by the Thlinkits: The black shining stems of the maidenhair fern, which are easily distinguished and which add a rich touch; the split stems of the brome-grass as an overlaying material for the white patterns of spruce-root baskets; for the same purpose, the split stem of bluejoint; the stem of wood reed-grass; the stem of tufted hair-grass; the stem of beech-rye; the root of horsetail, which works in a rich purple; wolf moss, boiled for canary-yellow dye; manna-grass; root of the Sitka spruce tree; juice of the blueberry for a purple dye.

The Attu weaver uses the stems and leaves of grass, having no trees and few plants. When she wants the grass white, it is cut in November and hung, points down, out-doors to dry; if yellow be desired, as it usually is, it is cut in July and the two youngest full-grown blades are cut out and split into three pieces, the middle one being rejected and the others hung up to dry out-doors; if green is wanted, the grass is prepared as for yellow, except that the first two weeks of curing is carried on in the heavy shade of thick grasses, then it is taken into the house and dried. Curing requires about a month, during which time the sun is never permitted to touch the grass.

Ornamentation by means of color is wrought by the use of materials which are naturally of a different color; by the use of dyed materials; by overlaying the weft and warp with strips of attractive material before weaving; by embroidering on the texture during the process of manufacture, this being termed "false" embroidery; by covering the texture with plaiting, called imbrication; by the addition of feathers, beads, shells, and objects of like nature.

Some otherwise fine specimens of Atkan basketry are rendered valueless, in my judgment, by the present custom of introducing flecks of gaily dyed wool, the matchless beauty of these baskets lying in their delicate, even weaving, and in their exquisite natural coloring—the faintest old rose, lavender, green, yellow and purple being woven together in one ravishing mist of elusive splendor. So enchanting to the real lover of basketry are the creations of those far lonely women's hands and brains, that they seem fairly to breathe out their loveliness upon the air, as a rose.

This basketry was first introduced to the world in 1874, by

William H. Dall, to whom Alaska and those who love Alaska owe so much. Warp and weft are both of beach grass or wild rye. One who has never seen a fine specimen of these baskets has missed one of the joys of this world.

The Aleuts perpetuate no story or myth in their ornamentation. With them it is art for art's sake; and this is, doubtless, one reason why their work draws the beholder spellbound.

The symbolism of the Thlinkit is charming. It is found not alone in their basketry, but in their carvings in stone, horn, and wood, and in Chilkaht blankets. The favorite designs are: shadow of a tree, water drops, salmon berry cut in half, the Arctic tern's tail, flaking of the flesh of a fish, shark's tooth, leaves of the fireweed, an eye, raven's tail, and the crossing. It must be confessed that only a wild imagination could find the faintest resemblance of the symbols woven into the baskets to the objects they represent. The symbol called "shadow of a tree" really resembles sunlight in moving water.

With the Haidah hats and Chilkaht blankets, it is very different. The head, feet, wings, and tail of the raven, for instance, are easily traced. In more recent basketry the swastika is a familiar design. Many Thlinkit baskets have "rattly" covers. Seeds found in the crops of quail are woven into these covers. They are "good spirits" which can never escape; and will insure good fortune to the owner. Woe be to him, however, should he permit his curiosity to tempt him to investigate; they will then escape and work him evil instead of good, all the days of his life.

In Central Alaska, the basketry is usually of the coiled variety, coarsely and very indifferently executed. Both spruce and willow are used. From Dawson to St. Michael, in the summer of 1907, stopping at every trading post and Indian village, I did not see a single piece of basketry that I would carry home. Coarse, unclean, and of slovenly workmanship, one could but turn away in pity and disgust for the wasted effort.

The Innuit in the Behring Sea vicinity make both coiled and twined basketry from dried grasses; but it is even worse than the Yukon basketry, being carelessly done,—the Innuit infinitely preferring the carving and decorating of walrus ivory to basket weaving. It is delicious to find an Innuit who never saw a glacier decorating a paper-knife with something that looks like a pond lily, and labelling it Taku Glacier, which is three thousand miles to the southeastward. I saw no attempt on the Yukon, nor on Behring Sea, at what Mr. Mason calls imbrication,—the beautiful ornamentation which the Indians of Columbia, Frazer, and Thompson rivers and of many Salish tribes of Northwestern Washington use to distinguish their coiled work. It resembles knife-plaiting before it is pressed flat. This imbrication is frequently of an exquisite, dull, reddish brown over an old soft yellow. Baskets adorned with it often have handles and flat covers; but papoose baskets and covered long baskets, almost as large as trunks, are common.

There was once a tide in my affairs which, not being taken at the flood, led on to everlasting regret.

One August evening several years ago I landed on an island in Puget Sound where some Indians were camped for the fishing season. It was Sunday; the men were playing the fascinating gambling game of slahal, the children were shouting at play, the women were gathered in front of their tents, gossiping.

In one of the tents I found a coiled, imbricated Thompson River basket in old red-browns and yellows. It was three and a half feet long, two and a half feet high, and two and a half wide, with a thick, close-fitting cover. It was offered to me for ten dollars, and—that I should live to chronicle it!—not knowing the worth of such a basket, I closed my eyes to its appealing and unforgettable beauty, and passed it by.

But it had, it has, and it always will have its silent revenge. It is as bright in my memory to-day as it was in my vision that August Sunday ten years ago, and more enchanting. My longing to see it again, to possess it, increases as the years go by. Never have I seen its equal, never shall I. Yet am I ever looking for that basket, in every Indian tent or hovel I may stumble upon—in villages, in camps, in out-of-the-way places. Sure am I that I should know it from all other baskets, at but a glance.

I knew nothing of the value of baskets, and I fancied the woman was taking advantage of my ignorance. While I hesitated, the steamer whistled. It was all over in a moment; my chance was gone. I did not even dream how greatly I desired that basket until I stood in the bow of the steamer and saw the little white camp fade from view across the sunset sea.

The original chaste designs and symbols of Thlinkit, Haidah, and Aleutian basketry are gradually yielding, before the coarse taste of traders and tourists, to the more modern and conventional designs. I have lived to see a cannery etched upon an exquisitely carved paper-knife; while the things produced at infinite labor and care and called cribbage-boards are in such bad taste that tourists buying them become curios themselves.

The serpent has no place in Alaskan basketry for the very good reason that there is not a snake in all Alaska, and the Indians and Innuit probably never saw one. A woman may wade through the swampiest place or the tallest grass without one shivery glance at her pathway for that little sinuous ripple which sends terror to most women's hearts in warmer climes. Indeed, it is claimed that no poisonous thing exists in Alaska.

The tourist must not expect to buy baskets farther north than Skaguay, where fine ones may be obtained at very reasonable prices. Having visited several times every place where basketry is sold, I would name first Dundas, then Yakutat, and then Sitka as the most desirable places for "shopping," so far as southeastern Alaska is concerned; out "to Westward," first Unalaska and Dutch Harbor, then Kodiak and Seldovia.

Eskimo in Walrus-skin Kamelayka

But the tourists who make the far, beautiful voyage out among the Aleutians to Unalaska might almost be counted annually upon one's fingers—so unexploited are the attractions of that region; therefore, I will add that fine specimens of the Attu and Atka work may be found at Wrangell, Juneau, Skaguay, and Sitka, without much choice, either in workmanship or price. But fortunate may the tourist consider himself who travels this route on a steamer that gathers the salmon catch in August or September, and is taken through Icy Strait to the Dundas cannery. There, while a cargo of canned salmon is being taken aboard, the passengers have time to barter with the good-looking and intelligent Indians for the superb baskets laid out in the immense warehouse. Nowhere in Alaska have I seen baskets of such beautiful workmanship, design, shape, and coloring as at Dundas—excepting always, of course, the Attu and Atka; nowhere have I seen them in such numbers, variety, and at such low prices.

My own visit to Dundas was almost pathetic. It was on my return from a summer's voyage along the coast of Alaska, as far westward as Unalaska. I had touched at every port between Dixon's Entrance and Unalaska, and at many places that were not ports; had been lightered ashore, rope-laddered and doried ashore, had waded ashore, and been carried ashore on sailors' backs; and then, with my top berth filled to the ceiling with baskets and things, with all my money spent and all my clothes worn out, I stood in the warehouse at Dundas and saw those dozens of beautiful baskets, and had them offered to me at but half the prices I had paid for inferior baskets. It was here that the summer hats and the red kimonos and the pretty collars were brought out, and were eagerly seized by the dark and really handsome Indian girls. A ten-dollar hat—at the end of the season!—went for a fifteen-dollar basket; a long, red woollen kimono,—whose warmth had not been required on this ideal trip, anyhow,—secured another of the same price; and may heaven forgive me, but I swapped one twenty-two-inch gold-embroidered belt for a three-dollar basket, even while I knew in my sinful heart that there was not a waist in that warehouse that measured less than thirty-five inches; and from that to fifty!

However, in sheer human kindness, I taught the girl to whom I swapped it how it might be worn as a garter, and her delight was so great and so unexpected that it caused me some apprehension as to the results. My very proper Scotch friend and travelling companion was so aghast at my suggestion that she took the girl aside and advised her to wear the belt for collars, cut in half, or as a gay decoration up the front plait of her shirt-waist, or as armlets; so that, with it all, I was at last able to retire to my stateroom and enjoy my bargains with a clear conscience, feeling that after some fashion the girl would get her basket's worth out of the belt.

CHAPTER VIII

Leaving Wrangell, the steamer soon passes, on the port side and at the entrance to Sumner Strait, Zarembo Island, named for that Lieutenant Zarembo who so successfully prevented the Britishers from entering Stikine River. Baron Wrangell bestowed the name, desiring in his gratitude and appreciation to perpetuate the name and fame of the intrepid young officer.

From Sumner Strait the famed and perilously beautiful Wrangell Narrows is entered. This ribbonlike water-way is less than twenty miles long, and in many places so narrow that a stone may be tossed from shore to shore. It winds between Mitkoff and Kupreanoff islands, and may be navigated only at certain stages of the tide. Deep-draught vessels do not attempt Wrangell Narrows, but turn around Cape Decision and proceed by way of Chatham Strait and Frederick Sound—a course which adds at least eighty miles to the voyage.

The interested voyager will not miss one moment of the run through the narrows, either for sleep or hunger. Better a sleepless night or a dinnerless day than one minute lost of this matchless scenic attraction.

The steamer pushes, under slow bell, along a channel which, in places, is not wider than the steamer itself. Its sides are frequently touched by the long strands of kelp that cover the sharp and dangerous reefs, which may be plainly seen in the clear water.

The timid passenger, sailing these narrows, holds his breath a good part of the time, and casts anxious glances at the bridge, whereon the captain and his pilots stand silent, stern, with steady, level gaze set upon the course. One moment's carelessness, ten seconds of inattention, might mean the loss of a vessel in this dangerous strait.

Intense silence prevails, broken only by the heavy, slow throb of the steamer and the swirl of the brown water in whirlpools over the rocks; and these sounds echo far.

The channel is marked by many buoys and other signals. The island shores on both sides are heavily wooded to the water, the branches spraying out over the water in bright, lacy green. The tree trunks are covered with pale green moss, and long moss-fringes hang from the branches, from the tips of the trees to the water's edge. The effect is the same as that of festal decoration.

Eagles may always be seen perched motionless upon the tall tree-tops or upon buoys.

The steamship *Colorado* went upon the rocks between Spruce and Anchor points in 1900, where her storm-beaten hull still lies as a silent, but eloquent, warning of the perils of this narrow channel.

The tides roaring in from the ocean through Frederick Sound on the north and Sumner Strait on the south meet near Finger Point in the narrows.

Sunrise and sunset effects in this narrow channel are justly famed. I once saw a mist blown ahead of my steamer at sunset that, in the vivid brilliancy of its mingled scarlets, greens, and purples, rivalled the coloring of a humming-bird.

At dawn, long rays of delicate pink, beryl, and pearl play through this green avenue, deepening in color, fading, and withdrawing like Northern Lights. When the scene is silvered and softened by moonlight, one looks for elves and fairies in the shadows of the moss-dripping spruce trees.

The silence is so intense and the channel so narrow, that frequently at dawn wild birds on the shores are heard saluting the sun with song; and never, under any other circumstances, has bird song seemed so nearly divine, so golden with magic and message, as when thrilled through the fragrant, green stillness of Wrangell Narrows at such an hour.

I was once a passenger on a steamer that lay at anchor all night in Sumner Strait, not daring to attempt the Narrows on account of storm and tide. A stormy sunset burned about our ship. The sea was like a great, scarlet poppy, whose every wave petal circled upward at the edges to hold a fleck of gold. Island upon island stood out through that riot of color in vivid, living green, and splendid peaks shone burnished against the sky.

There was no sleep that night. Music and the dance held sway in the cabins for those who cared for them, and for the others there was the beauty of the night. In our chairs, sheltered by the great smoke-stacks of the hurricane-deck, we watched the hours go by—each hour a different color from the others—until the burned-out red of night had paled into the new sweet primrose of dawn. The wind died, leaving the full tide "that, moving, seems asleep"; and no night was ever warmer and sweeter in any tropic sea than that.

Wrangell Narrows leads into Frederick Sound—so named by Whidbey and Johnstone, who met there, in 1794, on the birthday of Frederick, Duke of York.

Vancouver's expedition actually ended here, and the search for the "Strait of Anian" was finally abandoned.

Several glaciers are in this vicinity: Small, Patterson, Summit, and Le Conte. The Devil's Thumb, a spire-shaped peak on the mainland, rises more than two thousand feet above the level of the sea, and stands guard over Wrangell Narrows and the islands and glaciers of the vicinity.

On Soukhoi Island fox ranches were established about five years ago; they are said to be successful.

The Thunder Bay Glacier is the first on the coast that discharges bergs. The thunder-like roars with which the vast bulks of beautiful blue-white ice broke from the glacier's front caused the Indians to believe this bay to be the home of the thunder-bird, who always produces thunder by the flapping of his mighty wings.

Baird Glacier is in Thomas Bay, noted for its scenic charms,—glaciers, forestation, waterfalls, and sheer heights combining to give it a deservedly wide reputation among tourists. Elephant's Head, Portage Bay, Farragut Bay, and Cape Fanshaw are important features of the vicinity. The latter is a noted landmark and storm-point. It fronts the southwest, and the full fury of the fiercest storms beats mercilessly upon it. Light craft frequently try for days to make this point, when a wild gale is blowing from the Pacific.

Of the scenery to the south of Cape Fanshaw, Whidbey reported to Vancouver, on his final trip of exploration in August, 1794, that "the mountains rose abruptly to a prodigious height ... to the South, a part of them presented an uncommonly awful appearance, rising with an inclination towards the water to a vast height, loaded with an immense quantity of ice and snow, and overhanging their base, which seemed to be insufficient to bear the ponderous fabric it sustained, and rendered the view of the passage beneath it horribly magnificent."

At the Cape he encountered such severe gales that a whole day and night were consumed in making a distance of sixteen miles.

There are more fox ranches on "The Brothers" Islands, and soon after passing them Frederick Sound narrows into Stephens' Passage. Here, to starboard, on the mainland, is Mount Windham, twenty-five hundred feet in height, in Windham Bay.

Gold was discovered in this region in the early seventies, and mines were worked for a number of years before the Juneau and Treadwell excitement. The mountains abound in game.

Sumdum is a mining town in Sumdum, or Holkham, Bay. The fine, live glacier in this arm is more perfectly named than any other in Alaska—Sumdum, as the Indians pronounce it, more clearly describing the deep roar of breaking and falling ice, with echo, than any other syllables.

Large steamers do not enter this bay; but small craft, at slack-tide, may make their way among the rocks and icebergs. It is well worth the extra expense and trouble of a visit.

To the southwest of Cape Fanshaw, in Frederick Sound, is Turnabout Island, whose suggestive name is as forlorn as Turnagain Arm, in Cook Inlet, where Cook was forced to "turn again" on what proved to be his last voyage.

Stephens' Passage is between the mainland and Admiralty Island. This island barely escapes becoming three or four islands. Seymour Canal, in the eastern part, almost cuts off a large portion, which is called Glass Peninsula, the connecting strip of land being merely a portage; Kootznahoo Inlet cuts more than halfway across from west to east, a little south of the centre of the island; and at the northern end had Hawk Inlet pierced but a little farther, another island would have been formed. The scenery along these inlets, particularly Kootznahoo, where the lower wooded hills rise from sparkling blue waters to glistening snow peaks, is magnificent. Whidbey reported that although this island appeared to be composed of a rocky substance covered with but little soil, and that chiefly consisting of vegetables in an imperfect state of dissolution, yet it produced timber which he considered superior to any he had before observed on the western coast of America.

It is a pity that some steamship company does not run at least one or two excursions during the summer to the little-known and unexploited inlets of southeastern Alaska—to the abandoned Indian villages, graveyards, and totems; the glaciers, cascades, and virgin spruce glades; the roaring narrows and dim, sweet fiords, where the regular passenger and "tourist" steamers do not touch. A month might easily be spent on such a trip, and enough nature-loving, interested, and interesting people could be found to take every berth—without the bugaboo, the increasing nightmare of the typical tourist, to rob one of his pleasure.

At present an excursion steamer sails from Seattle, and from the hour of its sailing the steamer throbs through the most beautiful archipelago in the world, the least known, and the one most richly repaying study, making only five or six landings, and visiting two glaciers at most. It is quite true that every moment of this "tourist" trip of ten days is, nevertheless, a delight, if the weather be favorable; that the steamer rate is remarkably cheap, and that no one can possibly regret having made this trip if he cannot afford a longer one in Alaska. But this does not alter the fact that there are hundreds of people who would gladly make the longer voyage each summer, if transportation were afforded. Local transportation in Alaska is so expensive that few can afford to go from place to place, waiting for steamers, and paying for boats and guides for every side trip they desire to make.

Admiralty Island is rich in gold, silver, and other minerals. There are whaling grounds in the vicinity, and a whaling station was recently established on the southwestern end of the Island, near Surprise Harbor and Murder Cove. Directly across Chatham Strait from this station, on Baranoff Island, only twenty-five miles from Sitka, are the famous Sulphur Hot Springs.

There are fine marble districts on the western shores of Admiralty Island.

On the southern end are Woewodski Harbor and Pybas Bay.

Halfway through Stephens' Passage are the Midway Islands, and but a short distance farther, on the mainland, is Port Snettisham, a mining settlement on an arm whose northern end is formed by Cascades Glacier, and from whose southern arm musically and exquisitely leaps a cascade which is the only rival of Sarah Island in the affections of mariners—*Sweetheart Falls.*

Who so tenderly named this cascade, and for whom, I have not been able to learn; but those pale green, foam-crested waters shall yet give up their secret. Never would Vancouver be suspected of such naming. Had he so prettily and sentimentally named it, the very waters would have turned to stone in their fall, petrified by sheer amazement.

The scenery of Snettisham Inlet is the finest in this vicinity of fine scenic effects, with the single exception of Taku Glacier.

In Taku Harbor is an Indian village, called Taku, where may be found safe anchorage, which is frequently required in winter, on account of what are called "Taku winds." Passing Grand Island, which rises to a wooded peak, the steamer crosses the entrance to Taku Inlet and enters Gastineau Channel.

There are many fine peaks in this vicinity, from two to ten thousand feet in height.

The stretch of water where Stephens' Passage, Taku Inlet, Gastineau Channel, and the southeastern arm of Lynn Canal meet is in winter dreaded by pilots. A squall is liable to come tearing down Taku Inlet at any moment and meet one from some other direction, to the peril of navigation.

At times a kind of fine frozen mist is driven across by the violent gales, making it difficult to see a ship's length ahead. At such times the expressive faces on the bridge of a steamer are psychological studies.

In summer, however, no open stretch of water could be more inviting. Clear, faintly rippled, deep sapphire, flecked with the first glistening bergs floating out of the inlet, it leads the way to the glorious presence that lies beyond.

I had meant to take the reader first up lovely Gastineau Channel to Juneau; but now that I have unintentionally drifted into Taku Inlet, the glacier lures me on. It is only an hour's run, and the way is one of ever increasing beauty, until the steamer has pushed its prow through the hundreds of sparkling icebergs, under slow bell, and at last lies motionless. One feels as though in the presence of some living, majestic being, clouded in mystery. The splendid front drops down sheer to the water, from a height of probably three hundred feet. A sapphire mist drifts over it, without obscuring the exquisite tintings of rose, azure, purple, and green that flash out from the glistening spires and columns. The crumpled mass pushing down from the mountains strains against the front, and sends towered bulks plunging headlong into the sea, with a roar that echoes from peak to peak in a kind of "linked sweetness long drawn out" and ever diminishing.

There is no air so indescribably, thrillingly sweet as the air of a glacier on a fair day. It seems to palpitate with a fragrance that ravishes the senses. I saw a great, recently captured bear, chained on the hurricane deck of a steamer, stand with his nose stretched out toward the glacier, his nostrils quivering and a look of almost human longing and rebellion in his small eyes. The feeling of pain and pity with which a humane person always beholds a chained wild animal is accented in these wide and noble spaces swimming from snow mountain to snow mountain, where the very watchword of the silence seems to be "Freedom." The chained bear recognized the scent of the glacier and remembered that he had once been free.

In front of the glacier stretched miles of sapphire, sunlit sea, set with sparkling, opaline-tinted icebergs. Now and then one broke and fell apart before our eyes, sending up a funnel-shaped spray of color,—rose, pale green, or azure.

At every blast of the steamer's whistle great masses of ice came thundering headlong into the sea—to emerge presently, icebergs. Canoeists approach glaciers closely at their peril, never knowing when an iceberg may shoot to the surface and wreck their boat. Even larger craft are by no means safe, and tourists desiring a close approach should voyage with intrepid captains who sail safely through everything.

The wide, ceaseless sweep of a live glacier down the side of a great mountain and out into the sea holds a more compelling suggestion of power than any other action of nature. I have never felt the appeal of a mountain glacier—of a stream of ice and snow that, so far as the eye can discover, never reaches anywhere, although it keeps going forever. The feeling of forlornness with which, after years of anticipation, I finally beheld the renowned glacier of the Selkirks, will never be forgotten. It was the forlornness of a child who has been robbed of her Santa Claus, or who has found that her doll is stuffed with sawdust.

But to behold the splendid, perpendicular front of a live glacier rising out of a sea which breaks everlastingly upon it; to see it under the rose and lavender of sunset or the dull gold of noon; to see and hear tower, minaret, dome, go thundering down into the clear depths and pound them into foam—this alone is worth the price of a trip to Alaska.

We were told that the opaline coloring of the glacier was unusual, and that its prevailing color is an intense blue, more beautiful and constant than that of other glaciers; and that even the bergs floating out from it were of a more pronounced blue than other bergs.

But I do not believe it. I have seen the blue of the Columbia Glacier in Prince William Sound; and I have sailed for a whole afternoon among the intensely blue ice shallops that go drifting in an endless fleet from Glacier Bay out through Icy Straits to the ocean. If there be a more exquisite blue this side of heaven than I have seen in Icy Straits and in the palisades of the Columbia Glacier, I must see it to believe it.

There are three glaciers in Taku Inlet: two—Windham and Twin—which are at present "dead"; and Taku, the Beautiful, which is very much alive. The latter was named Foster, for the former Secretary of the Treasury; but the Indian name has clung to it, which is one more cause for thanksgiving.

The Inlet is eighteen miles long and about seven hundred feet wide. Taku River flows into it from the northeast, spreading out in blue ribbons over the brown flats; at high tide it may be navigated, with caution, by small row-boats and canoes. It was explored in early days by the Hudson Bay Company, also by surveyors of the Western Union Telegraph Company.

Whidbey, entering the Inlet in 1794, sustained his reputation for absolute blindness to beauty. He found "a compact body of ice extending some distance nearly all around." He found "frozen mountains," "rock sides," "dwarf pine trees," and "undissolving frost and snow." He lamented the lack of a suitable landing-place for boats; and reported the aspect in general to be "as dreary and inhospitable as the imagination can possibly suggest."

Alas for the poor chilly Englishman! He, doubtless, expected silvery-gowned ice maidens to come sliding out from under the glacier in pearly boats, singing and kissing their hands, to bear him back into their deep blue grottos and dells of ice, and refresh him with Russian tea from old brass samovars; he expected these maidens to be girdled and crowned with carnations and poppies, and to pluck winy grapes—with *dust* clinging to their bloomy roundness—from living vines for him to eat; and most of all, he expected to find in some remote corner of the clear and sparkling cavern a big fireplace, "which would remind him pleasantly of England;" and a brilliant fire on a well-swept hearth, with the smoke and sparks going up through a melted hole in the glacier.

About fifteen miles up Taku River, Wright Glacier streams down from the southeast and fronts upon the low and marshy lands for a distance of nearly three miles.

The mountains surrounding Taku Inlet rise to a height of four thousand feet, jutting out abruptly, in places, over the water.

CHAPTER IX

Gastineau Channel is more than a mile wide at the entrance, and eight miles long; it narrows gradually as it separates Douglas Island from the mainland, and, still narrowing, goes glimmering on past Juneau, like a silver-blue ribbon. Down this channel at sunset burns the most beautiful coloring, which slides over the milky waters, producing an opaline effect. At such an hour this scene—with Treadwell glittering on one side, and Juneau on the other, with Mount Juneau rising in one swelling sweep directly behind the town—is one of the fairest in this country of fair scenes.

The unique situation of Juneau appeals powerfully to the lover of beauty. There is an unforgettable charm in its narrow, crooked streets and winding, mossed stairways; its picturesque shops,—some with gorgeous totem-poles for signs,—where a small fortune may be spent on a single Attu or Atka basket; the glitter and the music of its streets and its "places," the latter open all night; its people standing in doorways and upon corners, eager to talk to strangers and bid them welcome; and its gayly clad squaws, surrounded by fine baskets and other work of their brown hands.

The streets are terraced down to the water, and many of the pretty, vine-draped cottages seem to be literally hung upon the side of the mountain. One must have good, strong legs to climb daily the flights of stairs that steeply lead to some of them.

In the heart of the town is an old Presbyterian Mission church, built of logs, with an artistic square tower, also of logs, at one corner. This church is now used as a brewery and soda-bottling establishment!

The lawns are well cared for, and the homes are furnished with refined taste, giving evidences of genuine comfort, as well as luxury.

My first sight of Juneau was at three o'clock of a dark and rainy autumn night in 1905. We had drifted slowly past the mile or more of brilliant electric lights which is Treadwell and Douglas; and turning our eyes to the north, discovered, across the narrow channel, the lights of Juneau climbing out of the darkness up the mountain from the water's edge. Houses and buildings we could not see; only those radiant lights, leading us on, like will-o'-the-wisps.

When we landed it seemed as though half the people of the town, if not the entire population, must be upon the wharf. It was then that we learned that it is always daytime in Alaskan towns when a steamer lands—even though it be three o'clock of a black night.

The business streets were brilliant. Everything was open for business, except the banks; a blare of music burst through the open door of every saloon and dance-hall; blond-haired "ladies" went up and down the streets in the rain and mud, bare-headed, clad in gauze and other airy materials, in silk stockings and satin slippers. They laughed and talked with men on the streets in groups; they were heard singing; they were seen dancing and inviting the young waiters and cabin-boys of our steamer into their dance halls.

"How'd you like Juneau?" asked my cabin-boy the next day, teetering in the doorway with a plate of oranges in his hand, and a towel over his arm.

"It seemed very lively," I replied, "for three o'clock in the morning."

"Oh, hours don't cut any ice in Alaska," said he. "People in Alaska keep their clo's hung up at the head of their beds, like the harness over a fire horse. When the boat whistles, it loosens the clo's from the hook; the people spring out of bed right under 'em; the clo's fall onto 'em—an' there they are on the wharf, all dressed, by the time the boat docks. They're all right here, but say! they can't hold a candle to the people of Valdez for gettin' to the dock. They just cork you at Valdez."

At Juneau I went through the most brilliant business transaction of my life. I was in the post-office when I discovered that I had left my pocket-book on the steamer. I desired a curling-iron; so I borrowed a big silver dollar of a friend, and hastened away to the largest dry-goods shop.

A sleepy clerk waited upon me. The curling-iron was thirty cents. I gave him the dollar, and he placed the change in my open hand. Without counting it, I went back to the post-office, purchased twenty-five cents' worth of stamps, and gave the balance to the friend from whom I had borrowed the dollar.

"Count it," said I, "and see how much I owe you."

She counted it.

"How much did you spend?" she asked presently.

"Fifty-five cents."

She began to laugh wildly.

"You have a thirty-cent curling-iron, twenty-five cents' worth of stamps, and you've given me back a dollar and sixty-five cents—all out of one silver dollar!"

I counted the money. It was too true.

With a burning face I took the change and went back to the store. My friend insisted upon going with me, although I would have preferred to see her lost on the Taku Glacier. I cannot endure people who laugh like children at everything.

Eskimo in Bidarka

The captain and several passengers were in the store. They heard my explanation; and they all gathered around to assist the polite but sleepy clerk.

One would say that it would be the simplest thing in the world to straighten out that change; but the postage stamps added complications. Everybody figured, explained, suggested, criticised, and objected. Several times we were quite sure we had it. Then, some one would titter—and the whole thing would go glimmering out of sight.

However, at the end of twenty minutes it was arranged to the clerk's and my own satisfaction. Several hours later, when we were well on our way up Lynn Canal, a calmer figuring up proved that I had not paid one cent for my curling-iron.

From the harbor Mount Juneau has the appearance of rising directly out of the town—so sheer and bold is its upward sweep to a height of three thousand feet. Down its many pale green mossy fissures falls the liquid silver of cascades.

It is heavily wooded in some places; in others, the bare stone shines through its mossy covering, giving a soft rose-colored effect, most pleasing to the eye.

Society in Juneau, as in every Alaskan town, is gay. Its watchword is hospitality. In summer, there are many excursions to glaciers and the famed inlets which lie almost at their door, and to see which other people travel thousands of miles. In winter, there is a brilliant whirl of dances, card parties, and receptions. "Smokers" to which ladies are invited are common—although they are somewhat like the pioneer dish of "potatoes-and-point."

When the pioneers were too poor to buy sufficient bacon for the family dinner, they hung a small piece on the wall; the family ate their solitary dish of potatoes and pointed at the piece of bacon.

So, at these smokers, the ladies must be content to see the men smoke, but they might, at least, be allowed to point.

Most of the people are wealthy. Money is plentiful, and misers are unknown. The expenditure of money for the purchase of pleasure is considered the best investment that an Alaskan can make.

Fabulous prices are paid for luxuries in food and dress.

"I have lived in Dawson since 1897," said a lady last summer, "and have never been ill for a day. I attribute my good health to the fact that I have never flinched at the price of anything my appetite craved. Many a time I have paid a dollar for a small cucumber; but I have never paid a dollar for a drug. I have always had fruit, regardless of the price, and fresh vegetables. No amount of time or money is considered wasted on flowers. Women of Alaska invariably dress well and present a smart appearance. Many wear imported gowns and hats—and I do not mean imported from 'the states,' either—and costly jewels and furs are more common than in any other section of America. We entertain lavishly, and our hospitality is genuine."

Every traveller in Alaska will testify to the truth of these assertions. If a man looks twice at a dollar before spending it, he is soon "jolted" out of the pernicious habit.

The worst feature of Alaskan social life is the "coming out" of many of the women in winter, leaving their husbands to spend the long, dreary winter months as they may. To this selfishness on the part of the women is due much of the intoxication and immorality of Alaska—few men being of sufficiently strong character to withstand the distilled temptations of the country.

That so many women go "out" in winter, is largely due to the proverbial kindness and indulgence of American husbands, who are loath to have their wives subjected to the rigors and the hardships of an Alaskan winter.

However, the winter exodus may scarcely be considered a feature of the society of Juneau, or other towns of southeastern Alaska. The climate resembles that of Puget Sound; there is a frequent and excellent steamship service to and from Seattle; and the reasons for the exodus that exist in cold and shut-in regions have no apparent existence here.

Every business—and almost every industry—is represented in Juneau. The town has excellent schools and churches, a library, women's clubs, hospitals, a chamber of commerce, two influential newspapers, a militia company, a brass band—and a good brass band is a feature of real importance in this land of little music—an opera-house, and, of course, electric lights and a good water system.

Juneau has for several years been the capital of Alaska; but not until the appointment of Governor Wilford B. Hoggatt, in 1906, to succeed Governor J. G. Brady, were the Executive Office and Governor's residence established here. So confident have the people of Juneau always been that it would eventually become the capital of Alaska, that an eminence between the town and the Auk village has for twenty years been called Capitol Hill. During all these years there has been a fierce and bitter rivalry between Juneau and Sitka.

Juneau was named for Joseph Juneau, a miner who came, "grub-staked," to this region in 1880. It was the fifth name bestowed upon the place, which grew from a single camp to the modern and independent town it is to-day—and the capital of one of the greatest countries in the world.

In its early days Juneau passed through many exciting and charming vicissitudes. Anything but monotony is welcomed by a town in Alaska; and existence in Juneau in the eighties was certainly not monotonous.

The town started with a grand stampede and rush, which rivalled that of the Klondike seventeen years later; the Treadwell discovery and attendant excitement came during the second year of its existence, and a guard of marines was necessary to preserve order, until, upon its withdrawal, a vigilance committee took matters into its own hands, with immediate beneficial results.

The population of Juneau is about two thousand, which—like that of all other northern towns—is largely increased each fall by the miners who come in from the hills and inlets to "winter."

In the middle eighties there were Chinese riots. The little yellow men were all driven out of town, and their quarters were demolished by a mob.

A recent attempt to introduce Hindu labor in the Treadwell mines resulted as disastrously.

Railroad Construction, Eyak Lake

CHAPTER X

Treadwell! Could any mine employing stamps have a more inspiring name, unless it be Stampwell? It fairly forces confidence and success.

Douglas Island, lying across the narrow channel from Juneau, is twenty-five miles long and from four to nine miles wide. On this island are the four famous Treadwell mines, owned by four separate companies, but having the same general managership.

Gold was first discovered on this island in 1881. Sorely against his will, John Treadwell was forced to take some of the original claims, having loaned a small amount upon them, which the borrower was unable to repay.

Having become possessed of these claims, a gambler's "hunch" impelled him to buy an adjoining claim from "French Pete" for four hundred dollars. On this claim is now located the famed "Glory Hole."

This is so deep that to one looking down into it the men working at the bottom and along the sides appear scarcely larger than flies. Steep stairways lead, winding, to the bottom of this huge quartz bowl; but visitors to the dizzy regions below are not encouraged, on account of frequent blasting and danger of accidents.

It is claimed that Treadwell is the largest quartz mine in the world, and that it employs the largest number of stamps—nine hundred. The ore is low grade, not yielding an average of more than two dollars to the ton; but it is so easily mined and so economically handled that the mines rank with the Calumet and Hecla, of Michigan; the Comstock Lode mines, of Nevada; the Homestake, of South Dakota; and the Portland, of Colorado.

The Treadwell is the pride of Alaska. Its poetic situation, romantic history, and admirable methods should make it the pride of America.

Its management has always been just and liberal. It has had fewer labor troubles than any other mine in America.

There are two towns on the island—Treadwell and Douglas. The latter is the commercial and residential portion of the community—for the towns meet and mingle together.

The entire population, exclusive of natives, is three thousand people—a population that is constantly increasing, as is the demand for laborers, at prices ranging from two dollars and sixty cents per day up to five dollars for skilled labor.

The island is so brilliantly lighted by electricity that to one approaching on a dark night it presents the appearance of a city six times its size.

The nine hundred stamps drop ceaselessly, day and night, with only two holidays in a year—Christmas and the Fourth of July. The noise is ferocious. In the stamp-mill one could not distinguish the boom of a cannon, if it were fired within a distance of twenty feet, from the deep and continuous thunder of the machinery.

In 1881 the first mill, containing five stamps, was built and commenced crushing ore that came from a streak twenty feet wide. This ore milled from eight to ten dollars a ton, proving to be of a grade sufficiently high to pay for developing and milling, and leave a good surplus.

It was soon recognized that the great bulk of the ore was extremely low grade, and that, consequently, a large milling capacity would be required to make the enterprise a success. A one-hundred-and-twenty-stamp-mill was erected and began crushing ore in June, 1885. At the end of three years the stamps were doubled. In another year three hundred additional stamps were dropping. Gradually the three other mines were opened up and the stamps were increased until nine hundred were dropping.

The shafts are from seven to nine hundred feet below sea level, and one is beneath the channel; yet very little water is encountered in sinking them. Most of the water in the mines comes from the surface and is caught up and pumped out, from the first level.

The net profits of these mines to their owners are said to be six thousand dollars a day; and mountains of ore are still in sight.

Our captain obtained permission to take us down into the mine. This was not so difficult as it was to elude the other passengers. At last, however, we found ourselves shut into a small room, lined with jumpers, slickers, and caps.

Shades of the things we put on to go under Niagara Falls!

"Get into this!" commanded the captain, holding a sticky and unclean slicker for me. "And make haste! There's no time to waste for you to examine it. Finicky ladies don't get two invitations into the Treadwell. Put in your arm."

My arm went in. When an Alaskan sea captain speaks, it is to obey. Who last wore that slicker, far be it from me to discover. Chinaman, leper, Jap, or Auk—it mattered not. I was in it, then, and curiosity was sternly stifled.

"Now put on this cap." Then beheld mine eyes a cap that would make a Koloshian ill.

"Must I put *that* on?"

I whispered it, so the manager would not hear.

"You must put this on. Take off your hat."

My hat came off, and the cap went on. It was pushed down well over my hair; down to my eyebrows in the front and down to the nape of my neck in the back.

"There!" said the captain, cheerfully. "You needn't be afraid of anything down in the mine now."

Alas! there was nothing in any mine, in any world, that I dreaded as I did what might be in that cap.

There were four of us, with the manager, and there was barely room on the rather dirty "lift" for us.

We stood very close together. It was as dark as a dungeon.

"Now—look out!" said the manager.

As we started, I clutched somebody—it did not matter whom.

I also drew one wild and amazed breath; before I could possibly let go of that one—to say nothing of drawing another—there was a bump, and we were in a level one thousand and eighty feet below the surface of the earth.

We stepped out into a brilliantly lighted station, with a high, glittering quartz ceiling. The swift descent had so affected my hearing that I could not understand a word that was spoken for fully five minutes. None of my companions, however, complained of the same trouble.

It has been the custom to open a level at every hundred and ten feet; but hereafter the distance between levels in the Treadwell mine will be one hundred and fifty feet.

At each level a station, or chamber, is cut out, as wide as the shaft, from forty to sixty feet in length, and having an average height of eight feet. A drift is run from the shaft for a distance of twenty-five feet, varying in height from fifteen feet in front to seven at the back. The main crosscut is then started at right angles to the station drift.

From east and west the "drifts" run into this crosscut, like little creeks into a larger stream.

No one has ever accused me of being shy in the matter of asking questions. It was the first time I had been down in one of the famous gold mines of the world, and I asked as many questions as a woman trying to rent a forty-dollar house for twenty dollars. Between shafts, stations, ore bins, crosscuts, stopes, drifts, levels, and *winzes*, it was less than fifteen minutes before I felt the cold moisture of despair breaking out upon my brow. Winzes proved to be the last straw. I could get a glimmering of what the other things were; but *winzes*!

The manager had been polite in a forced, friend-of-the-captain kind of way. He was evidently willing to answer every question once, but whenever I forgot and asked the same question twice, he balked instantly. Exerting every particle of intelligence I possessed, I could not make out the difference between a stope and a station, except that a stope had the higher ceiling.

"I have told you the difference *three times* already," cried the manager, irritably.

The captain, back in the shadow, grinned sympathetically.

"Nor'-nor'-west, nor'-by-west, a-quarter-nor'," said he, sighing. "She'll learn your gold mine sooner than she'll learn my compass."

Then they both laughed. They laughed quite a while, and my disagreeable friend laughed with them. For myself, I could not see anything funny anywhere.

I finally learned, however, that a station is a place cut out for a stable or for the passage of cars, or other things requiring space; while a stope is a room carried to the level of the top of the main crosscut. It is called a stope because the ore is "stoped" out of it.

But winzes! What winzes are is still a secret of the ten-hundred-and-eighty-foot level of the Treadwell mine.

Tram-cars filled with ore, each drawn by a single horse, passed us in every drift—or was it in crosscuts and levels? One horse had been in the mine seven years without once seeing sunlight or fields of green grass; without once sipping cool water from a mountain creek with quivering, sensitive lips; without once stretching his aching limbs upon the soft sod of a meadow, or racing with his fellows upon a hard road.

But every man passing one of these horses gave him an affectionate pat, which was returned by a low, pathetic whinny of recognition and pleasure.

"One old fellow is a regular fool about these horses," said the manager, observing our interest. "He's always carrying them down armfuls of green grass, apples, sugar, and everything a horse will eat. You'd ought to hear them nicker at sight of him. If they pass him in a drift, when he hasn't got a thing for them, they'll nicker and nicker, and keep turning their heads to look after him. Sometimes it makes me feel queer in my throat."

No one can by any chance know what noise is until he has stood at the head of a drift and heard three Ingersoll-Sergeant drills beating with lightning-like rapidity into the walls of solid quartz for the purpose of blasting.

Standing between these drills and within three feet of them, one suddenly is possessed of the feeling that his sense of hearing has broken loose and is floating around in his head in waves. This feeling is followed by one of suffocation. Shock succeeds shock until one's very mind seems to go vibrating away.

At a sign from the manager the silence is so sudden and so intense that it hurts almost as much as the noise.

There is a fascination in walking through these high-ceiled, brilliantly lighted stopes, and these low-ceiled, shadowy drifts. Walls and ceilings are gray quartz, glittering with gold. One is constantly compelled to turn aside for cars of ore on their way to the dumping-places, where their burdens go thundering to the levels below.

At last the manager paused.

"I suppose," said he, sighing, "you wouldn't care to see the—"

I did not catch the last word, and had no notion what it was, but I instantly assured him that I would rather see it than anything in the whole mine.

His face fell.

"Really—" he began.

"Of course we'll see it," said the captain; "we want to see everything."

The manager's face fell lower.

"All right," said he, briefly, "come on!"

We had gone about twenty steps when I, who was close behind him, suddenly missed him. He was gone.

Had he fallen into a dump hole? Had he gone to atoms in a blast? I blinked into the shadows, standing motionless, but could see no sign of him.

Then his voice shouted from above me—"Come on!"

I looked up. In front of me a narrow iron ladder led upward as straight as any flag-pole, and almost as high. Where it went, and why it went, mattered not. The only thing that impressed me was that the manager, halfway up this ladder, had commanded me to "come on."

I? to "come on!" up that perpendicular ladder whose upper end was not in sight!

But whatever might be at the top of that ladder, I had assured him that I would rather see it than anything in the whole mine. It was not for me to quail. I took firm hold of the cold and unclean rungs, and started.

When we had slowly and painfully climbed to the top, we worked our way through a small, square hole and emerged into another stope, or level, and in a very dark part of it. Each man worked by the light of a single candle. They were stoping out ore and making it ready to be dumped into lower levels—from which it would finally be hoisted out of the mine in skips.

The ceiling was so low that we could walk only in a stooping position. The laborers worked in the same position; and what with this discomfort and the insufficient light, it would seem that their condition was unenviable. Yet their countenances denoted neither dissatisfaction nor ill-humor.

"Well," said the manager, presently, "you can have it to say that you have been under the bay, anyhow."

"*Under the*—"

"Yes; under Gastineau Channel. That's straight. It is directly over us."

We immediately decided that we had seen enough of the great mine, and cheerfully agreed to the captain's suggestion that we return to the ship. We were compelled to descend by the perpendicular ladder; and the descent was far worse than the ascent had been.

On our way to the "lift" by which we had made our advent into the mine, we met another small party. It was headed by a tall and handsome man, whose air of delicate breeding would attract attention in any gathering in the world. His distinction and military bearing shone through his greasy slicker and greasier cap—which he instinctively fumbled, in a futile attempt to lift it, as we passed.

It was that brave and gallant explorer, Brigadier-General

Greely, on his way to the Yukon. He was on his last tour of inspection before retirement. It was his farewell to the Northern country which he has served so faithfully and so well.

Eyak Lake, near Cordova

One stumbles at almost every turn in Alaska upon some world-famous person who has answered Beauty's far, insistent call. The modest, low-voiced gentleman at one's side at the captain's table is more likely than not a celebrated explorer or geologist, writer or artist; or, at the very least, an earl.

"After we've seen our passengers eat their first meal," said the chief steward, "we know how to seat them. You can pick out a lady or a gentleman at the table without fail. A boor can fool you every place except at the table. We never assign seats until after the first meal; and oftener than you would suppose we seat them according to their manners at the first meal."

I smiled and smiled, then, remembering the first meal on our steamer. It was breakfast. We had been down to the dining room for something and, returning, found ourselves in a mob at the head of the stairs.

There were one hundred and sixty-five passengers on the boat, and fully one hundred and sixty of them were squeezed like compressed hops around that stairway. In two seconds I was a cluster of hops myself, simply that and nothing more. I do not know how the compressing of hops is usually accomplished; but in my particular case it was done between two immensely big and disagreeable men. They ignored me as calmly as though I were a little boy, and talked cheerfully over my head, although it soon developed that they were not in the least acquainted.

A little black-ringleted, middle-aged woman who seemed to be mounted on wires, suddenly squeezed her head in under their arms, simpering.

"Oh, Doctor!" twittered she, coquettishly. "You are talking to *my husband*."

"The deuce!" ejaculated the Doctor, but whether with evil intent or not, I could not determine from his face.

"Yes, truly. Doctor Metcalf, let me introduce my husband, Mr. Wildey."

They shook hands on my shoulder—but I didn't mind a little thing like that.

"On your honeymoon, eh?" chuckled the Doctor, amiably. The other big man grew red to his hair, and the lady's black ringlets danced up and down.

"Now, now, Doctor," chided she, shaking a finger at him,—she was at least fifty,—"no teasing. No steamer serenades, you know. I was on an Alaskan steamer once, and they pinned red satin hearts all over a bride's stateroom door. Just fancy getting up some morning and finding my stateroom door covered with red satin hearts!"

"I can smell mackerel," said a shrill tenor behind me; and alas! so could I. If there be anything that I like the smell of less than a mackerel, it is an Esquimau hut only.

Somebody sniffed delightedly.

"Fried, too," said a happy voice. "Can't you squeeze down

closer to the stairway?"

Almost at once the big man behind me was tipped forward into the big man in front of me—and, as a mere incident in passing, of course, into me as well. We all went tipping and bobbing and clutching toward the stairway.

Life does not hold many half-hours so rich and so full as the one that followed. As a revelation of the baser side of human nature, it was precious.

My friend was tall; and once, far down the saloon, I caught a glimpse of her handsome, well-carried head as the mob parted for an instant. The expression on her face was like that on the face of the Princess de Lamballe when Lorado Taft has finished with her.

Suddenly I began to move forward. Rather, I was borne forward without effort on my part. A great wave seemed to pick me up and carry me to the head of the stairway. I fairly floated down into the dining room. I fell into the first chair at the first table I came to; but the mob flowed by, looking for something better. Every woman was on a mad hunt for the captain's table. My table remained unpeopled until my friend came in and found me. Gradually and reluctantly the chairs were filled and we devoted ourselves to the mackerel.

In a far corner at the other end of the room, there was a table with flowers on it. With a sigh of relief I saw black ringlets dancing thereat.

"Thank heaven!" I said. "The bride is at the captain's table."

"Ho, no, ma'am," said the gentle voice of the waiter in my ear. "You're hat hit yourself, ma'am. You're hin the captain's hown seat, ma'am. 'E don't come down to the first meal, though, ma'am," he added hastily, seeing my look of horror. For the first, last, and, I trust, only, time in my life I had innocently seated myself at a captain's table, without an invitation.

After breakfast we hastened on deck and went through deep-breathing exercises for an hour, trying to work ourselves back to our usual proportions.

I should like to see a chief steward seat that mob.

I was greatly amused, by the way, at a young waiter's description of an earl.

"We have lots of earls goin' up," said he, easily. "Oh, yes; they go up to Cook Inlet and Kodiak to hunt big game. I always know an earl the first meal. He makes me pull his corks, and he gives me a quarter or a half for every cork I pull. Sometimes I make six bits or a dollar at a meal, just pulling one earl's corks. I'd rather wait on earls than anybody—except ladies, of course," he added, with a positive jerk of remembrance; whereupon we both smiled.

CHAPTER XI

Gastineau Channel northwest of Juneau is not navigable for craft drawing more than three feet of water, at high tide.

Coming out of the channel the steamer turns around the southern end of Douglas Island and heads north into Lynn Canal, with Admiralty Island on the port side and Douglas on the starboard.

Directly north of the latter island is Mendenhall Glacier, formerly known as the Auk. The Indians of this vicinity bear the same name, and have a village north of Juneau. They were a warlike offshoot of the Hoonahs, and bore a bad reputation for treachery and unreliability. Only a few now remain.

In the neighborhood of this glacier—at which the steamer does not call but which may be plainly seen streaming down—are several snow mountains, from five thousand to seven thousand feet in height. They seem hardly worthy of the name of mountain in Alaska; but they float so whitely and so beautifully above the deep blue waters of Lynn Canal that the voyager cannot mistake their mission.

Shelter Island, west of Mendenhall Glacier, forms two channels—Saginaw and Favorite. The latter, as indicated by its name, is the one followed by steamers going to Skaguay. Saginaw is taken by steamers going down Chatham Straits, or Icy Straits, to Sitka.

Indian Houses, Cordova

Sailing up Favorite Channel, Eagle Glacier is passed on the starboard side. It is topped by a great crag which so closely resembles in outline our national emblem that it was so named by Admiral Beardslee, in 1879. The glacier itself is not of great importance.

On Benjamin Island, a fair anchorage may be secured for vessels bound north which have unfortunately been caught in a strong northwest gale.

After the dangerous Vanderbilt Reef is passed, Point Bridget and Point St. Mary's are seen at the entrance to Berner's Bay, where is situated the rich gold mine belonging to Governor Hoggatt.

A light was established in 1905 on Point Sherman; also, on Eldred Rock, where the *Clara Nevada* went down, in 1898, with the loss of every soul on board. For ten years repeated attempts to locate this wreck have been made, on account of the rich treasure which the ship was supposed to carry; but not until 1908 was it discovered—when, upon the occurrence of a phenomenally low tide, it was seen gleaming in clear green depths for a few hours by the keeper of the lighthouse. There was a large loss of life.

There is a mining and mill settlement at Seward, in this vicinity.

William Henry Bay, lying across the canal from Berner's, is celebrated as a sportsman's resort, although this recommendation has come to bear little distinction in a country where it is so common. Enormous crabs, rivalling those to the far "Westward," are found here. Their meat is not coarse, as would naturally be supposed, because of their great size, but of a fine flavor.

Seduction Point, on the island bearing the same name, lies between Chilkaht Inlet on the west and Chilkoot Inlet on the east. For once, Vancouver rose to the occasion and bestowed a striking name, because at this point the treacherous Indians tried to lure Whidbey and his men up the inlet to their village. Upon his refusal to go, they presented a warlike front, and the sincerity of their first advances was doubted.

At the entrance to Chilkaht Inlet, Davidson Glacier is seen sweeping down magnificently from near the summit of the White Mountains. Although this glacier does not discharge bergs, nor rise in splendid tinted palisades straight from the water, as do Taku and Columbia, it is, nevertheless, very imposing—especially if seen from the entrance of the inlet at sunset of a clear day.

The setting of the glaciers of Lynn Canal is superb. The canal itself, named by Vancouver for his home in England, is the most majestic slender water-way in Alaska. From Puget Sound, fiord after fiord leads one on in ever increasing, ever changing splendor, until the grand climax is reached in Lynn Canal.

For fifty-five miles the sparkling blue waters of the canal push almost northward. Its shores are practically unbroken by inlets, and rise in noble sweeps or stately palisades, to domes and peaks of snow. Glaciers may be seen at every turn of the steamer. Not an hour—not one mile of this last fifty-five—should be missed.

In winter the snow descends to the water's edge and this stretch is exalted to sublimity. The waters of the canal take on deep tones of purple at sunset; fires of purest old rose play upon the mountains and glaciers; and the clear, washed-out atmosphere brings the peaks forward until they seem to overhang the steamer throbbing up between them.

Lynn Canal is really but a narrowing continuation of Chatham Strait. Together they form one grand fiord, two hundred miles in length, with scarcely a bend, extending directly north and south. From an average width of four or five miles, they narrow, in places, to less than half a mile.

In July, 1794, Vancouver, lying at Port Althorp, in Cross Sound, sent Mr. Whidbey to explore the continental shore to the eastward. Mr. Whidbey sailed through Icy Strait, seeing the glacier now known as the Brady Glacier, and rounding Point Couverden, sailed up Lynn Canal.

Here, as usual, he was simply stunned by the grandeur and magnificence of the scenery, and resorted to his pet adjectives.

"Both sides of this arm were bounded by *lofty, stupendous mountains, covered with perpetual ice and snow*, whilst the shores in this neighborhood appeared to be composed of cliffs of very fine slate, interspersed with beaches of very fine paving stone.... Up this channel the boats passed, and found the continental shore now take a direction N. 22 W., to a point where the arm narrowed to two miles across; from whence it extended ten miles further in a direction N. 30 W., where its navigable extent terminated in latitude 59° 12′, longitude 224° 33′. This station was reached in the morning of the 16th, after passing some islands and some rocks nearly in mid-channel." (It was probably on one of these that the *Clara Nevada* was wrecked a hundred years later.) "Above the northernmost of these (which lies four miles below the shoal that extends across the upper part of the arm, there about a mile in width) the water was found to be perfectly fresh. Along the edge of this shoal, the boats passed from side to side, in six feet water, and beyond it, the head of the arm extended about half a league, where a small opening in the land was seen, about the fourth of a mile wide, leading to the northwestward, from whence a rapid stream of fresh water rushed over the shoal" (this was Chilkaht River). "But this, to all appearance, was bounded at no great distance by a continuation of the same lofty ridge of snowy mountains so repeatedly mentioned, as stretching eastwardly from Mount Fairweather, and which, in every point of view they had hitherto been seen, appeared to be a firm and close-connected range of *stupendous mountains, forever doomed to support a burthen of undissolving ice and snow.*"

Here, it will be observed, Whidbey was so unconsciously wrought upon by the sublimity of the country that he was moved to fairly poetic utterance. He seemed, however, to be himself doomed to support forever a burthen of gloom and undissolving weariness as heavy as that borne by the mountains.

Up this river, or, as Whidbey called it, *brook*, the Indians informed him, eight chiefs of great consequence resided in a number of villages. He was urged to visit them. Their behavior was peaceable, civil, and friendly; but Mr. Whidbey declined the invitation, and returning, rounded, and named, Point Seduction, and passing into Chilkoot Inlet, discovered more "high, stupendous mountains, loaded with perpetual ice and snow."

After exploring Chilkoot Inlet, they returned down the canal, soon falling in with a party of friendly Indians, who made overtures of peace. Mr. Whidbey describes their chief as a tall, thin, elderly man. He was dressed superbly, and supported a degree of state, consequence, and personal dignity which had been found among no other Indians. His external robe was a very fine large garment that reached from his neck down to his heels, made of wool from the mountain goat—the famous Chilkaht blanket here described, for the first time, by the unappreciative Whidbey. It was neatly variegated with several colors, and edged and otherwise decorated with little tufts of woollen yarn, dyed of various colors. His head-dress was made of wood, resembling a crown, and adorned with bright copper and brass plates, whence hung a number of tails, or streamers, composed of wool and furworked together, dyed of various colors, and each terminating in a whole ermine skin.

His whole appearance, both as to dress and manner, was magnificent.

Mr. Whidbey was suspicious of the good intentions of these new acquaintances, and was therefore well prepared for the trouble that followed.

Headed by the splendid chief, the Indians attacked Whidbey's party in boats, and, being repulsed, followed for two days.

As the second night came on boisterously, Mr. Whidbey was compelled to seek shelter. The Indians, understanding his design, hastened to shore in advance, got possession of the only safe beach, drew up in battle array, and stood with spears couched, ready to receive the exploring party. (This was on the northern part of Admiralty Island.)

Here appears the most delicious piece of unintentional humor in all Vancouver's narrative.

"There was now no alternative but either to force a landing by firing upon them, or to remain at their oars all night. The latter Mr. Whidbey considered to be not only the most humane, but the most prudent to adopt, concluding that their habitations were not far distant, and believing them, from the number of smokes that had been seen during the day, to be a very numerous tribe."

They probably appeared more "stupendous" than any snow-covered mountain in poor Mr. Whidbey's startled eyes.

To avoid a "dispute" with these "troublesome people," Mr. Whidbey withdrew to the main canal and stopped "to take some rest" at a point which received the felicitous name of Point Retreat, on the northern part of Admiralty Island—a name which it still retains.

In the following month Mr. Whidbey was compelled to rest again upon his extremely humane spirit, to the southward in

Frederick Sound.

"The day being fair and pleasant," chronicles Vancouver, "Mr. Whidbey wished to embrace this opportunity of drying their wet clothes, putting their arms in order.... For this purpose the party landed on a commodious beach; but before they had finished their business a large canoe arrived, containing some women and children, and sixteen stout Indian men, well appointed with the arms of the country.... Their conduct afterward put on a very suspicious appearance; the children withdrew into the woods, and the rest fixed their daggers round their wrists, and exhibited other indications not of the most friendly nature. To avoid the chance of anything unpleasant taking place, Mr. Whidbey considered it most humane and prudent to withdraw"—which he did, with all possible despatch.

They were pursued by the Indians; this conduct "greatly attracting the observation of the party."

Mr. Whidbey did not scruple to fire into a fleeing canoe; nor did he express any sorrow when "most hideous and extraordinary noises" indicated that he had fired to good effect; but the instant the Indians lined up in considerable numbers with "couched spears" and warlike attitude, the situation immediately became "stupendous" and Whidbey's ever ready "humaneness" came to his relief.

CHAPTER XII

The Davidson Glacier was named for Professor George Davidson, who was one of its earliest explorers. A heavy forest growth covers its terminal moraine, and detracts from its lower beauty.

Pyramid Harbor, at the head of Chilkaht Inlet, has an Alaska Packers' cannery at the base of a mountain which rises as straight as an arrow from the water to a height of eighteen hundred feet. This mountain was named *Labouchere*, for the Hudson Bay Company's steamer which, in 1862, was almost captured by the Hoonah Indians at Port Frederick in Icy Strait.

Pyramid Harbor was named for a small pyramid-shaped island which now bears the same name, but of which the Indian name is Schlayhotch. The island is but little more than a tiny cone, rising directly from the water. Indians camp here, in large numbers in the summer-time, to work in the canneries. The women sell berries, baskets, Chilkaht blankets of deserved fame, and other curios.

It was this harbor which the Canadians in the Joint High Commission of 1898 unblushingly asked the United States to cede to them, together with Chilkaht Inlet and River, and a strip of land through the *lisière* owned by us.

The Chilkaht River flows into this inlet from the northwest. At its mouth it widens into low tide flats, over which, at low tide, the water flows in ribbonish loops. Here, during a "run," the salmon are taken in countless thousands.

The Chilkahts and Chilkoots are the great Indians of Alaska. They comprise the real aristocracy. They are a brave, bold, courageous race; saucy and independent, constantly carrying a "chip on the shoulder," or a "feather pointing forward" in the head-gear. They are looked up to and feared by the Thlinkits of inferior tribes.

Their villages are located up the Chilkaht and Chilkoot rivers; and their frequent mountain journeyings have developed their legs, giving them a well-proportioned, athletic physique, in marked contrast to the bowed- and scrawny-legged canoe dwellers to the southward and westward.

They are skilful in various kinds of work; but their fame will eventually endure in the exquisite dance-blankets, known as the Chilkaht blanket. These blankets are woven of the wool of the mountain goat, whose winter coat is strong and coarse. At shedding time in the spring, as the goat leaps from place to place, the wool clings to trees, rocks, and bushes in thick festoons. These the indolent Indians gather for the weaving of their blankets, rather than take the trouble of killing the goats.

This delicate and beautiful work is, like the Thlinkit and Chilkaht basket, in simple twined weaving. The warp hangs loose from the rude loom, and the wool is woven upward, as in Attu and Haidah basketry.

The owner of one of the old Chilkaht blankets possesses a treasure beyond price. The demand has cheapened the quality of those of the present day; but those of Baranoff's time were marvels of skill and coloring, considering that Indian women's dark hands were the only shuttles.

Black, white, yellow, and a peculiar blue are the colors most frequently observed in these blankets; and a deep, rich red is becoming more common than formerly. A wide black, or dark, band usually surrounds them, border-wise, and a fringe as wide as the blanket falls magnificently from the bottom; a narrower one from the sides.

The old and rare ones were from a yard and a half to two yards long. The modern ones are much smaller, and may be obtained as low as seventy-five dollars. The designs greatly resemble those of the Haidah hats and basketry.

The full face, with flaring nostrils, small eyes, and ferocious display of teeth, is the bear; the eye which appears in all places and in all sizes is that of the thunder-bird, or, with the Haidahs, the sacred raven.

There is an Indian mission, named Klukwan, at the head of the inlet.

The Chilkahts were governed by chiefs and sub-chiefs. At the time of the transfer "Kohklux" was the great chief of the region. He was a man of powerful will and determined character. He wielded a strong influence over his tribes, who believed that he bore a charmed life. He was friendly to Americans and did everything in his power to assist Professor George Davidson, who went to the head of Lynn Canal in 1869 to observe the solar total eclipse.

The Indians apparently placed no faith in Professor Davidson's announcement of approaching darkness in the middle of the day, however, and when the eclipse really occurred, they fled from him, as from a devil, and sought the safety of their mountain fastnesses.

The passes through these mountains they had held from time immemorial against all comers. The Indians of the vast interior regions and those of the coast could trade only through the Chilkahts—the scornful aristocrats and powerful autocrats of the country.

CHAPTER XIII

Coming out of Chilkaht Inlet and passing around Seduction Point into Chilkoot Inlet, Katschin River is seen flowing in from the northeast. The mouth of this river, like that of the Chilkaht, spreads into extensive flats, making the channel very narrow at this point.

Across the canal lies Haines Mission, where, in 1883, Lieutenant Schwatka left his wife to the care of Doctor and Mrs. Willard, while he was absent on his exploring expedition down the Yukon.

The Willards were in charge of this mission, which was maintained by the Presbyterian Board of Missions, until some trouble arose with the Indians over the death of a child, to whom the Willards had administered medicines.

"Crossing the Mission trail," writes Lieutenant Schwatka, "we often traversed lanes in the grass, which here was fully five feet high, while, in whatever direction the eye might look, wild flowers were growing in the greatest profusion. Dandelions as big as asters, buttercups twice the usual size, and violets rivalling the products of cultivation in lower latitudes were visible around. It produced a singular and striking contrast to raise the eyes from this almost tropical luxuriance, and allow them to rest on Alpine hills, covered halfway down their shaggy sides with the snow and glacier ice, and with cold mist condensed on their crowns.... Berries and berry blossoms grew in a profusion and variety which I have never seen equalled within the same limits in lower latitudes."

This was early in June. Here the lieutenant first made the acquaintance of the Alaska mosquito and gnat, neither of which is to be ignored, and may be propitiated by good red blood only; also, the giant devil's-club, which he calls devil's-sticks. He was informed that this nettle was formerly used by the shamans, or medicine-men, as a prophylactic against witchcraft, applied externally.

The point of this story will be appreciated by all who have come in personal contact with this plant, so tropical in appearance when its immense green leaves are spread out flat and motionless in the dusk of the forest.

From Chilkoot Inlet the steamer glides into Taiya Inlet, which leads to Skaguay. Off this inlet are many glaciers, the finest of which is Ferebee.

Chilkoot Inlet continues to the northwestward. Chilkoot River flows from a lake of the same name into the inlet. There are an Indian village and large canneries on the inlet.

Taiya Inlet leads to Skaguay and Dyea. It is a narrow water-way between high mountains which are covered nearly to their crests with a heavy growth of cedar and spruce. They are crowned, even in summer, with snow, which flows down their fissures and canyons in small but beautiful glaciers, while countless cascades foam, sparkling, down to the sea, or drop sheer from such great heights that the beholder is bewildered by their slow, never ceasing fall.

Here,—at the mouth of the Skaguay River, with mountains rising on all sides and the green waters of the inlet pushing restlessly in front; with its pretty cottages climbing over the foot-hills, and with well-worn, flower-strewn paths enticing to the heights; with the Skaguay's waters winding over the grassy flats like blue ribbons; with flower gardens beyond description and boxes in every window scarlet with bloom; with cascades making liquid and most sweet music by day and irresistible lullabies by night, and with snow peaks seeming to float directly over the town in the upper pearl-pink atmosphere—is Skaguay, the romantic, the marvellous, the town which grew from a dozen tents to a city of fifteen thousand people almost in a night, in the golden year of ninety-eight.

I could not sleep in Skaguay for the very sweetness of the July night. A cool lavender twilight lingered until eleven o'clock, and then the large moon came over the mountains, first outlining their dark crests with fire; then throbbing slowly on from peak to peak—bringing irresistibly to mind the lines:—

"Like a great dove with silver wingsStretched, quivering o'er the sea,The moon her glistening plumage bringsAnd hovers silently."

The air was sweet to enchantment with flowers; and all night long through my wide-open window came the far, dreamy, continuous music of the waterfalls.

On all the Pacific Coast there is not a more interesting, or a more profitable, place in which to make one's headquarters for the summer, than Skaguay. More side trips may be made, with less expenditure of time and money, from this point than from any other. Launches may be hired for expeditions down Lynn Canal and up the inlets,—whose unexploited splendors may only be seen in this way; to the Mendenhall, Davidson, Denver, Bertha, and countless smaller glaciers; to Haines, Fort Seward, Pyramid Harbor, and Seduction Point; while by canoe, horse, or his own good legs, one may get to the top of Mount Dewey and to Dewey Lake; up Face Mountain; to Dyea; and many hunting grounds where mountain sheep, bear, goat, ptarmigan, and grouse are plentiful.

The famous White Pass railway—which was built in eighteen months by the "Three H's," Heney, Hawkins, and Hislop, and which is one of the most wonderful engineering feats of the world—may be taken for a trip which is, in itself, worth going a thousand miles to enjoy. Every mile of the way is historic ground—not only to those who toiled over it in 'ninety-seven and 'ninety-eight, bent almost to the ground beneath their burdens, but to the whole world, as well. The old Brackett wagon road; White Pass City; the "summit"; Bennett Lake; Lake Lindeman;

White Horse Rapids; Grand Canyon; Porcupine Ridge—to whom do these names not stand for tragedy and horror and broken hearts?

The town of Skaguay itself is more historic than any other point. Here the steamers lightered or floated ashore men, horses, and freight. "You pay your money and you take your chance," the paraphrase went in those days. Many a man saw every dollar he had in provisions—and often it was a grubstake, at that—sink to the bottom of the canal before his eyes. Others saw their outfits soaked to ruin with salt water. For those who landed safely, there were horrors yet to come.

And here, between these mountains, in this wind-racked canyon, the town of Skaguay grew; from one tent to hundreds in a day, from hundreds to thousands in a week; from tents to shacks, from shacks to stores and saloons. Here "Soapy" Smith and his gang of outlaws and murderers operated along the trail; here he was killed; here is his dishonored grave, between the mountains which will not endure longer than the tale of his desperate crimes, and his desperate expiation.

Not the handsome style of man that one would expect of such a bold and daring robber was "Soapy." No flashing black eyes, heavy black hair, and long black mustache made him "a living flame among women," as Rex Beach would put it. Small, spare, insignificant in appearance, it has been said that he looked more like an ill-paid frontier minister than the head of a lawless and desperate gang of thieves.

His "spotters" were scattered along the trail all the way to Dawson. They knew what men were "going in," what ones "coming out," "heeled." Such men were always robbed; if not on the road, then after reaching Skaguay; when they could not safely, or easily, be robbed alive, they were robbed dead. It made no difference to "Soapy" or his gang of men and women. It was a reign of terror in that new, unknown, and lawless land.

There is nothing in Skaguay to-day—unless it be the sinking grave of "Soapy" Smith, which is not found by every one—to suggest the days of the gold rush, to the transient visitor. It is a quiet town, where law and order prevail. It is built chiefly on level ground, with a few very long streets—running out into the alders, balms, spruces, and cottonwoods, growing thickly over the river's flats.

In all towns in Alaska the stores are open for business on Sunday when a steamer is in. If the door of a curio-store, which has tempting baskets or Chilkaht blankets displayed in the window, be found locked, a dozen small boys shout as one, "Just wait a minute, lady. Propri'tor's on the way now. He just stepped out for breakfast. Wait a minute, lady."

We arrived at Skaguay early on a Sunday morning, and were directed to the "'bus" of the leading hotel. We rode at least a mile before reaching it. We found it to be a wooden structure, four or five stories in height; the large office was used as a kind of general living-room as well. The rooms were comfortable and the table excellent. The proprietress grows her own vegetables and flowers, and keeps cows, chickens, and sheep, to enrich her table.

About ten o'clock in the forenoon we went to the station to have our trunks checked to Dawson. The doors stood open. We entered and passed from room to room. There was no one in sight. The square ticket window was closed.

We hammered upon it and upon every closed door. There was no response. We looked up the stairway, but it had a personal air. There are stairways which seem to draw their steps around them, as a duchess does her furs, and to give one a look which says, "Do not take liberties with me!"—while others seem to be crying, "Come up; come up!" to every passer-by. I have never seen a stairway that had the duchess air to the degree that the one in the station at Skaguay has it. If any one doubts, let him saunter around that station until he finds the stairway and then take a good look at it.

We went outside, and I, being the questioner of the party, asked a man if the ticket office would be open that day.

He squared around, put his hands in his pockets, bent his

wizened body backward, and gave a laugh that echoed down the street.

"God bless your soul, lady," said he, "*on Sunday!* Only an extry goes out on Sundays, to take round-trip tourists to the summit and back while the steamer waits. To-day's extry has gone."

"Yes," said I, mildly but firmly, "but we are going to Dawson to-morrow. Our train leaves at nine o'clock, and there will be so many to get tickets signed and baggage checked—"

He gave another laugh.

"Don't you worry, lady. Take life easy, the way we do here. If we miss one train, we take the next—unless we miss it, too!" He laughed again.

At that moment, bowing and smiling in the window of the ticket office, appeared a man—the nicest man!

"Will you see him bow!" gasped my friend. "Is he bowing at *us*? Why—are you *bowing back*?"

"Of course I am."

"What on earth does he want?"

"He wants to be nice to us," I replied; and she followed me inside.

The nice face was smiling through the little square window.

"I was upstairs," he said—ah, he had descended by way of the "Duchess," "and I heard you rapping on windows and doors"—the smile deepened, "so I came down to see if I could serve you."

We related our woes; we got our tickets signed and our baggage checked; had all our questions answered—and they were not few—and the following morning ate our breakfast at our leisure and were greatly edified by our fellow-travellers' wild scramble to get their bills paid and to reach the station in time to have their baggage checked.

Valdez

CHAPTER XIV

Sailing down Lynn Canal, Chatham Strait, and the narrow, winding Peril Strait, the sapphire-watered and exquisitely islanded Bay of Sitka is entered from the north. Six miles above the Sitka of to-day a large wooden cross marks the site of the first settlement, the scene of the great massacre.

On one side are the heavily and richly wooded slopes of Baranoff Island, crested by many snow-covered peaks which float in the higher primrose mist around the bay; on the other, water avenues—growing to paler, silvery blue in the distance—wind in and out among the green islands to the far sea, glimpses of which may be had; while over all, and from all points for many miles, the round, deeply cratered dome of Edgecumbe shines white and glistening in the sunlight. It is the superb feature of the landscape; the crowning glory of a scene that would charm even without it.

Mount Edgecumbe is the home of Indian myth and legend—as is Nass River to the southeastward. In appearance, it is like no other mountain. It is only eight thousand feet in height, but it is so round and symmetrical, it is so white and sparkling, seen either from the ocean or from the inner channels, and its crest is sunken so evenly into an unforgettable crater, that it instantly impresses upon the beholder a kind of personality among mountains.

In beauty, in majesty, in sublimity, it neither approaches nor compares with twenty other Alaskan mountains which I have seen; but, like the peerless Shishaldin, to the far westward, it stands alone, distinguished by its unique features from all its sister peaks.

Not all the streams of lava that have flowed down its sides for hundreds of years have dulled its brilliance or marred its graceful outlines.

I have searched Vancouver's chronicles, expecting to fined Edgecumbe described as "a mountain having a very elegant hole in the top,"—to match his "elegant fork" on Mount Olympus of Puget Sound.

Peril Strait is a dangerous reach leading in sweeping curves from Chatham Strait to Salisbury Sound. It is the watery dividing line between Chichagoff and Baranoff islands. It has two narrows, where the rapids at certain stages of the tides are most dangerous.

Upon entering the strait from the east, it is found to be wide and peaceful. It narrows gradually until it finally reaches, in its forty-mile windings, a width of less than a hundred yards.

There are several islands in Peril Strait: Fairway and Trader's at the entrance; Broad and Otstoi on the starboard; Pouverstoi, Elovoi, Rose, and Kane. Between Otstoi and Pouverstoi islands is Deadman's Reach. Here are Peril Point and Poison Cove, where Baranoff lost a hundred Aleuts by their eating of poisonous mussels in 1799. For this reason the Russians gave it the name, Pogibshi, which, interpreted, means "Destruction," instead of the "Pernicious" or "Peril" of the present time.

Deadman's Reach is as perilous for its reefs as for its mussels. Hoggatt Reef, Dolph Rock, Ford Rock, Elovoi Island, and Krugloi Reef are all dangerous obstacles to navigation, making this reach as interestingly exciting as it is beautiful.

Fierce tides race through Sergius Narrows, and steamers going to and from Sitka are guided by the careful calculation of their masters, that they may arrive at the narrows at the favorable stage of the tides. Bores, racing several feet high, terrific whirlpools, and boiling geysers make it impossible for vessels to approach when the tides are at their worst. This is one of the most dangerous reaches in Alaska.

Either Rose or Adams Channel may be used going to Sitka, but the latter is the favorite.

Kakul Narrows leads into Salisbury Sound; but the

Sitkan steamers barely enter this sound ere they turn to the southeastward into Neva Strait. It was named by Portlock for the Marquis of Salisbury.

Entrance Island rises between Neva Strait and St. John the Baptist Bay. There are both coal and marble in the latter bay.

Halleck Island is completely surrounded by Nakwasina Passage and Olga Strait, joining into one grand canal of uniform width.

All these narrow, tortuous, and perilous water-ways wind around the small islands that lie between Baranoff Island on the east and Kruzoff Island on the west. Baranoff is one hundred and thirty miles long and as wide as thirty miles in places. Kruzoff Island is small, but its southern extremity, lying directly west of Sitka, shelters that favored place from the storms of the Pacific.

Whitestone Narrows in the southern end of Neva Strait is extremely narrow and dangerous, owing to sunken rocks. Deep-draught vessels cannot enter at low tide, but must await the favorable half-hour.

Sitka Sound is fourteen miles long and from five to eight wide. It is more exquisitely islanded than any other bay in the world; and after passing the site of Baranoff's first settlement and Old Sitka Rocks, the steamer's course leads through a misty emerald maze. Sweeping slowly around the green shore of one island, a dozen others dawn upon the beholder's enraptured vision, frequently appearing like a solid wall of green, which presently parts to let the steamer slide through,—when, at once, another dazzling vista opens to the view.

Before entering Sitka Sound, Halleck, Partoffs-Chigoff, and Krestoff are the more important islands; in Sitka Sound, Crow, Apple, and Japonski. The latter island is world-famous. It is opposite, and very near, the town; it is about a mile long, and half as wide; its name, "Japan," was bestowed because, in 1805, a Japanese junk was wrecked near this island, and the crew was forced to dwell upon it for weeks. It is greenly and gracefully draped with cedar and spruce trees, and is an object of much interest to tourists.

Around Japonski cluster more than a hundred small islands of the Harbor group; in the whole sound there are probably a thousand, but some are mere green or rocky dots floating upon the pale blue water.

A magnetic and meteorological observatory was established on Japonski by the Russians and was maintained until 1867.

An Alaskan Road House

CHAPTER XV

The Northwest Coast of America extended from Juan de Fuca's Strait to the sixtieth parallel of north latitude. Under the direction of the powerful mind of Peter the Great explorations in the North Pacific were planned. He wrote the following instructions with his own hand, and ordered the Chief Admiral, Count Fedor Apraxin, to see that they were carried into execution:—

First.—One or two boats, with decks, to be built at Kamchatka, or at any other convenient place, with which

Second.—Inquiry should be made in relation to the northerly coasts, to see whether they were not contiguous with America, since their end was not known. And this done, they should

Third.—See whether they could not somewhere find an harbor belonging to Europeans, or an European ship. They should likewise set apart some men who were to inquire after the name and situation of the coasts discovered. Of all this an exact journal should be kept, with which they should return to St. Petersburg.

Before these instructions could be carried out, Peter the Great died.

His Empress, Catherine, however, faithfully carried out his plans.

The first expedition set out in 1725, under the command of Vitus Behring, a Danish captain in the Russian service, with Lieutenants Spanberg and Chirikoff as assistants. They carried several officers of inferior rank; also seamen and ship-builders. Boats were to be built at Kamchatka, and they started overland through Siberia on February the fifth of that year. Owing to many trials and hardships, it was not until 1728 that Behring sailed along the eastern shore of the peninsula, passing and naming St. Lawrence Island, and on through Behring Strait. There, finding that the coast turned westward, his natural conclusion was that Asia and America were not united, and he returned to Kamchatka. In 1734, under the patronage of the Empress Elizabeth, Peter the Great's daughter, a second expedition made ready; but owing to insurmountable difficulties, it was not until September, 1740, that Behring and Chirikoff set sail in the packet-boats *St. Peter* and *St. Paul*—Behring commanding the former—from Kamchatka. They wintered at Avatcha on the Kamchatkan Peninsula, where a few buildings, including a church, were hastily erected, and to which the name of Petropavlovsk was given.

On June 4, 1741, the two ships finally set sail on their eventful voyage—how eventful to us of the United States we are only, even now, beginning to realize. They were accompanied by Lewis de Lisle de Croyere, professor of astronomy, and Georg Wilhelm Steller, naturalist.

Müller, the historian, and Gmelin, professor of chemistry and natural history, also volunteered in 1733 to accompany the expedition; but owing to the long delay, and ill-health arising from arduous labors in Kamchatka, they were compelled to permit the final expedition to depart without them.

On the morning of June 20, the two ships became separated in a gale and never again sighted one another. Chirikoff took an easterly course, and to him, on the fifteenth of July, fell, by chance, the honor of the first discovery of land on the American continent, opposite Kamchatka, in 55° 21′. Here he lost two boatloads of seamen whom he sent ashore for investigation, and whose tragic fate may only be guessed from the appearance of savages later, upon the shore.

That the first Russians landing upon the American continent should have met with so horrible a fate as theirs is supposed to have been, has been considered by the superstitious as an evil omen. The first boat sent ashore contained ten armed sailors and was commanded by the mate, Abraham Mikhailovich Dementief. The latter is described as a capable young man, of distinguished family, of fine personal appearance, and of kind heart, who, having suffered from an unfortunate love affair, had offered himself to serve his country in this most hazardous expedition. They were furnished with provisions and arms, including a small brass cannon, and given a code of signals by Chirikoff, by which they might communicate with the ship. The boat reached the shore and passed behind a point of land. For several days signals which were supposed to indicate that the party was alive and well, were observed rising at intervals. At last, however, great anxiety was experienced by those on board lest the boat should have sustained damage in some way, making it impossible for the party to return. On the fifth day another boat was sent ashore with six men, including a carpenter and a calker. They effected a landing at the same place, and shortly afterward a great smoke was observed, pushing its dark curls upward above the point of land behind which the boats had disappeared.

The following morning two boats were discovered putting off from the shore. There was great rejoicing on the ship, for the night had been passed in deepest anxiety, and without further attention to the boats, preparations were hastily made for immediate sailing. Soon, however, to the dread and horror of all, it was discovered that the boats were canoes filled with savages, who, at sight of the ship, gave unmistakable signs of astonishment, and shouting "Agaï! Agaï!" turned hastily back to the shore.

Silence and consternation fell upon all. Chirikoff, humane and kind-hearted, bitterly bewailed the fate of his men. A wind soon arising, he was forced to make for the open sea. He remained in the vicinity, and as soon as it was possible, returned to his anchorage; but no signs of the unfortunate sailors were ever discovered.

Without boats, and without sufficient men, no attempt at a rescue could be made; nor was further exploration possible; and heavy-hearted and discouraged, notwithstanding his brilliant success, Chirikoff again weighed anchor and turned his ship homeward.

He and his crew were attacked by scurvy; provisions and

water became almost exhausted; Chirikoff was confined to his berth, and many died; some islands of the chain now known as the Aleutians were discovered; and finally, on the 8th of October, 1741, after enduring inexpressible hardships, great physical and mental suffering, and the loss of twenty-one men, they arrived on the coast of Kamchatka near the point of their departure.

In the meantime, on the day following Chirikoff's discovery of land, Commander Behring, far to the northwestward, saw, rising before his enraptured eyes, the splendid presence of Mount St. Elias, and the countless, and scarcely less splendid, peaks which surround it, and which, stretching along the coast for hundreds of miles, whitely and silently people this region with majestic beauty. Steller, in his diary, claims to have discovered land on the fifteenth, but was ridiculed by his associates, although it was clearly visible to all in the same place on the following day.

They effected a landing on an island, which they named St. Elias, in honor of the day upon which it was discovered. It is now known as Kayak Island, but the mountain retains the original name. Having accomplished the purpose of his expedition, Behring hastily turned the *St. Peter* homeward.

For this haste Behring has been most severely criticised. But when we take into consideration the fact that preparations for this second expedition had begun in 1733; that during all those years of difficult travelling through Siberia, of boat building and the establishment of posts and magazines for the storing of provisions, he had been hampered and harassed almost beyond endurance by the quarrelling, immorality, and dishonesty of his subordinates; that for all dishonesty and blunders he was made responsible to the government; and that so many complaints of him had been forwarded to St. Petersburg by officers whom he had reprimanded or otherwise punished that at last, in 1739, officers had been sent to Ohkotsk to investigate his management of the preparations; that he had now discovered that portion of the American continent which he had set out to discover, had lost Chirikoff, upon whose youth and hopefulness he had been, perhaps unconsciously, relying; and—most human of all—that he had a young and lovely wife and two sons in Russia whom he had not seen for years (and whom he was destined never to see again); when we take all these things into consideration, there seems to be but little justice in these harsh criticisms.

To-day, there is no portion of the Alaskan coast more unreliable, nor more to be dreaded by mariners, than that in the vicinity of Behring's discovery. Even in summer violent winds and heavy seas are usually encountered. Steamers cannot land at Kayak, and passengers and freight are lightered ashore; and when this is accomplished without disaster or great difficulty, the trip is spoken of as an exceptional one. Yet Behring remained in this dangerous anchorage five days. Several landings were made on the two Kayak Islands, and on various smaller ones. Some Indian huts, without occupants, were found and entered. They were built of logs and rough bark and roofed with tough dried grasses. There were, also, some sod cellars, in which dried salmon was found. In one of the cabins were copper implements, a whetstone, some arrows, ropes, and cords made of sea-weed, and rude household utensils; also herbs which had been prepared according to Kamchatkan methods.

Returning, Behring discovered and named many of the Aleutian Islands and exchanged presents with the friendly natives. They were, however, overtaken by storms and violent illness; they suffered of hunger and thirst; so many died that barely enough remained to manage the ship. Finally on November 5, in attempting to land, the *St. Peter* was wrecked on a small island, where, on the 8th of December, in a wretched hut, half covered with sand which sifted incessantly through the rude boards that were his only roof, and after suffering unimaginable agonies, the illustrious Dane, Vitus Behring, died the most miserable of deaths. The island was named for him, and still retains the name, being the larger of the Commander Islands.

The survivors of the wreck remaining on Behring Island dragged out a wretched existence until spring, in holes dug in the sand and roofed with sails. Water they had; but their food consisted chiefly of the flesh of sea-otters and seals. In May, weak, emaciated, and hopeless though they were, and with their brave leader gone, they began building a boat from the remnants of the *St. Peter.* It was not completed until August; when, with many fervent prayers, they embarked, and, after nine days of mingled dread and anxiety in a frail and leaking craft, they arrived safely on the Kamchatkan shore.

All hope of their safety had long been abandoned, and there was great rejoicing upon their return. Out of their own deep gratitude a memorial was placed in the church at Petropavlovsk, which is doubtless still in existence, as it was in a good state of preservation a few years ago.

Russian historians at first seemed disposed to depreciate Behring's achievement, and to over-exalt the Russian, Chirikoff. They made the claim that the latter was a man of high intellectual attainments, courageous, hopeful, and straightforward; kind-hearted, and giving thought to and for others. He was instructor of the marines of the guard, but after having been recommended to Peter the Great as a young man highly qualified to accompany the expedition under Behring, he was promoted to a lieutenancy and accompanied the latter on his first expedition in 1725; and on the second, in 1741, he was made commander of the *St. Pevril*, or *St. Paul*, "not by seniority but on account of superior knowledge and worth." Despite the fact that Behring was placed by the emperor in supreme command of both expeditions, the Russians looked upon Chirikoff as the real hero. He was a favorite with all, and in the accounts of quarrels and dissensions among the heads of the various detachments of scientists and naval officers of the expedition, the name of Chirikoff does not appear. His wife and daughter accompanied him to Siberia.

Captain Vitus Behring—or Ivan Ivanovich, as the Russians called him—is described as a man of intelligence, honesty, and irreproachable conduct, but rather inclined in his later years to vacillation of purpose and indecision of character, yielding easily to an irritable and capricious temper. Whether these facts were due to age or disease is not known; but that they seriously affected his fitness for the command of an exploration is not denied, even by his admirers. Even so sane and conscientious an historian as Dall calls him timid, hesitating, and indolent, and refers to his "characteristic imbecility," "utter incapacity," and "total incompetency." It is incredible, however, that a man of such gross faults should have been given the command of this brilliant expedition by so wise and great a monarch as Peter. Behring died,—old, discouraged, in indescribable anguish; suspicious of every one, doubting even Steller, the naturalist who accompanied the expedition and who was his faithful friend. Chirikoff returned, young, flushed with success, popular and in favor with all, from the Empress down to his subordinates. Favored at the outset by youth and a cheerful spirit, his bright particular star guided him to the discovery of land a few hours in advance of Behring. This was his good luck and his good luck only. Vitus Behring, the Dane in the Russian service, was in supreme command of the expedition; and to him belongs the glory. One cannot to-day sail that magnificent sweep of purple water between Alaska and Eastern Siberia without a thrill of thankfulness that the fame and the name of the illustrious Dane are thus splendidly perpetuated.

To-day, his name is heard in Alaska a thousand times where Chirikoff's is heard once. The glory of the latter is fading, and Behring is coming to his own—Russians speaking of him with a pride that approaches veneration.

Kow-Ear-Nuk and his Drying Salmon

Captain Martin Petrovich Spanberg, the third in command of the expedition, was also a Dane. He is everywhere described as an illiterate, coarse, cruel man; grasping, selfish, and unscrupulous in attaining ends that made for his own advancement. In his study of the character of Spanberg, Bancroft—who has furnished the most complete and painstaking description of these expeditions—makes comment which is, perhaps unintentionally, humorous. After describing Spanberg as exceedingly avaricious and cruel, and stating that his bad reputation extended over all Siberia, and that his name appears in hundreds of complaints and petitions from victims of his licentiousness, cruelty, and avarice, Bancroft näively adds, "He was just the man to become rich." Wealthy people may take such comfort as they can out of the comment.

CHAPTER XVI

Inspired by the important discoveries of this expedition and by the hope of a profitable fur trade with China, various Russian traders and adventurers, known as "promyshleniki," made voyages into the newly discovered regions, pressing eastward island by island, and year by year; beginning that long tale of cruelty and bloodshed in the Aleutian Islands which has not yet reached an end. Men as harmless as the pleading, soft-eyed seals were butchered as heartlessly and as shamelessly, that their stocks of furs might be appropriated and their women ravished. In 1745 Alexeï Beliaief and ten men inveigled fifteen Aleutians into a quarrel with the sole object of killing them and carrying off their women. In 1762, the crew of the *Gavril* persuaded twenty-five young Aleutian girls to accompany them "to pick berries and gather roots for the ship's company." On the Kamchatkan coast several of the crew and sixteen of these girls were landed to pick berries. Two of the girls made their escape into the hills; one was killed by a sailor; and the others cast themselves into the sea and were drowned. Gavril Pushkaref, who was in command of the vessel, ordered that all the remaining natives, with the exception of one boy and an interpreter, should be thrown overboard and drowned.

These are only two instances of the atrocious outrages perpetrated upon these innocent and childlike people by the brutal and licentious traders who have frequented these far beautiful islands from 1745 to the present time. From year to year now dark and horrible stories float down to us from the far northwestward, or vex our ears when we sail into those pale blue water-ways. Nor do they concern "promyshleniki" alone. Charges of the gravest nature have been made against men of high position who spend much time in the Aleutian Islands. That these gentle people have suffered deeply, silently, and shamefully, at the hands of white men of various nationalities, has never been denied, nor questioned. It is well known to be the simple truth. From 1760 to about 1766 the natives rebelled at their treatment and active hostilities were carried on. Many Russians were killed, some were tortured. Solovief, upon arriving at Unalaska and learning the fate of some of his countrymen, resolved to avenge them. His designs were carried out with unrelenting cruelty. By some writers, notably Berg, his crimes have been palliated, under the plea that nothing less than extreme brutality could have so soon reduced the natives to the state of fear and humility in which they have ever since remained—failing to take into consideration the atrocities perpetrated upon the natives for years before their open revolt.

In 1776 we find the first mention of Grigor Ivanovich Shelikoff; but it was not until 1784 that he succeeded in making the first permanent Russian settlement in America, on Kodiak Island,—forty-three dark and strenuous years after Vitus Behring saw Mount St. Elias rising out of the sea. Shelikoff was second only to Baranoff in the early history of Russian America, and is known as "the founder and father of Russian colonies in America." His wife, Natalie, accompanied him upon all his voyages. She was a woman of very unusual character, energetic and ambitious, and possessed of great business and executive ability. After her husband's death, her management for many years of not only her own affairs, but those of the Shelikoff Company as well, reflected great credit upon herself.

It was the far-sighted Shelikoff who suggested and carried out the idea of a monopoly of the fur trade in Russian America under imperial charter. As a result of his forceful presentation of this scheme and the able—and doubtless selfish—assistance of General Jacobi, the governor-general of Eastern Siberia, the Empress became interested. In 1788 an imperial ukase was issued, granting to the Shelikoff Company exclusive control of the territory already occupied by them. Assistance from the public coffers was at that time withheld; but the Empress graciously granted to Shelikoff and his partner, Golikof, swords and medals containing her portrait. The medals were to be worn around their necks, and bore inscriptions explaining that they "had been conferred for services rendered to humanity by noble and bold deeds."

Although Shelikoff greatly preferred the pecuniary assistance from the government, he nevertheless accepted with a good grace the honor bestowed, and bided his time patiently.

In accordance with commands issued by the commander at Ohkotsk and by the Empress herself, Shelikoff adopted a policy of humanity in his relations with the natives, although it is suspected that this was on account of his desire to please the Empress and work out his own designs, rather than the result of his own kindness of heart.

Steamer "Resolute"

With the clearness of vision which distinguished his whole career, Shelikoff selected Alexander Baranoff as his agent in the territory lying to the eastward of Kodiak. In Voskressenski, or Sunday, Harbor—now Resurrection Bay, on which the town of Seward is situated—Baranoff built in 1794 the first vessel to glide into the waters of Northwestern America—the *Phœnix*. At the request of Shelikoff a colony of two hundred convicts, accompanied by twenty priests, were sent out by imperial ukase, and established at Yakutat Bay, under Baranoff. During the years that followed many complaints were entered by the clergy against Baranoff for cruelty, licentiousness, and mismanagement of the company's affairs. But, whatever his faults may have been, it is certain that no man could have done so much for the promotion of the company's interests at that time as Baranoff; nor could any other so efficiently have conducted its affairs.

It was during his governorship that the rose of success bloomed brilliantly for the Russian-American Company in the colonies. He was a shrewd, tireless, practical business man. His successors were men distinguished in army and navy circles, haughty and patrician, but absolutely lacking in business ability, and ignorant of the unique conditions and needs of the country.

After Baranoff's resignation and death, the revenues of the company rapidly declined, and its vast operations were conducted at a loss.

It was in 1791 that Baranoff assumed command of all the establishments on the island of the Shelikoff Company which, under imperial patronage, had already secured a partial monopoly of the American fur trade. Owing to competition by independent traders, the large company, after the death of Shelikoff, united with its most influential rival, under the name of the Shelikoff United Company. The following year this company secured an imperial ukase which granted to it, under the name of the Russian-American Company, "full privileges, for a period of twenty years, on the coast of Northwestern America, beginning from latitude fifty-five degrees North, and including the chain of islands extending from Kamchatka northward to America and southward to Japan; the exclusive right to all enterprises, whether hunting, trading, or building, and to new discoveries,with strict prohibition from profiting by any of these pursuits, not only to all parties who might engage in them on their own responsibility, but also to those who formerly had ships and establishments there, except those who have united with the new company."

In the same year a fort was established by Baranoff, on what is now Sitka Sound. This was destroyed by natives; and in 1804 another fort was erected by Baranoff, near the site of the former one, which he named Fort Archangel Michael. This fort is the present Sitka. Its establishment enabled the Russian-American Company to extend its operations to the islands lying southward and along the continental shore.

We now come to the most fascinating portion of the history of Alaska. Not even the wild and romantic days of gold excitement in the Klondike can equal Baranoff's reign at Sitka for picturesqueness and mysterious charm. The strength and personality of the man were such that to-day one who is familiar with his life and story, entering Sitka, will unconsciously feel his presence; and will turn, with a sigh, to gaze upon the commanding height where once his castle stood.

There were many dark and hopeless days for Baranoff during his first years with the company, and it was while in a state of deep discouragement and hopelessness that he received the news of his appointment as chief manager of the newly organized Russian-American Company. Most of his plans and undertakings had failed; many Russians and natives had been lost on hunting voyages; English and American traders had superseded him at every point to the eastward of Kodiak; many of his Aleutian hunters had been killed in conflict with the savage Thlinkits; he had lost a sloop which had been constructed at Voskressenski Bay; and finally, he had returned to Kodiak enduring the agonies of inflammatory rheumatism, only to be reproached by the subordinates, who were suffering of actual hunger—so long had they been without relief from supply ships.

In this dark hour the ship arrived which carried not only good tidings, but plentiful supplies as well. Baranoff's star now shone brightly, leading him on to hope and renewed effort.

In the spring of the following year, 1799, Baranoff, with two vessels manned by twenty-two Russians, and three hundred and fifty canoes, set sail for the eastward. Many of the natives were lost by foundering of the canoes, and many more by slaughter at the hands of the Kolosh, but finally they arrived at a point now known as Old Sitka, six miles north of the present Sitka, and bartered with the chief of the natives for a site for a settlement. Captain Cleveland, whose ship *Caroline*, of Boston, was then lying in the harbor, describes the Indians of the vicinity as follows: "A more hideous set of beings in the form of men and women, I had never before seen. The fantastic manner in which many of the faces were painted was probably intended to give them a more ferocious appearance; and some groups looked really as if they had escaped from the dominions of Satan himself. One had a perpendicular line dividing the two sides of the face, one side of which was painted red, the other black, with the hair daubed with grease and red ochre, and filled with the down of birds. Another had the face divided with a horizontal line in the middle, and painted black and white. The visage of a third was painted in checkers, etc. Most of them had little mirrors, before the acquisition of which they must have been dependent on each other for those correct touches of the pencil which are so much in vogue, and which daily require more time than the toilet of a Parisian belle."

These savages were known to be treacherous and dangerous, but they pretended to be friendly, and fears were gradually allayed by continued peace. The story of the great massacre and destruction of the fort is of poignant interest, as simply and pathetically told by one of the survivors, a hunter: "In this present year 1802, about the twenty-fourth of June—I do not remember the exact date, but it was a holiday—about two o'clock in the afternoon, I went to the river to look for our calves, as I had been detailed by the commander of the fort, Vassili Medvednikof, to take care of the cattle. On returning soon after, I noticed at the fort a great multitude of Kolosh people, who had not only surrounded the barracks below, but were already climbing over the balcony and to the roof with guns and cannon; and standing upon a little knoll in front of the out-houses, was the Sitka toyon, or chief, Mikhail, giving orders to those who were around the barracks, and shouting to some people in canoes not far away, to make haste and assist in the fight. In answer to his shouts sixty-two canoes emerged from behind the points of rocks." (One is inclined to be sceptical concerning the exact number of canoes; the frightened hunter would scarcely pause to count the war canoes as they rounded the point.) "Even if I had reached the barracks, they were already closed and barricaded, and there was no safety outside; therefore, I rushed away to the cattle yard, where I had a gun. I only waited to tell a girl who was employed in the yard to take her little child and fly to the woods, when, seizing my gun, I closed up the shed. Very soon after this four Kolosh came to the door and knocked three times. As soon as I ran out of the shed, they seized me by the coat and took my gun from me. I was compelled to leave both in their hands, and jumping through a window, ran past the fort and hid in the thick underbrush of the forest, though two Kolosh ran after me, but could not find me in the woods. Soon after, I emerged from the underbrush, and approached the barracks to see if the attack had been repulsed, but I saw that not only the barracks, but the ship recently built, the warehouse and the sheds, the cattle sheds, bath house and other small buildings, had been set on fire and were already in full blaze. The sea-otter skins and other property of the company, as well as the private property of Medvednikof and the hunters, the savages were throwing from the balcony to the ground on the water side, while others seized them and carried them to the canoes, which were close to the fort.... All at once I saw two

Kolosh running toward me armed with guns and lances, and I was compelled to hide again in the woods. I threw myself down among the underbrush on the edge of the forest, covering myself with pieces of bark. From there I saw Nakvassin drop from the upper balcony and run toward the woods; but when nearly across the open space he fell to the ground, and four warriors rushed up and carried him back to the barracks on the points of their lances and cut off his head. Kabanof was dragged from the barracks into the street, where the Kolosh pierced him with their lances; but how the other Russians who were there came to their end, I do not know. The slaughter and incendiarism were continued by the savages until the evening, but finally I stole out among the ruins and ashes, and in my wanderings came across some of our cows, and saw that even the poor dumb animals had not escaped the bloodthirsty fiends, having spears stuck in their sides. Exercising all my strength, I was barely able to pull out some of the spears, when I was observed by two Kolosh, and compelled to leave the cows to their fate and hide again in the woods.

"I passed the night not far from the ruins of the fort. In the morning I heard the report of a cannon and looked out of the brush, but could see nobody, and not wishing to expose myself again to further danger, went higher up in the mountain through the forest. While advancing cautiously through the woods, I met two other persons who were in the same condition as myself,—a girl from the Chiniatz village, Kodiak, with an infant on her breast, and a man from the Kiliuda village, who had been left behind by the hunting party on account of sickness. I took them both with me to the mountain, but each night I went with my companions to the ruins of the fort and bewailed the fate of the slain. In this miserable condition we remained for eight days, with nothing to eat and nothing but water to drink. About noon of the last day we heard from the mountain two cannon-shots, which raised some hopes in me, and I told my companions to follow me at a little distance, and then went down toward the river through the woods to hide myself near the shore and see whether there was a ship in the bay."

He discovered, to his unspeakable joy, an English ship in the bay. Shouting to attract the attention of those on board, he was heard by six Kolosh, who made their way toward him and had almost captured him ere he saw them and made his escape in the woods. They forced him to the shore at a point near the cape, where he was able to make himself heard by those on the vessel. A boat put off at once, and he was barely able to leap into it when the Kolosh, in hot pursuit, came in sight again. When they saw the boat, they turned and fled.

When the hunter had given an account of the massacre to the commander of the vessel, an armed boat was sent ashore to rescue the man and girl who were in hiding. They were easily located and, with another Russian who was found in the vicinity, were taken aboard and supplied with food and clothing.

The commander himself then accompanied them, with armed men, to the site of the destroyed fort, where they examined and buried the dead. They found that all but Kabanof had been beheaded.

Three days later the chief, Mikhail, went out to the ship, was persuaded to go aboard, and with his nephew was held until all persons captured during the massacre and still living had been surrendered. The prisoners were given up reluctantly, one by one; and when it was believed that all had been recovered, the chief and his nephew were permitted to leave the ship.

The survivors were taken to Kodiak, where the humane captain of the ship demanded of Baranoff a compensation of fifty thousand roubles in cash. Baranoff, learning that the captain's sole expense had been in feeding and clothing the prisoners, refused to pay this exorbitant sum; and after long wrangling it was settled for furs worth ten thousand roubles.

Accounts of the massacre by survivors and writers of that time vary somewhat, some claiming that the massacre was occasioned by the broken faith and extreme cruelty of the Russians in their treatment of the savages; others, that the Sitkans had been well treated and that Chief Mikhail had falsely pretended to be the warm and faithful friend of Baranoff, who had placed the fullest confidence in him.

Baranoff was well-nigh broken-hearted by his new and terrible misfortune. The massacre had been so timed that the most of the men of the fort were away on a hunting expedition; and Baranoff himself was on Afognak Island, which is only a few hours' sail from Kodiak. Several Kolosh women lived at the fort with Russian men; and these women kept their tribesmen outside informed as to the daily conditions within the garrison. On the weakest day of the fort, a holiday, the Kolosh had, therefore, suddenly surrounded it, armed with guns, spears, and daggers, their faces covered with masks representing animals.

About this time Krusenstern and Lisiansky sailed from Kronstadt, in the hope—which was fulfilled—of being the first to carry the Russian flag around the world. Lisiansky arrived at Kodiak, after many hardships, only to receive a written request from Baranoff to proceed at once to Sitka and assist him in subduing the savages and avenging the officers and men lost in the fearful massacre. On the 15th of August, 1804, he therefore sailed to eastward, and on the twentieth of the same month entered Sitka Sound. The day must have been gloomy and Lisiansky's mood in keeping with the day, for he thus describes a bay which is, under favorable conditions, one of the most idyllically beautiful imaginable: "On our entrance into Sitka Sound to the place where we now were, there was not to be seen on the shore the least vestige of habitation. Nothing presented itself to our view but impenetrable woods reaching from the water-side to the very tops of the mountains. I never saw a country so wild and gloomy; it appeared more adapted for the residence of wild beasts than of men."

Shortly afterward Baranoff arrived in the harbor with several hundred Aleutians and many Russians, after a tempestuous and dangerous voyage from Yakutat, the site of the convict settlement.

He learned that the savages had taken up their position on a bluff a few miles distant, where they had fortified themselves. This bluff was the noble height upon which Baranoff's castle was afterward erected, and which commands the entire bay upon which the Sitka of to-day is located. Lisiansky, in his "Voyage around the World," describes the Indians' fort as "an irregular polygon, its longest side facing the sea. It was protected by a breastwork two logs in thickness, and about six feet high. Around and above it tangled brushwood was piled. Grape-shot did little damage, even at the distance of a cable's length. There were two embrasures for cannon in the side facing the sea, and two gates facing the forest. Within were fourteen large huts, or, as they were called then, and are called at the present time by the natives, barabaras. Judging from the quantity of provisions and domestic implements found there, it must have contained at least eight hundred warriors."

An envoy from the Kolosh fort came out with friendly overtures, but was informed that peace conditions could only be established through the chiefs. He departed, but soon returned and delivered a hostage.

Baranoff made plain his conditions; agreement with the chiefs in person, the delivery of two more hostages, and permanent possession of the fortified bluff.

The chiefs did not appear, and the conditions were not accepted. Then, on October 1, after repeated warnings, Baranoff gave the order to fire upon the fort. Immediately afterward, Baranoff, Lieutenant Arlusof, and a party of Russians and Aleutians landed with the intention of storming the fort. They were repulsed, the panic-stricken Aleutians stampeded, and Baranoff was left almost without support. In this condition, he could do nothing but retreat to the boats,—which they were barely able to reach before the Kolosh were upon them. They saved their field-pieces, but lost ten men. Twenty-six were wounded, including Baranoff himself. Had not their retreat at this point been covered by the guns of the ship, the loss of life would have been fearful.

The following day Lisiansky was placed in command. He

opened a rapid fire upon the fort, with such effect that soon after noon a peace envoy arrived, with promise of hostages. His overtures were favorably received, and during the following three days several hostages were returned to the Russians. The evacuation of the fort was demanded; but, although the chief consented, no movements in that direction could be discovered from the ships. Lisiansky moved his vessel farther in toward the fort and sent an interpreter to ascertain how soon the occupants would be ready to abandon their fortified and commanding position. The reply not being satisfactory, Lisiansky again fired repeatedly upon the stronghold of the Kolosh. On the 3d of October a white flag was hoisted, and the firing was discontinued. Then arose from the rocky height and drifted across the water until far into the night the sound of a mournful, wailing chant.

When dawn came the sound had ceased. Absolute silence reigned; nor was there any living object to be seen on the shore, save clouds of carrion birds, whose dark wings beat the still air above the fort. The Kolosh had fled; the fort was deserted by all save the dead. The bodies of thirty Kolosh warriors were found; also those of many children and dogs, which had been killed lest any cry from them should betray the direction of their flight.

The fort was destroyed by fire, and the construction of magazines, barracks, and a residence for Baranoff was at once begun. A stockade surrounded these buildings, each corner fortified with a block-house. The garrison received the name of Novo Arkangelsk, or New Archangel. The tribal name of the Indians in that locality was Sitkah—pronounced Seetkah—and this short and striking name soon attached itself permanently to the place.

Immense houses were built solidly and with every consideration for comfort and safety, and many families lived in each. They ranged in size from one hundred to one hundred and fifty feet in length, and about eighty in width, and were from one to three stories high with immense attics. They were well finished and richly papered. The polished floors were covered with costly rugs and carpets, and the houses were furnished with heavy and splendid furniture, which had been brought from St. Petersburg. The steaming brass samovar was everywhere a distinctive feature of the hospitality and good cheer which made Sitka famous.

To the gay and luxurious life, the almost prodigal entertainment of guests by Sitkans from this time on to 1867, every traveller, from writers and naval officers down to traders, has enthusiastically testified. At the first signal from a ship feeling its way into the dark harbor, a bright light flashed a welcome across the water from the high cupola on Baranoff's castle, and fires flamed up on Signal Island to beacon the way.

The officers were received as friends, and entertained in a style of almost princely magnificence during their entire stay—the only thing asked in return being the capacity to eat like gluttons, revel like roisterers, and drink until they rolled helplessly under the table; and, in Baranoff's estimation, these were small returns, indeed, to ask of a guest for his ungrudging and regal hospitality.

Visions of those high revels and glittering banquets of a hundred years ago come glimmering down to us of to-day. Beautiful, gracious, and fascinating were the Russian ladies who lived there,—if we are to believe the stories of voyagers to the Sitka of Baranoff's and Wrangell's times. Baranoff's furniture was of specially fine workmanship and exceeding value; his library was remarkable, containing works in nearly all European languages, and a collection of rare paintings—the latter having been presented to the company at the time of its organization.

Baranoff had left a wife and family in Russia. He never saw them again, although he sent allowances to them regularly. He was not bereft of woman's companionship, however, and we have tales of revelry by night when Baranoff alternately sang and toasted everybody, from the Emperor down to the woman upon his knee with whom he shared every sparkling glass. He had a beautiful daughter by a native woman, and of her he was exceedingly careful. A governess whom he surprised in the act of drinking a glass of liquor was struck in sudden blind passion and turned out of the house. The following day he sent for her, apologized, and reinstalled her with an increased salary, warning her, however, that his daughter must never see her drink a drop of liquor. When in his most gloomy and hopeless moods, this daughter could instantly soothe and cheer him by playing upon the piano and singing to him songs very different from those sung at his drunken all-night orgies.

That there was a very human and tender side to Baranoff's nature cannot be doubted by those making a careful study of his tempestuous life. He was deeply hurt and humiliated by the insolent and supercilious treatment of naval officers who considered him of inferior position, notwithstanding the fact that he was in supreme command of all the Russian territory in America. From time to time the Emperor conferred honors upon him, and he was always deeply appreciative; and it is chronicled that when a messenger arrived with the intelligence that he had been appointed by the Emperor to the rank of Collegiate Councillor, Baranoff, broken by the troubles, hardships, and humiliations of his stormy life, was suddenly and completely overcome by joy. He burst into tears and gave thanks to God.

"I am a nobleman!" he exclaimed. "I am the equal in position and the superior in ability of these insolent naval officers."

In 1812 Mr. Wilson P. Hunt, of the Pacific Fur Company, sailed from Astoria for Sitka on the *Beaver* with supplies for the Russians. By that time Baranoff had risen to the title and pomp of governor, and was living in splendid style befitting his position and his triumph over the petty officers, whose names are now insignificant in Russian history.

Mr. Hunt found this hyperborean veteran ensconced in a fort which crested the whole of a high, rocky promontory. It mounted one hundred guns, large and small, and was impregnable to Indian attack unaided by artillery. Here the old governor lorded it over sixty Russians, who formed the corps of the trading establishment, besides an indefinite number of Indian hunters of the Kodiak tribe, who were continually coming and going, or lounging and loitering about the fort like so many hounds round a sportsman's hunting quarters. Though a loose liver among his guests, the governor was a strict disciplinarian among his men, keeping them in perfect subjection and having seven guards on duty night and day.

Besides those immediate serfs and dependents just mentioned, the old Russian potentate exerted a considerable sway over a numerous and irregular class of maritime traders, who looked to him for aid and munitions, and through whom he may be said to have, in some degree, extended his power along the whole Northwest Coast. These were American captains of vessels engaged in a particular department of trade. One of the captains would come, in a manner, empty-handed, to New Archangel. Here his ship would be furnished with about fifty canoes and a hundred Kodiak hunters, and fitted out with provisions and everything necessary for hunting the sea-otter on the coast of California, where the Russians had another establishment. The ship would ply along the California coast, from place to place, dropping parties of otter hunters in their canoes, furnishing them only with water, and leaving them to depend upon their own dexterity for a maintenance. When a sufficient cargo was collected, she would gather up her canoes and hunters and return with them to Archangel, where the captain would render in the returns of his voyage and receive one-half of the skins as his share.

Over these coasting captains the old governor exerted some sort of sway, but it was of a peculiar and characteristic kind; it was the tyranny of the table. They were obliged to join in his "prosnics" or carousals and his heaviest drinking-bouts. His carousals were of the wildest and coarsest, his tempers violent, his language strong. "He is continually," said Mr. Hunt, "giving entertainment by way of parade; and if you do not drink raw rum, and boiling punch as strong as sulphur, he will insult you as soon as he gets drunk, which is very shortly after sitting down at table."

A "temperance captain" who stood fast to his faith and kept his sobriety inviolate might go elsewhere for a market; he was not a man after the governor's heart. Rarely, however, did any captain made of such unusual stuff darken the doors of Baranoff's high-set castle. The coasting captains knew too well his humor and their own interests. They joined with either real or well-affected pleasure in his roistering banquets; they ate much and drank more; they sang themselves hoarse and drank themselves under the table; and it is chronicled that never was Baranoff satisfied until the last-named condition had come to pass. The more the guests that lay sprawling under the table, upon and over one another, the more easily were trading arrangements effected with Baranoff later on.

Mr. Hunt relates the memorable warning to all "flinchers" which occurred shortly after his arrival. A young Russian naval officer had recently been sent out by the Emperor to take command of one of the company's vessels. The governor invited him to one of his "prosnics" and plied him with fiery potations. The young officer stoutly maintained his right to resist—which called out all the fury of the old ruffian's temper, and he proceeded to make the youth drink, whether he would or not. As the guest began to feel the effect of the burning liquors, his own temper rose to the occasion. He quarrelled violently with his almost royal host, and expressed his young opinion of him in the plainest language—if Russian language ever can be plain. For this abuse of what Baranoff considered his magnificent hospitality, he was given seventy-nine lashes when he was quite sober enough to appreciate them.

With all his drinking and prodigal hospitality, Baranoff always managed to get his own head clear enough for business before sobriety returned to any of his guests, who were not so accustomed to these wild and constant revels of their host's; so that he was never caught napping when it came to bargaining or trading. His own interests were ever uppermost in his mind, which at such times gave not the faintest indication of any befuddlement by drink or by licentiousness of other kinds.

For more than twenty years Baranoff maintained a princely and despotic sway over the Russian colonies. His own commands were the only ones to receive consideration, and but scant attention was given by him to orders from the Directory itself. Complaints of his rulings and practices seldom reached Russia. Tyrannical, coarse, shrewd, powerful, domineering, and of absolutely iron will, all were forced to bow to his desires, even men who considered themselves his superiors in all save sheer brute force of will and character. Captain Krusenstern, a contemporary, in his account of Baranoff, says: "None but vagabonds and adventurers ever entered the company's services as Promishléniks;"—uneducated Russian traders, whose inferior vessels were constructed usually of planks lashed to timbers and calked with moss; they sailed by dead reckoning, and were men controlled only by animal instincts and passions;—"it was their invariable destiny to pass a life of wretchedness in America." "Few," adds Krusenstern, "ever had the good fortune to touch Russian soil again."

In the light of present American opinion of the advantages and joys of life in Russia, this naïve remark has an almost grotesque humor. Like many of the brilliantly successful, but unscrupulous, men of the world, Baranoff seemed to have been born under a lucky star which ever led him on. Through all his desperate battles with Indians, his perilous voyages by sea, and the plottings of subordinates who hated him with a helpless hate, he came unharmed.

During his later years at Sitka, Baranoff, weighed down by age, disease, and the indescribable troubles of his long and faithful service, asked frequently to be relieved. These requests were ignored, greatly to his disappointment.

When, finally, in 1817, Hagemeister was sent out with instructions to assume command in Baranoff's place, if he deemed it necessary, the orders were placed before the old governor so suddenly and so unexpectedly that he was completely prostrated. He was now failing in mind, as well as body; and in this connection Bancroft adds another touch of ironical humor, whether intentional or accidental it is impossible to determine. "One of his symptoms of approaching imbecility," writes Bancroft, "being in his sudden attachment to the church. He kept constantly about him the priest who had established the first church at Sitka, and, urged by his spiritual adviser, made large donations for religious purposes."

The effect of the unexpected announcement is supposed to have shortened Baranoff's days. Lieutenant Yanovsky, of the vessel which had brought Hagemeister, was placed in charge by the latter as his representative. Yanovsky fell in love with Baranoff's daughter and married her. It was, therefore, to his own son-in-law that the old governor at last gave up the sceptre.

"Obleuk," an Eskimo Girl in Parka

By strength of his unbreakable will alone, he arose from a bed of illness and painfully and sorrowfully arranged all the affairs of his office, to the smallest and most insignificant detail, preparatory to the transfer to his successor.

It was in January, 1818, that Hagemeister had made known his appointment to the office of governor; it was not until September that Baranoff had accomplished his difficult task and turned over the office.

There was then, and there is to-day, halfway between the site of the castle and Indian River, a gray stone about three feet high and having a flat, table-like surface. It stands on the shore beside the hard, white road. The lovely bay, set with a thousand isles, stretches sparkling before it; the blue waves break musically along the curving shingle; the wooded hills rise behind it; the winds murmur among the tall trees.

The name of this stone is the "blarney" stone. It was a favorite retreat of Baranoff's and there, when he was sunken in one of his lonely or despondent moods, he would sit for hours, staring out over the water. What his thoughts were at such times, only God and he knew,—for not even his beloved daughter dared to approach him when one of his lone moods was upon him.

In the first hour that he was no longer governor of the country he had ruled so long and so royally, he walked with bowed head along the beach until he reached his favorite retreat. There he sat himself down and for hours remained in silent communion with his own soul. He had longed for relief from his arduous duties, but it had come in a way that had broken his heart. His government had at last listened to complaints against him, and, ungrateful for his long and faithful service, had finally relieved him with but scant consideration; with an abruptness and a lack of courtesy that had sorely wounded him.

Nearly thirty of his best years he had devoted to the company. He had conquered the savages and placed the fur trade upon a highly profitable basis; he had built many vessels and had established trading relations with foreign countries; forts, settlements, and towns had risen at his indomitable will. Sitka, especially, was his own; her storied splendor, whose fame has endured through all the years, she owed entirely to him; she was the city of his heart. He was her creator; his life-blood, his very heart beats, were in her; and now that the time had really come to give her up forever, he found the hour of farewell the hardest of his hard life. No man, of whatsoever material he may be made, nor howsoever insensible to the influence of beauty he may deem himself to be, could dwell for twenty years in Sitka without finding, when it came to leaving her, that the tendrils of her loveliness had twined themselves so closely about his heart that their breaking could only be accomplished by the breaking of the heart itself.

Of his kin, only a brother remained. The offspring of his connection with a Koloshian woman was now married and settled comfortably. A son by the same mistress had died. He had first thought of going to his brother, who lived in Kamchatka; but Golovnin was urging him to return to Russia, which he had left forty years before. This he had finally decided to do, it having been made clear to him that he could still be of service to his country and his beloved colonies by his experience and advice. Remain in the town he had created and ruled so tyrannically, and which he still loved so devotedly, he could not. The mere thought of that was unendurable.

All was now in readiness for his departure, but the old man—he was now seventy-two—had not anticipated that the going would be so hard. The blue waves came sparkling in from the outer sea and broke on the curving shingle at his feet; the white and lavender wings of sea-birds floated, widespread, upon the golden September air; vessels of the fleet he had built under the most distressing difficulties and disadvantages lay at anchor under the castle wherein he had banqueted every visitor of any distinction or position for so many years, and the light from whose proud tower had guided so many worn voyagers to safety at last; the yellow, red-roofed buildings, the great ones built of logs, the chapel, the significant block-houses—all arose out of the wilderness before his sorrowful eyes, taking on lines of beauty he had never discovered before.

From this hour Baranoff failed rapidly from day to day. His time was spent in bidding farewell to the Russians and natives—to many of whom he was sincerely attached—and to places which had become endeared to him by long association. He was frequently found in tears. Those who have seen fair Sitka rising out of the blue and islanded sea before their raptured eyes may be able to appreciate and sympathize with the old governor's emotion as, on the 27th of November, 1818, he stood in the stern of the *Kutusof* and watched the beloved city of his creation fade lingeringly from his view. He was weeping, silently and hopelessly, as the old weep, when, at last, he turned away.

Baranoff never again saw Sitka. In March the *Kutusof* landed at Batavia, where it remained more than a month. There he was very ill; and soon after the vessel had again put to sea, he died, like Behring, a sad and lonely death, far from friends and home. On the 16th of April, 1819, the waters of the Indian Ocean received the body of Alexander Baranoff.

Notwithstanding his many and serious faults, or, possibly because of their existence in so powerful a character—combined as they were with such brilliant talent and with so many admirable and conscientious qualities—Baranoff remains through all the years the most fascinating figure in the history of the Pacific Coast. None is so well worth study and close investigation; none is so rich in surprises and delights; none has the charm of so lone and beautiful a setting. There was no littleness, no niggardliness, in his nature. "He never knew what avarice was," wrote Khlebnikof, "and never hoarded riches. He did not wait until his death to make provision for the living, but gave freely to all who had any claim upon him."

He spent money like a prince. He received ten shares of stock in the company from Shelikoff and was later granted twenty more; but he gave many of these to his associates who were not so well remunerated for their faithful services. He provided generously during his life for his family; and for the families in Russia of many who lost their lives in the colonies, or who were unable through other misfortunes to perform their duties in this respect.

Born of humble parentage in Kargopal, Eastern Russia, in 1747, he had, at an early age, drifted to Moscow, where he was engaged as a clerk in retail stores until 1771, when he established himself in business.

Not meeting with success, he four years later emigrated to Siberia and undertook the management of a glass factory at Irkutsk. He also interested himself in other industries; and on account of several valuable communications to the Civil Economical Society on the subject of manufacture he was in 1789 elected a member of the society.

A Northern Madonna

His life here was a humdrum existence, of which his restless spirit soon wearied. Acquainting himself with the needs, resources, and possibilities of Kamchatka, he set out to the eastward with an assortment of goods and liquors, which he sold to the savages of that and adjoining countries.

At first his operations were attended by success; but when, in 1789, two of his caravans were captured by Chuckchi, he found himself bankrupt, and soon yielded to Shelikoff's urgent entreaties to try his fortunes in America.

Such is the simple early history of this remarkable man. Not one known descendant of his is living to-day. But men like Baranoff do not need descendants to perpetuate their names.

Bancroft is the highest authority on the events of this period, his assistant being Ivan Petroff, a Russian, who was well-informed on the history of the colonies.

Many secret reasons have been suspected for the sale of the magnificent country of Alaska to the United States for so paltry a sum.

The only revenue, however, that Russia derived from the colonies was through the rich fur trade; and when, after Baranoff's death, this trade declined and its future seemed hopeless, the country's vast mineral wealth being unsuspected, Russia found herself in humor to consider any offer that might be of immediate profit to herself. For seven millions and two hundred thousands of dollars Russia cheerfully, because unsuspectingly, yielded one of the most marvellously rich and beautiful countries of the world—its valleys yellow with gold, its mountains green with copper and thickly veined with coal, its waters alive with fish and fur-bearing animals, its scenery sublime—to the scornful and unappreciative United States.

As early as the fifties it became rumored that Russia, foreseeing the entire decline of the fur trade, considered Alaska a white elephant upon its hands, and that an offer for its purchase would not meet with disfavor. The matter was discussed in Washington at various times, but it was not until 1866 that it was seriously considered. The people of the present state of Washington were among those most desirous of its purchase; and there was rumor of the organization of a trading company of the Pacific Coast for the purpose of purchasing the rights of the Russian-American Company and acquiring the lease of the *lisière* which was to expire in 1868. The Russian-American Company was then, however, awaiting the reply of the Hudson Bay Company concerning a renewal of the lease; and the matter drifted on until, in the spring of 1867, the Russian minister opened negotiations for the purchase of the country with Mr. Seward. There was some difficulty at first over the price, but the matter was one presenting so many mutual advantages that this was soon satisfactorily arranged.

On Friday evening, March 25, 1867, Mr. Seward was playing whist with members of his family when the Russian minister was announced. Baron Stoeckl stated that he had received a despatch from his government by cable, conveying the consent of the Emperor to the cession.

"To-morrow," he added, "I will come to the department, and we can enter upon the treaty."

With a smile of satisfaction, Seward replied:—

"Why wait till to-morrow? Let us make the treaty to-night."

"But your department is closed. You have no clerks, and my secretaries are scattered about the town."

"Never mind that," said Seward; "if you can muster your legation together before midnight, you will find me awaiting you at the department."

By four o'clock on the following morning the treaty was engrossed, sealed, and ready for transmission by the President to the Senate. The end of the session was approaching, and there was need of haste in order to secure action upon it.

Leutze painted this historic scene. Mr. Seward is seen sitting at his table, pen in hand, listening to the Russian minister. The gaslight, streaming down on the table, illuminates the outline of "the great country."

When, immediately afterward, the treaty was presented for consideration in the Senate, Charles Sumner delivered his famous and splendid oration which stands as one of the masterpieces of history, and which revealed an enlightened knowledge and understanding of Alaska that were remarkable at that time—and which probably surpassed those of Seward. Among other clear and beautiful things he said:—

"The present treaty is a visible step in the occupation of the whole North American Continent. As such it will be recognized by the world and accepted by the American people. But the treaty involves something more. By it we dismiss one more monarch from this continent. One by one they have retired; first France, then Spain, then France again, and now Russia—all giving way to that absorbing unity which is declared in the national motto: *E pluribus unum.*"

There is yet one more monarch to be retired, in all kindness and good-will, from our continent; and that event will take place when our brother-Canadians unite with us in deed as they already have in spirit.

For years the purchase was unpopular, and was ridiculed by the press and in conversation. Alaska was declared to be a "barren, worthless, God-forsaken region," whose only products were "icebergs and polar bears"; vegetation was "confined to mosses"; and "Walrussia" was wittily suggested as an appropriate name for our new possession—as well as "Icebergia"; but in the face of all the opposition and ridicule, those two great Americans, Seward and Sumner, stood firmly for the acquisition of this splendid country. They looked through the mist of their own day and saw the day that is ours.

CHAPTER XVII

Since Sitka first dawned upon my sight on a June day, in her setting of vivid green and glistening white, she has been one of my dearest memories. Four times in all have the green islands drifted apart to let her rise from the blue sea before my enchanted eyes; and with each visit she has grown more dear, and her memory more tormenting.

Something gives Sitka a different look and atmosphere from any other town. It may be her whiteness, glistening against the rich green background of forest and hill, with the whiteness of the mountains shining in the higher lights; or it may be the severely white and plain Greek church, rising in the centre of the main street, not more than a block from the water, that gives Sitka her chaste and immaculate appearance.

No buildings obstruct the view of the church from the water. There it is, in the form of a Greek cross, with its green roof, steeple, and bulbous dome.

This church is generally supposed to be the one that Baranoff built at the beginning of the century; but this is not true. Baranoff did build a small chapel, but it was in 1848 that the foundation of the present church was laid—almost thirty years after the death of Baranoff. It was under the special protection of the Czar, who, with other members of the imperial family, sent many costly furnishings and ornaments.

Veniaminoff—who was later made Archpriest, and still later the Archbishop of Kamchatka, and during the last years of his noble life, the Metropolitan of Moscow—sent many of the rich vestments, paintings, and furnishings. The chime of silvery bells was also sent from Moscow.

Upon landing at Sitka, one is confronted by the old log storehouse of the Russians. This is an immense building, barricading the wharf from the town. A narrow, dark, gloomy passage-way, or alley, leads through the centre of this building. It seems as long as an ordinary city square to the bewildered stranger groping through its shadows.

In front of this building, and inside both ends of the passage as far as the light reaches, squat squaws, young and old, pretty and hideous, starry-eyed and no-eyed, saucy and kind, arrogant and humble, taciturn and voluble, vivacious and weary-faced. Surely no known variety of squaw may be asked for and not found in this long line that reaches from the wharf to the green-roofed church.

There is no night so wild and tempestuous, and no hour of any night so late, or of any morning so early, that the passenger hastening ashore is not greeted by this long line of dark-faced women. They sit like so many patient, noiseless statues, with their tempting wares clustered around the flat, "toed-in" feet of each.

Not only is this true of Sitka, but of every landing-place on the whole coast where dwells an Indian or an Aleut that has something to sell. Long before the boat lands, their gay shawls by day, or their dusky outlines by night, are discovered from the deck of the steamer.

How they manage it, no ship's officer can tell; for the whistle is frequently not blown until the boat is within a few yards of the shore. Yet there they are, waiting!

Sometimes, at night, they appear simultaneously, fluttering down into their places, swiftly and noiselessly, like a flock of birds settling down to rest for a moment in their flight.

Some of these women are dressed in skirts and waists, but the majority are wrapped in the everlasting gay blankets. No lip or nose ornaments are seen, even in the most aged. Two or three men are scattered down the line, to guard the women from being cheated.

These tall and lordly creatures strut noiselessly and superciliously about, clucking out guttural advice to the squaws, as well as, to all appearances, the frankest criticism of the persons examining their wares with a view to purchasing.

The women are very droll, and apparently have a keen sense of humor; and one is sure to have considerable fun poked at one, going down the line.

Mild-tempered people do not take umbrage at this ridicule; in fact, they rather enjoy it. Being one of them, I lost my temper only once. A young squaw offered me a wooden dish, explaining in broken English that it was an old eating dish.

It had a flat handle with a hole in it; and as cooking and eating utensils are never washed, it had the horrors of ages encrusted within it to the depth of an inch or more.

This, of course, only added to its value. I paid her a dollar for it, and had just taken it up gingerly and shudderingly with the tips of my fingers, when, to my amazement and confusion, the girl who had sold it to me, two older women who were squatting near, and a tall man leaning against the wall, all burst simultaneously into jeering and uncontrollable laughter.

As I gazed at them suspiciously and with reddening face, the young woman pointed a brown and unclean finger at me; while, as for the chorus of chuckles and duckings that assailed my ears—I hope I may never hear their like again.

To add to my embarrassment, some passengers at that moment approached.

"Hello, Sally," said one; "what's the matter?"

Laughing too heartily to reply, she pointed at the wooden dish, which I was vainly trying to hide. They all looked, saw, and laughed with the Indians.

For a week afterward they smiled every time they looked at me; and I do believe that every man, woman, and child on the steamer came, smiling, to my cabin to see my "buy." But the ridicule of my kind was as nothing compared to that of the Indians themselves. To be "taken in" by the descendant of a Koloshian, and then jeered at to one's very face!

The only possession of an Alaskan Indian that may not be

purchased is a rosary. An attempt to buy one is met with glances of aversion.

"It has been *blessed*!" one woman said, almost in a whisper.

But they have most beautiful long strings of big, evenly cut, sapphire-blue beads. They call them Russian beads, and point out certain ones which were once used as money among the Indians.

Their wares consist chiefly of baskets; but there are also immense spoons carved artistically out of the horns of mountain sheep; richly beaded moccasins of many different materials; carved and gayly painted canoes and paddles of the fragrant Alaska cedar or Sitka pine; totem-poles carved out of dark gray slate stone; lamps, carved out of wood and inlaid with a fine pearl-like shell. These are formed like animals, with the backs hollowed to hold oil. There are silver spoons, rings, bracelets, and chains, all delicately traced with totemic designs; knives, virgin charms, Chilkaht blankets, and now and then a genuine old spear, or bow and arrow, that proves the dearest treasure of all.

Eskimo Lad in Parka and Mukluks

Old wooden, or bone, gambling sticks, finely carved, polished to a satin finish, and sometimes inlaid with fragments of shell, or burnt with totemic designs, are also greatly to be desired.

The main features of interest in Sitka are the Greek-Russian church and the walk along the beach to Indian River Park.

A small admission fee is charged at the church door. This goes to the poor-fund of the parish. It is the only church in Alaska that charges a regular fee, but in all the others there are contribution boxes. When one has, with burning cheeks, seen his fellow-Americans drop dimes and nickels into the boxes of these churches, which have been specially opened at much inconvenience for their accommodation, he is glad to see the fifty-cent fee at the door charged.

There are no seats in the church. The congregation stands or kneels during the entire service. There are three sanctuaries and as many altars. The chief sanctuary is the one in the middle, and it is dedicated to the Archi-Strategos Michael.

The sanctuary is separated from the body of the church by a screen—which has a "shaky" look, by the way—adorned with twelve ikons, or images, in costly silver and gold casings, artistically chased.

The middle door leading into the sanctuary is called the Royal Gates, because through it the Holy Sacrament, or Eucharist, is carried out to the faithful. It is most beautifully carved and decorated. Above it is a magnificent ikon, representing the Last Supper. The heavy silver casing is of great value. The casings alone of the twelve ikons on the screen cost many thousands of dollars.

An interesting story is attached to the one of the patron saint of the church, the Archangel Michael. The ship *Neva*, on her way to Sitka, was wrecked at the base of Mount Edgecumbe. A large and valuable cargo was lost, but the ikon was miraculously cast upon the beach, uninjured.

Many of the ikons and other adornments of the church were presented by the survivors of wrecked vessels; others by illustrious friends in Russia. One that had paled and grown dim was restored by Mrs. Emmons, the wife of Lieutenant Emmons, whose work in Alaska was of great value.

When the Royal Gates are opened the entire sanctuary—or Holy of Holies, in which no woman is permitted to set foot, lest it be defiled—may be seen.

To one who does not understand the significance of the various objects, the sanctuary proves a disappointment until the splendid old vestments of cloth of gold and silver are brought out. These were the personal gifts of the great Baranoff. They are exceedingly rich and sumptuous, as is the bishop's stole, made of cloth woven of heavy silver threads.

The left-hand chapel is consecrated to "Our Lady of Kazan." It is adorned with several ikons, one of which, "The Mother of God," is at once the most beautiful and the most valuable object in the church. An offer of fifteen thousand dollars was refused for it. The large dark eyes of the madonna are so filled with sorrowful tenderness and passion that they cannot be forgotten. They follow one about the chapel; and after he has gone out into the fresh air and the sunlight he still feels them upon him. Those mournful eyes hold a message that haunts the one who has once tried to read it. The appeal which the unknown Russian artist has painted into them produces an effect that is enduring.

But most precious of all to me were those objects, of whatsoever value, which were presented by Innocentius, the Metropolitan of Moscow, the Noble and the Devoted. If ever a man went forth in search of the Holy Grail, it was he; and if ever a man came near finding the Holy Grail, it was, likewise, he.

From Sitka to Unalaska, and up the Yukon so far as the Russian influence goes, his name is still murmured with a veneration that is almost adoration.

Historians know him and praise him, without a dissenting voice, as Father Veniaminoff; for it was under this simple and unassuming title that the pure, earnest, and devout young Russian came to the colonies in 1823, carrying the high, white light of his faith to the wretched natives, among whom his life work was to be, from that time on, almost to the end.

No man has ever done as much for the natives of Alaska as he, not even Mr. Duncan. His heart being all love and his nature all tenderness, he grew to love the gentle Aleutians and Sitkans, and so won their love and trust in return.

In the Sitka church is a very costly and splendid vessel, used for the Eucharist, which was once stolen, but afterward returned. There are censers of pure silver and chaste design, which tinkle musically as they swing.

A visit to the building of the Russian Orthodox Mission is also interesting. There will be found some of the personal belongings of Father Veniaminoff—his clock, a writing-desk which was made by his own hands, of massive and enduring workmanship, and several articles of furniture; also the ikon which once adorned his cell—a gift of Princess Potemkin.

Sir George Simpson describes an Easter festival at Sitka in 1842. He found all the people decked in festal attire upon his arrival at nine o'clock in the morning. They were also, men and women, quite "tipsy."

Upon arriving at Governor Etholin's residence, he was ushered into the great banqueting room, where a large party was rising from breakfast. This party was composed of the bishop and priests, the Lutheran clergyman, the naval officers, the secretaries, business men, and masters and mates of vessels,—numbering in all about seventy,—all arrayed in uniforms or, at the least, in elegant dress.

From morning till night Sir George was compelled to "run a gantlet of kisses." When two persons met, one said, "Christ is risen"—and this was a signal for prolonged kissing. "Some of them," adds Sir George, naïvely, "were certainly pleasant enough; but many, even when the performers were of the fair sex, were perhaps too highly flavored for perfect comfort."

He was likewise compelled to accept many hard-boiled, gilded eggs, as souvenirs.

During the whole week every bell in the chimes of the church rang incessantly—from morning to night, from night to morning; and poor Sir George found the jangling of "these confounded bells" harder to endure than the eggs or the kisses.

Sir George extolled the virtues of the bishop—Veniaminoff. His appearance impressed the Governor-in-Chief with awe; his talents and attainments seemed worthy of his already exalted station; while the gentleness which characterized his every word and deed insensibly moulded reverence into love.

Whymper visited Sitka in 1865, and found Russian hospitality under the administration of Matsukoff almost as lavish as during Baranoff's famous reign.

Scales and Summit of Chilkoot Pass in 1898

"Russian hospitality is proverbial," remarks Whymper, "and we all somewhat suffered therefrom. The first phrase of their language acquired by us was 'petnatchit copla'—fifteen drops."

This innocently sounding phrase really meant a good half-tumbler of some undiluted liquor, ranging from cognac to raw vodhka, which was pressed upon the visitors upon every available occasion. A refusal to drink meant an insult to their host; and they were often sorely put to it to carry gracefully the burden of entertainment which they dared not decline.

The big brass samovar was in every household, and they were compelled to drink strong Russian tea, served by the tumblerful. Balls, banquets, and fêtes in the gardens of the social clubs were given in their honor; while their fleet of four vessels in the harbor was daily visited by large numbers of Russian ladies and gentlemen from the town.

At all seasons of the year the tables of the higher classes were supplied with game, chickens, pork, vegetables, berries, and every luxury obtainable; while the food of the common laborers was, in summer, fresh fish, and in winter, salt fish.

Sir George Simpson attended a Koloshian funeral at Sitka, or New Archangel, in 1842. The body of the deceased, arrayed in the gayest of apparel, lay in state for two or three days, during which time the relatives fasted and bewailed their loss. At the end of this period, the body was placed on a funeral pyre, round which the relatives gathered, their faces painted black and their hair covered with eagles' down. The pipe was passed around several times; and then, in obedience to a secret sign, the fire was kindled in several places at once. Wailings and loud lamentations, accompanied by ceaseless drumming, continued until the pyre was entirely consumed. The ashes were, at last, collected into an ornamental box, which was elevated on a scaffold. Many of these monuments were seen on the side of a neighboring hill.

A wedding witnessed at about the same time was quite as interesting as the funeral, presenting several unique features. A good-looking Creole girl, named Archimanditoffra, married the mate of a vessel lying in port.

Attended by their friends and the more important residents of Sitka, the couple proceeded at six o'clock in the evening to the church, where a tiresome service, lasting an hour and a half, was solemnized by a priest.

The bridegroom then led his bride to the ballroom. The most startling feature of this wedding was of Russian, rather than savage, origin. The person compelled to bear all the expense of the wedding was chosen to give the bride away; and no man upon whom this honor was conferred ever declined it.

This custom might be followed with beneficial results to-day, a bachelor being always honored, until, in sheer self-defence, many a young man would prefer to pay for his own wedding to constantly paying for the wedding of some other man. It is more polite than the proposed tax on bachelors.

At this wedding the beauty and fashion of Sitka were assembled. The ladies were showily attired in muslin dresses, white satin shoes, silk stockings, and kid gloves; they wore flowers and carried white fans.

The ball was opened by the bride and the highest officer present; and quadrille followed waltz in rapid succession until daylight.

The music was excellent; and the unfortunate host and paymaster of the ceremonies carried out his part like a prince. Tea, coffee, chocolate, and champagne were served generously, varied with delicate foods, "petnatchit coplas" of strong liquors, and expensive cigars.

According to the law of the church, the bridesmaids and bridesmen were prohibited from marrying each other; but, owing to the limitations in Sitka, a special dispensation had been granted, permitting such marriages.

From the old Russian cemetery on the hill, a panoramic view is obtained of the town, the harbor, the blue water-ways winding among the green islands to the ocean, and the snow mountains floating above the pearly clouds on all sides. In a quiet corner of the cemetery rests the first Princess Matsukoff, an Englishwoman, who graced the "Castle on the Rock" ere she died, in the middle sixties. Her successor was young, beautiful, and gay; and her reign was as brilliant as it was brief. She it was who, through bitter and passionate tears, dimly beheld the Russian flag lowered from its proud place on the castle's lofty flagstaff and the flag of the United States sweeping up in its stead. But the first proud Princess Matsukoff slept on in her quiet resting-place beside the blue and alien sea, and grieved not.

From all parts of the harbor and the town is seen the kekoor, the "rocky promontory," from which Baranoff and Lisiansky drove the Koloshians after the massacre, and upon which Baranoff's castle later stood.

It rises abruptly to a height of about eighty feet, and is ascended by a long flight of wooden steps.

The first castle was burned; another was erected, and was destroyed by earthquake; was rebuilt, and was again destroyed—the second time by fire. The eminence is now occupied by the home of Professor Georgeson, who conducts the government agricultural experimental work in Alaska.

The old log trading house which is on the right side of the street leading to the church is wearing out at last. On some of the old buildings patches of modern weather-boarding mingle with the massive and ancient logs, producing an effect that is almost grotesque.

In the old hotel Lady Franklin once rested with an uneasy heart, during the famous search for her husband.

The barracks and custom-house front on a vivid green parade ground that slopes to the water. Slender gravelled roads lead across this well-kept green to the quarters and to the building formerly occupied by Governor Brady as the Executive Offices. His residence is farther on, around the bay, in the direction of the Indian village.

There are fine fur and curio stores on the main street.

The homes of Sitka are neat and attractive. The window boxes and carefully tended gardens are brilliant with bloom in summer.

Passing through the town, one soon reaches the hard, white road that leads along the curving shingle to Indian River. The road curves with the beach and goes glimmering on ahead, until it disappears in the green mist of the forest.

Surely no place on this fair earth could less deserve the offensive name of "park" than the strip of land bordering Indian River,—five hundred feet wide on one bank, and two hundred and fifty feet on the other, between the falls and the low plain where it pours into the sea,—which in 1890 was set aside for this purpose.

It has been kept undefiled. There is not a sign, nor a painted seat, nor a little stiff flower bed in it. There is not a striped paper bag, nor a peanut shell, nor the peel of an orange anywhere.

It must be that only those people who live on beauty, instead of food, haunt this beautiful spot.

The spruce, the cedar, and the pine grow gracefully and luxuriantly, their lacy branches spreading out flat and motionless upon the still air, tapering from the ground to a fine point. The hard road, velvet-napped with the spicy needles of centuries, winds through them and under them, the branches often touching the wayfarer's bared head.

The devil's-club grows tall and large; there are thickets of salmon-berry and thimbleberry; there are banks of velvety green, and others blue with violets; there are hedges of wild roses, the bloom looking in the distance like an amethyst cloud floating upon the green.

The Alaskan thimbleberry is the most delicious berry that grows. Large, scarlet, velvety, yet evanescent, it scarcely touches the tongue ere its ravishing flavor has become a memory.

The vegetation is all of tropical luxuriance, and, owing to its constant dew and mist baths, it is of an intense and vivid green that is fairly dazzling where the sun touches it. One of the chief charms of the wooded reserve is its stillness—broken only by the musical rush of waters and the lyrical notes of birds. A kind of lavender twilight abides beneath the trees, and, with the narrow, spruce-aisled vistas that open at every turn, gives one a sensation as of being in some dim and scented cathedral.

Enticing paths lead away from the main road to the river, where the voices of rapids and cataracts call; but at last one comes to an open space, so closely walled round on all sides by the forest that it may easily be passed without being seen—and to which one makes his way with difficulty, pushing aside branches of trees and tall ferns as he proceeds.

Here, producing an effect that is positively uncanny, are several great totems, shining out brilliantly from their dark green setting.

One experiences that solemn feeling which every one has known, as of standing among the dead; the shades of Baranoff, Behring, Lisiansky, Veniaminoff, Chirikoff,—all the unknown murdered ones, too,—go drifting noiselessly, with reproachful faces, through the dim wood.

It was on the beach near this grove of totems that Lisiansky's men were murdered by Koloshians in 1804, while obtaining water for the ship.

The Sitka Industrial Training School was founded nearly thirty years ago by ex-Governor Brady, who was then a missionary to the Indians of Alaska.

It was first attended by about one hundred natives, ranging from the very young to the very old. This school was continued, with varied success, by different people—including Captain Glass, of the *Jamestown*—until Dr. Sheldon Jackson became interested, and, with Mr. Brady and Mr. Austin, sought and obtained aid from the Board of Home Missions of the Presbyterian Church.

A building was erected for a Boys' Home, and this was followed, a year later, by a Girls' Home.

The girls were taught to speak the English language, cook, wash, iron, sew, mend, and to become cleanly, cheerful, honest, honorable women.

The boys were taught to speak the English language; the trades of shoemaking, coopering, boat-building, carpentry, engineering, rope-making, and all kinds of agricultural work. The rudiments of bricklaying, painting, and paper-hanging are also taught.

During the year 1907 a Bible Training Department was added for those among the older boys and girls who desired to obtain knowledge along such lines, or who aspired to take up missionary work among their people.

Twelve pupils took up the work, and six continued it throughout the year. The work in this department is, of course, voluntary on the part of the student.

The Sitka Training School is not, at present, a government school. During the early nineties it received aid from the government, under the government's method of subsidizing denominational schools, where they were already established, instead of incurring the extra expense of establishing new government schools in the same localities.

When the government ceased granting such subsidies, the Sitka School—as well as many other denominational schools—lost this assistance.

The property of the school has always belonged to the Presbyterian Board of Home Missions.

For many years it was customary to keep pupils at the schools from their entrance until their education was finished.

In the summer of 1905 the experiment was tried of permitting a few pupils to go to their homes during vacation. All returned in September cheerfully and willingly; and now, each summer, more than seventy boys and girls return to their homes to spend the time of vacation with their families.

In former years, it would have been too injurious to the child to be subjected to the influence of its parents, who were but slightly removed from savagery. To-day, although many of the old heathenish rites and customs still exist, they have not so deep a hold upon the natives; and it is hoped, and expected, that the influence of the students for good upon their people will far exceed that of their people for ill upon them.

During the past year ninety boys and seventy-four girls were enrolled—or as many as can be accommodated at the schools.

They represent the three peoples into which the Indians of southeastern Alaska are now roughly divided—the Thlinkits, the Haidahs, and the Tsimpsians. They come from Katalla, Yakutat, Skagway, Klukwan, Haines, Douglas, Juneau, Kasa-an, Howkan, Metlakahtla, Hoonah—and, indeed, from almost every point in southeastern Alaska where a handful of Indians are gathered together.

CHAPTER XVIII

The many people who innocently believe that there are no birds in Alaska may be surprised to learn that there are, at least, fifty different species in the southeastern part of that country.

Among these are the song sparrow, the rufous humming-bird, the western robin, of unfailing cheeriness, the russet-backed thrush, the barn swallow, the golden-crowned kinglet, the Oregon Junco, the winter wren, and the bird that, in liquid clearness and poignant sweetness of note, is second only to the western meadow-lark—the poetic hermit thrush.

He that has heard the impassioned notes of this shy bird rising from the woods of Sitka will smile at the assertion that there are no birds in Alaska.

On the way to Indian River is the museum, whose interesting and valuable contents were gathered chiefly by Sheldon Jackson, and which still bears his name.

Dr. Jackson has been the general Agent of Education in Alaska since 1885, and the Superintendent of Presbyterian Missions since 1877. His work in Alaska in early years was, undoubtedly, of great value.

The museum stands in an evergreen grove, not far from the road. Here may be found curios and relics of great value. It is to be regretted, however, that many of the articles are labelled with the names of collectors instead of those of the real donors—at least, this is the information voluntarily given me by some of the donors.

In the collection is an interesting war bonnet, which was donated by Chief Kath-le-an, who planned and carried out the siege of 1878.

It was owned by one of Kath-le-an's ancestors. It is made of wood, carved into a raven's head. It has been worked and polished until the shell is more like velvet than wood, and is dyed black.

It was many years ago a polite custom of the Thlinkits to paint and oil the face of a visitor, as a matter of hospitality and an indication of friendly feeling and respect.

A visitor from another tribe to Sitka fell ill and died, shortly after having been so oiled and honored, and his people claimed that the oil was rancid,—or that some evil spell had been oiled into him,—and a war arose.

The Sitka tribe began the preparation of the raven war bonnet and worked upon it all summer, while actual hostilities were delayed.

As winter came on, Kath-le-an's ancestor one day addressed his young men, telling them that the new war bonnet on his head would serve as a talisman to carry them to a glorious victory over their enemies.

Through the battle that followed, the war bonnet was everywhere to be seen in the centre of the most furious fighting. Only once did it go down, and then only for a moment, when the chief struggled to his feet; and as his young men saw the symbol of victory rising from the dust, the thrill of renewed hope that went through them impelled them forward in one splendid, simultaneous movement that won the day.

In 1804 Kath-le-an himself wore the hat when his people were besieged for many days by the Russians.

On this occasion the spell of the war bonnet was broken; and upon his utter defeat, Kath-le-an, feeling that it had lost its charm for good luck, buried the unfortunate symbol in the woods.

Many years afterward Kath-le-an exhumed the hat and presented it to the museum.

"We will hereafter dwell in peace with the white people," he said; "so my young men will never again need the war bonnet."

Kath-le-an has to this day kept his word. He is still alive, but is nearly ninety years old.

Interesting stories and myths are connected with a large

number of the relics in the museum—to which the small admission fee of fifty cents is asked.

One of the early picturesque block-houses built by the Russians still stands in a good state of preservation on a slight eminence above the town, on the way to the old cemetery.

The story of the lowering of the Russian flag, and the hoisting of the American colors at Sitka, is fraught with significance to the superstitious.

The steamship *John L. Stevens*, carrying United States troops from San Francisco, arrived in Sitka Harbor on the morning of October 9, 1867. The gunboats *Jamestown* and *Resaca* had already arrived and were lying at anchor. The *Ossipee* did not enter the harbor until the morning of the eighteenth.

At three o'clock of the same day the command of General Jefferson C. Davis, about two hundred and fifty strong, in full uniform, armed and handsomely equipped, were landed, and marched to the heights where the famous Governor's Castle stood. Here they were met by a company of Russian soldiers who took their place upon the left of the flagstaff.

The command of General Davis formed on the right. The United States flag, which was to float for the first time in possession of Sitka, was in the care of a color guard—a lieutenant, a sergeant, and ten men.

Besides the officers and troops, there were present the Prince and Princess Matsukoff, many Russian and American residents, and some interested Indians.

It was arranged by Captain Pestchouroff and General Lovell N. Rosseau, Commissioner for the United States, that the United States should lead in firing the first salute, but that there should be alternate guns from the American and Russian batteries—thus giving the flag of each nation a double national salute.

The ceremony was begun by the lowering of the Russian flag—which caused the princess to burst into passionate weeping, while all the Russians gazed upon their colors with the deepest sorrow and regret marked upon their faces.

As the battery of the *Ossipee* led off in the salute and the deep peals crashed upon Mount Verstovi and reverberated across the bay, an accident occurred which has ever been considered an omen of misfortune.

The Russian flag became entangled about the ropes, owing to a high wind, and refused to be lowered.

The staff was a native pine, about ninety feet in height. Russian soldiers, who were sailors as well, at once set out to climb the pole. It was so far to the flag, however, that their strength failed ere they reached it.

A "boatswain's chair" was hastily rigged of rope, and another Russian soldier was hoisted to the flag. On reaching it, he untangled it and then made the mistake of dropping it to the ground, not understanding Captain Pestchouroff's energetic commands to the contrary.

It fell upon the bayonets of the Russian soldiers—which was considered an ill omen for Russia.

The United States flag was then slowly hoisted by George Lovell Rosseau, and the salutes were fired as before, the Russian water battery leading this time.

The hoisting of the flag was so timed that at the exact instant of its reaching its place, the report of the last big gun of the *Ossipee* roared out its final salute.

Upon the completion of the salutes, Captain Pestchouroff approached the commissioner and said:—

"General Rosseau, by authority of his Majesty, the Emperor of Russia, I transfer to the United States the Territory of Alaska."

The transfer was simply accepted, and the ceremony was at an end.

No one understanding the American spirit can seriously condemn the Americans present for the three cheers which burst spontaneously forth; yet there are occasions upon which an exhibition of good taste, repression, and consideration for the people of other nationalities present is more admirable and commendable than a spread-eagle burst of patriotism.

The last trouble caused by the Sitkan Indians was in 1878. The sealing schooner *San Diego* carried among its crew seven men of the Kake-sat-tee clan. The schooner was wrecked and six of the Kake-sat-tees were drowned. Chief Kath-le-an demanded of Colonel M. D. Ball, collector of customs and, at that time, the only representative of the government in Sitka, one thousand blankets for the life of each man drowned.

Colonel Ball, appreciating the gravity of the situation, and desiring time to prepare for the attack which he knew would be made upon the town, promised to write to the company in San Francisco and to the government in Washington.

After a long delay a reply to his letter arrived from the company, which refused, as he had expected, to allow the claim, and stated that no wages, even, were due the men who were drowned.

The government—which at that time had a vague idea that Alaska was a great iceberg floating between America and Siberia—paid no attention to the plea for assistance.

When Chief Kath-le-an learned that payment in blankets would not be made, he demanded the lives of six white men. This, also, being refused, he withdrew to prepare for battle.

Then hasty preparations were made in the settlement to meet the hourly expected attack. All the firearms were made ready for action, and a guard kept watch day and night. The Russian women and children were quartered in the home of Father Nicolai Metropolsky; the Americans in the custom-house.

The Indians held their war feast many miles from Sitka. On their way to attack the village they passed the White Sulphur Hot Springs, on the eastern shore of Baranoff Island, and murdered the man in charge.

They then demanded the lives of five white men, and when their demand was again refused, they marched stealthily upon the settlement.

However, Sitka possessed a warm and faithful friend in the person of Anna-Hoots, Chief of the Kak-wan-tans. He and his men met the hostile party and, while attempting to turn them aside from their murderous purpose, a general fight among the two clans was precipitated.

Before the Kake-sat-tees could again advance, a mail-boat arrived, and the war passion simmered.

When the boat sailed, a petition was sent to the British authorities at Esquimault, asking, for humanity's sake, that assistance be sent to Sitka.

Kath-le-an had retreated for reënforcement; and on the eve of his return to make a second attack, H.M.S. *Osprey* arrived in the harbor.

The appeal to another nation for aid, and the bitter newspaper criticism of its own indifference, had at last aroused the United States government to a realization of its responsibilities. The revenue cutter *Wolcott* dropped anchor in the Sitka Harbor a few days after the *Osprey*; and from that time on Sitka was not left without protection.

Along the curving road to Indian River stands the soft gray Episcopal Church, St. Peter's-by-the-Sea. Built of rough gray stone and shingles, it is an immediate pleasure and rest to the eye.

"Its doors stand open to the sea,The wind goes thro' at will,And bears the scent of brine and blueTo the far emerald hill."

Any stranger may enter alone, and passing into any pew, may kneel in silent communion with the God who has created few things on this earth more beautiful than Sitka.

No admission is asked. The church is free to the prince and the pauper, the sinner and the saint; to those of every creed, and to those of no creed at all.

The church has no rector, but is presided over by P. T. Rowe, the Bishop of All Alaska and the Beloved of All Men; him who carries over land and sea, over ice and everlasting snow, over far tundra wastes and down the lone and mighty Yukon in his solitary canoe or bidarka, by dog team and on foot, to white people and dark, and to whomsoever needs—the simple, sweet, and blessed message of Love.

It was in 1895 that Reverend P. T. Rowe, Rector of St. James' Church, Sault Sainte Marie, was confirmed as Bishop of Alaska. He went at once to that far and unknown land; and of him and his work there no words are ever heard save those of love and praise. He is bishop, rector, and travelling missionary; he is doctor, apothecary, and nurse; he is the hope and the comfort of the dying and the pall-bearer of the dead. He travels many hundreds of miles every year, by lone and perilous ways, over the ice and snow, with only an Indian guide and a team of huskies, to carry the word of God into dark places. He is equally at ease in the barabara and in the palace-like homes of the rich when he visits the large cities of the world.

Bishop Rowe is an exceptionally handsome man, of courtly bearing and polished manners. The moment he enters a church his personality impresses itself upon the people assembled to hear him speak.

On a gray August Sunday in Nome—three thousand miles from Sitka—I was surprised to see so many people on their way to midday service, Alaska not being famed for its church-going qualities.

"Oh, it is the Bishop," said the hotel clerk, smiling. "Bishop Rowe," he added, apparently as an after-thought. "Everybody goes to church when he comes to town."

I had never seen Bishop Rowe, and I had planned to spend the day alone on the beach, for the surf was rolling high and its musical thunder filled the town. Its lonely, melancholy spell was upon me, and its call was loud and insistent; and my heart told me to go.

But I had heard so much of Bishop Rowe and his self-devoted work in Alaska that I finally turned my back upon temptation and joined the narrow stream of humanity wending its way to the little church.

When Bishop Rowe came bending his dark head through the low door leading from the vestry, clad in his rich scarlet and purple and gold-embroidered robes, I thought I had never seen so handsome a man.

But his appearance was forgotten the moment he began to speak. He talked to us; but he did not preach. And we, gathered there from so many distant lands—each with his own hopes and sins and passions, his own desires and selfishness—grew closer together and leaned upon the words that were spoken there to us. They were so simple, and so earnest, and so sweet; they were so seriously and so kindly uttered.

And the text—it went with us, out into the sea-sweet, surf-beaten streets of Nome; and this was it, "Love me; and tell me so." Like the illustrious Veniaminoff, Bishop Rowe, of a different church and creed, and working in a later, more commercial age, has yet won his hold upon northern hearts by the sane and simple way of Love. The text of his sermon that gray day in the surf-beaten, tundra-sweet city of Nome is the text that he is patiently and cheerfully working out in his noble life-work.

Mr. Duncan, at Metlakahtla, has given his life to the Indians who have gathered about him; but Bishop Rowe, of All Alaska, has given his life to dark men and white, wherever they might be. Year after year he has gone out by perilous ways to find them, and to scatter among them his words of love—as softly and as gently as the Indians used to scatter the white down from the breasts of sea-birds, as a message of peace to all men.

The White Sulphur Hot Springs, now frequently called the Sitka Hot Springs, are situated on Hot Springs Bay on the eastern shore of Baranoff Island, almost directly east of Sitka.

The bay is sheltered by many small green islands, with lofty mountains rising behind the sloping shores. It is an ideally beautiful and desirable place to visit, even aside from the curative qualities of the clear waters which bubble from pools and crevices among the rocks. These springs have been famous since their discovery by Lisiansky in 1805. Sir George Simpson visited them in 1842; and with every year that has passed their praises have been more enthusiastically sung by the fortunate ones who have voyaged to that dazzlingly green and jewelled region.

Summit of Chilkoot Pass, 1898

The main spring has a temperature of one hundred and fifty-three degrees Fahrenheit, its waters cooking eggs in eight minutes. From this spring the baths are fed, their waters, flowing down to the sea, being soon reduced in temperature to one hundred and thirty degrees.

Filmy vapors float over the vicinity of the springs and rise in funnel-shaped columns which may be seen at a considerable distance, and which impart an atmosphere of mystery and unreality to the place.

Vegetation is of unusual luxuriance, even for this land of tropical growth; and in recent years experiments with melons and vegetables which usually mature in tropic climes only, have been entirely successful in this steamy and balmy region.

There are four springs, in whose waters the Indians, from the time of their discovery, have sought to wash away the ills to which flesh is heir. They came hundreds of miles and lay for hours at a time in the healing baths with only their heads visible. The bay was neutral ground where all might come, but where none might make settlement or establish claims.

The waters near abound in fish and water-fowl, and the forests with deer, bears, and other large game.

The place is coming but slowly to the recognition of the present generation. When the tropic beauty of its location and the curative powers of its waters are more generally known, it will be a Mecca for pilgrims.

The main station of Government Agricultural Experimental work in Alaska is located at Sitka. Professor C. C. Georgeson is the special agent in charge of the work, which has been very successful. It has accomplished more than anything else in the way of dispelling the erroneous impressions which people have received of Alaska by reading the descriptions of early explorers who fancied that every drift of snow was a living glacier and every feather the war bonnet of a savage.

In 1906, at Coldfoot, sixty miles north of the Arctic Circle, were grown cucumbers eight inches long, nineteen-inch rhubarb, potatoes four inches long, cabbages whose matured heads weighed eight pounds, and turnips weighing sixteen pounds—all of excellent quality.

At Bear Lake, near Seward and Cook Inlet, were grown good potatoes, radishes, lettuce, carrots, beets, rhubarb, strawberries, raspberries, Logan berries, blackberries; also, roses, lilacs, and English ivy. In this locality cows and chickens thrive and are profitable investments for those who are not too indolent to take care of them.

Alaskan lettuce must be eaten to be appreciated. During the hot days and the long, light hours of the nights it grows so rapidly that its crispness and delicacy of flavor cannot be imagined.

Everything in Alaska is either the largest, the best, or most beautiful, in the world, the people who live there maintain; and this soon grows to be a joke to the traveller. But when the assertion that lettuce grown in Alaska is the most delicious in the world is made, not a dissenting voice is heard.

Along the coast, sea-weed and fish guano are used as fertilizers; and soil at the mouth of a stream where there is silt is most desirable for vegetables.

In southeastern Alaska and along the coast to Kodiak, at Fairbanks and Copper Centre, at White Horse, Dawson, Rampart, Tanana, Council City, Eagle, and other places on the Yukon, almost all kinds of vegetables, berries, and flowers grow luxuriantly and bloom and bear in abundance. One turnip, of fine flavor, has been found sufficient for several people.

In the vicinity of the various hot springs, even corn, tomatoes, and muskmelons were successful to the highest degree.

On the Yukon cabbages form fine white, solid heads; cauliflower is unusually fine and white; beets grow to a good size, are tender, sweet, and of a bright red; peas are excellent; rhubarb, parsley, and celery were in many places successful. Onions seem to prove a failure in nearly all sections of the country; and potatoes, turnips, and lettuce are the prize vegetables.

Grain growing is no longer attempted. The experiment made by the government, in the coast region, proved entirely unsatisfactory. It will usually mature, but August, September, and October are so rainy that it is not possible to save the crop. It is, however, grown as a forage crop, for which purpose it serves excellently.

The numerous small valleys, coves, and pockets afford desirable locations for gardens, berries, and some varieties of fruit trees.

In the interior encouraging success has been obtained with grain. The experiments at Copper Centre have not been so satisfactory as at Rampart, three and a half degrees farther north, on the Yukon.

At Copper Centre heavy frosts occur as early as August 14; while at Rampart no "killing" frosts have been known before the grain had ripened, in the latter part of August.

Rampart is the loveliest settlement on the Yukon, with the exception of Tanana. Across the river from Rampart, the green fields of the Experimental Station slope down to the water. The experiments carried on here by Superintendent Rader, under the general supervision of Professor Georgeson—who visits the stations yearly—have been very satisfactory.

Experimental work was begun at Rampart in 1900, and grain has matured there every year, while at Copper Centre only one crop of four has matured. In 1906, owing to dry weather, the growth was slow until the middle of July; from that date on to the latter part of August there were frequent rains, causing a later growth of grain than usual. The result of these conditions was that when the first "killing" frost occurred, the grain was still growing, and all plats, save those seeded earliest, were spoiled for the finer purposes. The frosted grain was, however, immediately cut for hay, twenty tons of which easily sold for four thousand, one hundred and fifty-two dollars.

These results prove that even where grain cannot be grown to the best advantage, it may be profitably grown for hay. For the latter purpose larger growing varieties would be sown, which would produce a much heavier yield and bring larger profits. At present all the feed consumed in the interior by the horses of pack trains and of travellers is hauled in from tide-water,—a hundred miles, at least, and frequently two or three times as far,—and two hundred dollars a ton for hay is a low price. The actual cost of hauling a ton of hay from Valdez to Copper Centre, one hundred miles, is more than two hundred dollars.

Road-house keepers advertise "specially low" rates on hay at twenty cents a pound, the ordinary retail price at that distance from tide-water being five hundred dollars a ton.

The most serious drawback to the advancement of agriculture in Alaska is the lack of interest on the part of the inhabitants.

Probably not fifty people could be found in the territory who went there for the purpose of making homes. Now and then a lone dreamer of dreams may be found who lives there—or who would gladly live there, if he might—only for the beauty of it, which can be found nowhere else; and which will soon vanish before the brutal tread of civilization.

The others go for gold. If they do not expect to dig it out of the earth themselves, they plan and scheme to get it out of those who have so acquired it. There is no scheme that has not been worked upon Alaska and the real workers of Alaska.

The schemers go there to get gold; honestly, if possible, but to get gold; to live "from hand to mouth," while they are there, and to get away as quickly as possible and spend their gold far from the country which yielded it. They have neither the time nor the desire to do anything toward the development of the country itself.

Ex-Governor John G. Brady is one of the few who have devoted their lives to the interest and the up-building of Alaska.

Thirty years ago he went to Alaska and established his home at Sitka. There he has lived all these years with his large and interesting family; there he still lives.

He has a comfortable home, gardens and orchards that leave little to be desired, and has demonstrated beyond all doubt that the man who wishes to establish a modern, comfortable—even luxurious—home in Alaska, can accomplish his purpose without serious hardship to his family, however delicate the members thereof may be.

The Bradys are enthusiasts and authorities on all matters pertaining to Alaska.

Governor Brady has been called the "Rose Governor" of Alaska, because of his genuine admiration for this flower. He can scarcely talk five minutes on Alaska without introducing the subject of roses; and no enthusiast has ever talked more simply and charmingly of the roses of any land than he talks of the roses of Alaska,—the cherished ones of the garden, and the big pink ones of Unalaska and the Yukon.

As missionary and governor, Mr. Brady has devoted many years to this splendid country; and the distressful troubles into which he has fallen of late, through no fault of his own, can never make a grateful people forget his unselfish work for the up-building and the civilization of Alaska.

To-day, Sitka is idyllic. Her charm is too poetic and too elusive to be described in prose. A greater contrast than she presents to such hustling, commercial towns as Juneau, Valdez, Cordova, and Katalla, could scarcely be conceived. To drift into the harbor of Sitka is like entering another world.

The Russian influence is still there, after all these years—as it is in Kodiak and Unalaska.

CHAPTER XIX

In rough weather, steamers bound for Sitka from the westward frequently enter Cross Sound and proceed by way of Icy Straits and Chatham to Peril.

Icy Straits are filled, in the warmest months, with icebergs floating down from the many glaciers to the north. Of these Muir has been the finest, and is a world-famous glacier, owing to the charming descriptions written of it by Mr. John Muir. For several years it was the chief object of interest on the "tourist" trip; but early in 1900 an earthquake shattered its beautiful front and so choked the bay with immense bergs that the steamer *Spokane* could not approach closer than Marble Island, thirteen miles from the front. The bergs were compact and filled the whole bay. Since that time excursion steamers have not attempted to enter Glacier Bay.

In the summer of 1907, however, a steamer entered the bay and, finding it free of ice, approached close to the famed glacier—only to find it resembling a great castle whose towers and turrets have fallen to ruin with the passing of years. Where once shone its opaline palisades is now but a field of crumpled ice.

There are no less than seven glaciers discharging into Glacier Bay and sending out beautiful bergs to drift up and down Icy Straits with the tides and winds. Rendu, Carroll, Grand Pacific, Johns Hopkins, Hugh Miller, and Geikie front on the bay or its narrow inlets.

Brady Glacier has a three-mile frontage on Wimbledon, or Taylor, Bay, which opens into Icy Straits.

When, on her mid-June voyage from Seattle in 1905, the *Santa Ana* drew out and away from Sitka, and turning with a wide sweep, went drifting slowly through the maze of green islands and set her prow "to Westward," one of the dreams of my life was "come true."

I was on my way to the far, lonely, and lovely Aleutian Isles,—the green, green isles crested with fire and snow that are washed on the north by the waves of Behring Sea.

It was a violet day. There were no warm purple tones anywhere; but the cool, sparkling violet ones that mean the nearness of mountains of snow. One could almost feel the crisp *ting* of ice in the air, and smell the sunlight that opalizes, without melting, the ice.

Round and white, with the sunken nest of the thunder-bird on its crest, Mount Edgecumbe rose before us; the pale green islands leaned apart to let us through; the sea-birds, white and lavender and rose-touched, floated with us; the throb of the steamer was like a pulse beating in one's own blood; there were words in the violet light that lured us on, and a wild sweet song in the waves that broke at our prow.

"There can be nothing more beautiful on earth," I said; but I did not know. An hour came soon when I stood with bared head and could not speak for the beauty about me; when the speech of others jarred upon me like an insult, and the throb of the steamer, which had been a sensuous pleasure, pierced my exaltation like a blow.

The long violet day of delight wore away at last, and night came on. A wild wind blew from the southwest, and the mood of the North Pacific Ocean changed. The ship rolled heavily; the waves broke over our decks. We could see them coming—black, bowing, rimmed with white. Then came the shock—followed by the awful shudder and struggle of the boat. The wind was terrific. It beat the breath back into the breast.

It was terrible and it was glorious. Those were big moments on the texas of the *Santa Ana;* they were worth living, they were worth while. But on account of the storm, darkness fell at midnight; and as the spray was now breaking in sheets over the bridge and texas, I was assisted to my cabin—drenched, shivering, happy.

"Shut your door," said the captain, "or you will be washed out of your berth; and wait till to-morrow."

I wondered what he meant, but before I could ask him, before he could close my cabin door, a great sea towered and poised for an instant behind him, then bowed over him and carried him into the room. It drenched the whole room and everything and everybody in it; then swept out again as the ship rolled to starboard.

My travelling companion in the middle berth uttered such sounds as I had never heard before in my life, and will probably never hear again unless it be in the North Pacific Ocean in the vicinity of Yakutat or Katalla. She made one attempt to descend to the floor; but at sight of the captain who was struggling to take a polite departure after his anything but polite entrance, she uttered the most dreadful sound of all and fell back into her berth.

I have never seen any intoxicated man teeter and lurch as he did, trying to get out of our cabin. I sat upon the stool where I had been washed and dashed by the sea, and laughed.

He made it at last. He uttered no apologies and no adieux; and never have I seen a man so openly relieved to escape from the presence of ladies.

I closed the window. Disrobing was out of the question. I could neither stand nor sit without holding tightly to something with both hands for support; and when I had lain down, I found that I must hold to both sides of the berth to keep myself in.

"Serves you right," complained the occupant of the middle berth, "for staying up on the texas until such an unearthly hour. I'm glad you can't undress. Maybe you'll come in at a decent hour after this!"

It is small wonder that Behring and Chirikoff disagreed and drifted apart in the North Pacific Ocean. It is my belief that two angels would quarrel if shut up in a stateroom in a "Yakutat blow"—than which only a "Yakataga blow" is worse; and it comes later.

I am convinced, after three summers spent in voyaging along the Alaskan coast to Nome and down the Yukon, that quarrelling with one's room-mate on a long voyage aids digestion. My room-mate and I have never agreed upon any other subject; but upon this, we are as one.

Neither effort nor exertion is required to begin a quarrel. It is only necessary to ask with some querulousness, "Are you going to stand before that mirror *all day?*" and hey, presto! we are instantly at it with hammer and tongs.

Toward daylight the storm grew too terrible for further quarrelling; too big for all little petty human passions. A coward would have become a man in the face of such a conflict. I have never understood how one can commit a cowardly act during a storm at sea. One may dance a hornpipe of terror on a public street when a man thrusts a revolver into one's face and demands one's money. That is a little thing, and inspires to little sensations and little actions. But when a ship goes down into a black hollow of the sea, down, down, so low that it seems as though she must go on to the lowest, deepest depth of all—and then lies still, shudders, and begins to mount, higher, higher, higher, to the very crest of a mountainous wave; if God put anything at all of courage and of bravery into the soul of the human being that experiences this, it comes to the front now, if ever.

In that most needlessly cruel of all the ocean disasters of the Pacific Coast, the wreck of the *Valencia* on Seabird Reef of the rock-ribbed coast of Vancouver Island, more than a hundred people clung to the decks and rigging in a freezing storm for thirty-six hours. There was a young girl on the ship who was travelling alone. A young man, an athlete, of Victoria, who had never met her before, assisted her into the rigging when the decks were all awash, and protected her there. On the last day before the ship went to pieces, two life-rafts were successfully launched. Only a few could go, and strong men were desired to manage the rafts. The young man in the rigging might have been saved, for the ones who did go on the raft were the only ones rescued. But when summoned, he made simple answer:—

"No; I have some one here to care for. I will stay."

Better to be that brave man's wave-battered and fish-eaten corpse, than any living coward who sailed away and left those desperate, struggling wretches to their awful fate.

The storm died slowly with the night; and at last we could sleep.

It was noon when we once more got ourselves up on deck. The sun shone like gold upon the sea, which stretched, dimpling, away for hundreds upon hundreds of miles, to the south and west. I stood looking across it for some time, lost in thought, but at last something led me to the other side of the ship.

All unprepared, I lifted my eyes—and beheld before me the glory and the marvel of God. In all the splendor of the drenched sunlight, straight out of the violet, sparkling sea, rose the magnificent peaks of the Fairweather Range and towered against the sky. No great snow mountains rising from the land have ever affected me as did that long and noble chain glistening out of the sea. They seemed fairly to thunder their beauty to the sky.

From Mount Edgecumbe there is no significant break in the mountain range for more than a thousand miles; it is a stretch of sublime beauty that has no parallel. The Fairweather Range merges into the St. Elias Alps; the Alps are followed successively by the Chugach Alps, the Kenai and Alaskan ranges,—the latter of which holds the loftiest of them all, the superb Mount McKinley,—and the Aleutian Range, which extends to the end of the Aliaska Peninsula. The volcanoes on the Aleutian and Kurile islands complete the ring of snow and fire that circles around the Pacific Ocean.

CHAPTER XX

Our ship having been delayed by the storm, it was mid-afternoon when we reached Yakutat. A vast plateau borders the ocean from Cross Sound, north of Baranoff and Chicagoff islands, to Yakutat; and out of this plateau rise four great snow peaks—Mount La Pérouse, Mount Crillon, Mount Lituya, and Mount Fairweather—ranging in height from ten thousand to fifteen thousand nine hundred feet.

In all this stretch there are but two bays of any size, Lituya and Dry, and they have only historical importance.

Lituya Bay was described minutely by La Pérouse, who spent some time there in 1786 in his two vessels, the *Astrolabe* and *Boussole.*

The entrance to this bay is exceedingly dangerous; the tide enters in a bore, which can only be run at slack tide. La Pérouse lost two boatloads of men in this bore, on the eve of his departure,—a loss which he describes at length and with much feeling.

Before finally departing, he caused to be erected a monument to the memory of the lost officers and crew on a small island which he named Cénotaphe, or Monument, Isle. A bottle containing a full account of the disaster and the names of the twenty-one men was buried at the foot of the monument.

La Pérouse named this bay Port des Français.

The chronicles of this modest French navigator seem, somehow, to stand apart from those of the other early voyagers. There is an appearance of truth and of fine feeling in them that does not appear in all.

He at first attempted to enter Yakutat Bay, which he called the Bay of Monti, in honor of the commandant of an exploring expedition which he sent out in advance; but the sea was breaking with such violence upon the beach that he abandoned the attempt.

He described the savages of Lituya Bay as treacherous and thievish. They surrounded the ships in canoes, offering to exchange fresh fish and otter skins for iron, which seemed to be the only article desired, although glass beads found some small favor in the eyes of the women.

La Pérouse supposed himself to be the first discoverer of this bay. The Russians, however, had been there years before.

The savages appeared to be worshippers of the sun. La Pérouse pronounced the bay itself to be the most extraordinary spot on the whole earth. It is a great basin, the middle of which is unfathomable, surrounded by snow peaks of great height. During all the time that he was there, he never saw a puff of wind ruffle the surface of the water, nor was it ever disturbed, save by the fall of masses of ice which were discharged from five different glaciers with a thunderous noise which reëchoed from the farthest recesses of the surrounding mountains. The air was so tranquil and the silence so undisturbed that the human voice and the cries of sea-birds lying among the rocks were heard at the distance of half a league.

The climate was found to be "infinitely milder" than that of Hudson Bay of the same latitude. Vegetation was extremely vigorous, pines measuring six feet in diameter and rising to a height of one hundred and forty feet.

Celery, sorrel, lupines, wild peas, yarrow, chicory, angelica, violets, and many varieties of grass were found in abundance, and were used in soups and salads, as remedies for scurvy.

Strawberries, raspberries, gooseberries, the elder, the willow, and the broom were found then as they are to-day. Trout and salmon were taken in the streams, and in the bay, halibut.

It is to be feared that La Pérouse was not strong on birds; for in the copses he heard singing "linnets, *nightingales*, blackbirds, and water quails," whose songs were very agreeable. It was July, which he called the "pairing-time." He found one very fine blue jay; and it is surprising that he did not hear it sing.

For the savages—especially the women—the fastidious Frenchman entertained feelings of disgust and horror. He could discover no virtues or traits in them to praise, conscientiously though he tried.

They lived in the same kind of habitations that all the early explorers found along the coast of Alaska: large buildings consisting of one room, twenty-five by twenty feet, or larger. Fire was kindled in the middle of these rooms on the earth floor. Over it was suspended fish of several kinds to be smoked. There was always a large hole in the roof—when there was a roof at all—to receive the smoke.

About twenty persons of both sexes dwelt in each of these houses. Their habits, customs, and relations were indescribably disgusting and indecent.

Their houses were more loathsome and vile of odor than the den of any beast. Even at the present time in some of the native villages—notably Belkoffski on the Aliaskan Peninsula—all the most horrible odors ever experienced in civilization, distilled into one, could not equal the stench with which the natives and their habitations reek. As their customs are somewhat cleanlier now than they were a hundred and thirty years ago, and as upon this one point all the early navigators forcibly agree, we may well conclude that they did not exaggerate.

The one room was used for eating, sleeping, cooking, smoking fish, washing their clothes—in their cooking and eating wooden utensils, by the way, which are never cleansed—and for the habitation of their dogs.

The men pierced the cartilage of the nose and ears for the wearing of ornaments of shell, iron, or other material. They filed their teeth down even with the gums with a piece of rough stone. The men painted their faces and other parts of their bodies in a "frightful manner" with ochre, lamp-black, and black lead, mixed with the oil of the "sea-wolf." Their hair was frequently greased and dressed with the down of sea-birds; the women's, also. A plain skin covered the shoulders of the men, while the rest of the body was left entirely naked.

The women filled the Frenchman with a lively horror. The labret in the lower lip, or ladle, as he termed it, wore unbearably upon his fine nerves. He considered that the whole world would not afford another custom equally revolting and disgusting. When the ornament was removed, the lower lip fell down upon the chin, and this second picture was more hideous than the first.

The gallant Captain Dixon, on his voyage a year later, was more favorably impressed with the women. He must have worn rose-colored glasses. He describes their habits and habitations almost as La Pérouse did, but uses no expression of disgust or horror. He describes the women as being of medium size, having straight, well-shaped limbs. They painted their faces; but he prevailed upon one woman by persuasion and presents to wash her face and hands. Whereupon "her countenance had all the cheerful glow of an English milkmaid's; and the healthy red which suffused her cheeks was even beautifully contrasted with the white of her neck; her eyes were black and sparkling; her eyebrows of the same color *and most beautifully arched*; her forehead so remarkably clear that the translucent veins were seen meandering even in their minutest branches—in short, she would be considered handsome even in England." The worst adjectives he applied to the labret were "singular" and "curious."

Pine Falls, Atlin

Don Maurello and other navigators found now and then a woman who might compete with the beauties of Spain and other lands; but none shared the transports of Dixon, who idealized their virtues and condoned their faults.

Tebenkof located two immense glaciers in the bay of Lituya, one in each arm, describing them briefly:—

"The icebergs fall from the mountains and float over the waters of the bay throughout the year. Nothing disturbs the deep silence of this *terribly grand* gorge of the mountains but the thunder of the falling icebergs."

La Pérouse found enormous masses of ice detaching themselves from five different glaciers. The water was covered with icebergs, and nearness to the shore was exceedingly dangerous. His small boat was upset half a mile from shore by a mass of ice falling from a glacier.

Mr. Muir describes La Pérouse Glacier as presenting grand ice bluffs to the open ocean, into which it occasionally discharged bergs.

All agree that the appearance and surroundings of the bay are extraordinary.

Yakutat Bay is two hundred and fifteen miles from Sitka. It was called Behring Bay by Cook and Vancouver, who supposed it to be the bay in which the Dane anchored in 1741. It was named Admiralty Bay by Dixon, and the Bay of Monti by La Pérouse. The Indian name is the only one which has been preserved.

It is so peculiarly situated that although several islands lie in front of it, the full force of the North Pacific Ocean sweeps into it. At most seasons of the year it is full of floating ice which drifts down from the glaciers of Disenchantment Bay.

At the point on the southern side of the bay which Dixon named Mulgrave, and where there is a fine harbor, Baranoff established a colony of Siberian convicts about 1796. His instructions from Shelikoff for the laying-out of a city in such a wilderness make interesting reading.

"And now it only remains for us to hope that, having selected on the mainland a suitable place, you will lay out the settlement with some taste and with due regard for beauty of construction, in order that when visits are made by foreign ships, as cannot fail to happen, it may appear more like a town than a village, and that the Russians in America may live in a neat and orderly way, and not, as in Ohkotsk, in squalor and misery, caused by the absence of nearly everything necessary to civilization. Use taste as well as practical judgment in locating the settlement. Look to beauty, as well as to convenience of material and supplies. On the plans, as well as in reality, leave room for spacious squares for public assemblies. Make the streets not too long, but wide, and let them radiate from the squares. If the site is wooded, let trees enough stand to line the streets and to fill the gardens, in order to beautify the place and preserve a healthy atmosphere. Build the houses along the streets, but at some distance from each other, in order to increase the extent of the town. The roofs should be of equal height, and the architecture as uniform as possible. The gardens should be of equal size and provided with good fences along the streets. Thanks be to God that you will at least have no lack of timber."

In the same letter poor Baranoff was reproached for exchanging visits with captains of foreign vessels, and warned that he might be carried off to California or some other "desolate" place.

The colony of convicts had been intended as an "agricultural" settlement; but the bleak location at the foot of Mount St. Elias made a farce of the undertaking. The site had been chosen by a mistake. A post and fortifications were erected, but it is not chronicled that Shelikoff's instructions were carried out. There was great mortality among the colonists and their families, and constant danger of attack by the Kolosh. Finally, in 1805, the fort and settlement were entirely destroyed by their cruel and revengeful enemies.

The new town of Yakutat is three or four miles from the old settlement. There is a good wharf at the foot of a commanding plateau, which is a good site for a city. On the wharf are a saw-mill and cannery. A stiff climb along a forest road brings one to a store, several other business houses, and a few residences.

There are good coal veins in the vicinity. The Yakutat and Southern Railway leads several miles into the interior, and handles a great deal of timber.

In 1794 Puget sailed the *Chatham* through the narrow channel between the mainland and the islands, leading to Port Mulgrave—where Portoff was established in a tent with nine of his countrymen and several hundred Kadiak natives. He found the channel narrow and dangerous; his vessel grounded, but was successfully floated at returning tide. Passage to Mulgrave was found easy, however, by a channel farther to the westward and southward.

In this bay, as in nearly all other localities on the Northwest Coast, the Indians coming out to visit them paddled around the ship two or three times singing a ceremonious song, before offering to come aboard. They gladly exchanged bows, arrows, darts, spears, fish-gigs—whatever they may be—kamelaykas, or walrus-gut coats, and needlework for white shirts, collars, cravats, and other wearing apparel.

An Indian chief stole Mr. Puget's gold watch chain and seals from his cabin; but it was discovered by Portoff and returned.

The cape extending into the ocean south of the town was the Cape Phipps of the Russians. It has long been known, however, as Ocean Cape. Cape Manby is on the opposite side of the bay.

Sailing up Yakutat Bay, the Bay of Disenchantment is entered and continues for sixty miles, when it merges into Russell Fiord, which bends sharply to the south and almost reaches the ocean.

Enchantment Bay would be a more appropriate name. The scenery is of varied, magnificent, and ever increasing beauty. The climax is reached in Russell Fiord—named for Professor Russell, who explored it in a canoe in 1891.

From Yakutat Bay to the very head of Russell Fiord supreme splendor of scenery is encountered, surpassing the most vaunted of the Old World. Within a few miles, one passes from luxuriant forestation to lovely lakes, lacy cascades, bits of green valley; and then, of a sudden, all unprepared, into the most sublime snow-mountain fastnesses imaginable, surrounded by glaciers and many of the most majestic mountain peaks of the world.

Cascades spring, foaming, down from misty heights, and flowers bloom, large and brilliant, from the water to the line of snow.

Malaspina, an Italian in the service of Spain, named Disenchantment Bay. Turner Glacier and the vast Hubbard Glacier discharge into this bay; and from the reports of the Italian, Tabenkoff, and Vancouver, it has been considered possible that the two glaciers may have reached, more than a hundred years ago, across the narrowest bend at the head of Yakutat Bay.

The fiord is so narrow that the tops of the high snow mountains have the appearance of overhanging their bases; and to the canoeist floating down the slender, translucent water-way, this effect adds to the austerity of the scene.

Captains of regular steamers are frequently offered good prices to make a side trip up Yakutat Bay to the beginning of Disenchantment; but owing to the dangers of its comparatively uncharted waters, they usually decline with vigor.

One who would penetrate into this exquisitely beautiful, lone, and enchanted region must trust himself to a long canoe voyage and complete isolation from his kind. But what recompense—what life-rememberable joy!

Each country has its spell; but none is so great as the spell of this lone and splendid land. It is too sacred for any light word of pen or lip. The spell of Alaska is the spell of God; and it holds all save the basest, whether they acknowledge it or deny. Here are sphinxes and pyramids built of century upon century's snow; the pale green thunder of the cataract; the roar of the avalanche and the glacier's compelling march; the flow of mighty rivers; the unbroken silences that swim from snow mountain to snow mountain; and the rose of sunset whose petals float and fade upon mountain and sea.

As one sails past these mountains days upon days, they seem to lean apart and withdraw in pearly aloofness, that others more beautiful and more remote may dawn upon the enraptured beholder's sight. For hundreds of miles up and down the coast, and for hundreds into the interior, they rise in full view from the ocean which breaks upon the nearer ones. At sunrise and at sunset each is wrapped in a different color from the others, each in its own light, its own glory—caused by its own peculiar shape and its position among the others.

While the steamer lies at Yakutat passengers may, if they desire, walk through the forest to the old village, where there is an ancient Thlinkit settlement. There is a new one at the new town. The tents and cabins climb picturesquely among the trees and ferns from the water up a steep hill.

In 1880 there was a great gold excitement at Yakutat. Gold was discovered in the black-sand beaches. A number of mining camps were there until the late 'eighties, and by the use of rotary hand amalgamators, men were able to clean up forty dollars a day.

The bay was flooded by a tidal wave which left the beach covered with fish. The oil deposited by their decay prevented the action of the mercury, and the camp was abandoned.

The sea is now restoring the black sand, and a second Nome may one day spring up on these hills in a single night.

As I have said elsewhere, the Yakutat women are among the finest basket weavers of the coast. A finely twined Yakutat basket, however small it may be, is a prize; but the bottom should be woven as finely and as carefully as the body of the basket. Some of the younger weavers make haste by weaving the bottom coarsely, which detracts from both its artistic and commercial value.

The instant the end of the gangway touches the wharf at Yakutat, the gayly-clad, dark-eyed squaws swarm aboard.

They settle themselves noiselessly along the promenade decks, disposing their baskets, bracelets, carved horn spoons, totem-poles, inlaid lamps, and beaded moccasins about them.

If, during the hours of animated barter that follow, one or two of the women should disappear, the wise woman-passenger will saunter around the ship and take a look into her stateroom, to make sure that all is well; else, when she does return to it, she may miss silver-backed mirrors, bottles of lavender water, bits of jewellery that may have been carelessly left in sight, pretty collars—and even waists and hats—to say nothing of the things which she may later on find.

These poor dark people were born thieves; and neither the little education they have received, nor the treatment accorded them by the majority of white people with whom they have been brought into contact, has served to wean them entirely from the habits and the instincts of centuries.

At Yakutat, no matter how much good sound sense he may possess, the traveller parts with many large silver dollars. He thinks of Christmas, and counts his friends on one hand, then on the other; then over again, on both.

When the steamer has whistled for the sixth time to call in the wandering passengers, and the captain is on the bridge; when the last squaw has pigeon-toed herself up the gangway, flirting her gay shawl around her and chuckling and clucking over the gullibility of the innocent white people; when the last strain from the phonograph in the big store on the hill has died across the violet water widening between the shore and the withdrawing ship—the spendthrift passenger retires to his cabin and finds the berths overflowing and smelling to heaven with Indian things. Then—too late—he sits down, anywhere, and reflects.

The western shore of Yakutat Bay is bounded by the largest glacier in the world—the Malaspina. It has a sea-frontage of more than sixty miles extending from the bay "to Westward"; and the length of its splendid sweep from its head to the sea at the foot of Mount St. Elias is ninety miles.

For one whole day the majestic mountain and its beautiful companion peaks were in sight of the steamer, before the next range came into view. The sea breaks sheer upon the ice-palisades of the glacier. Icebergs, pale green, pale blue, and rose-colored, march out to meet and, bowing, pass the ship.

One cannot say that he knows what beauty is until he has cruised leisurely past this glacier, with the mountains rising behind it, on a clear day, followed by a moonlit night.

On one side are miles on miles of violet ocean sweeping away into limitless space, a fleck of sunlight flashing like a fire-fly in every hollowed wave; on the other, miles on miles of glistening ice, crowned by peaks of softest snow.

At sunset warm purple mists drift in and settle over the glacier; above these float banks of deepest rose; through both, and above them, glimmer the mountains pearlily, in a remote loveliness that seems not of earth.

But by moonlight to see the glacier streaming down from the mountains and out into the ocean, into the midnight—silent, opaline, majestic—is worth ten years of dull, ordinary living.

It is as if the very face of God shone through the silence and the sublimity of the night.

CHAPTER XXI

There is an open roadstead at Yaktag, or Yakataga. The ship anchors several miles from shore—when the fierce storms which prevail in this vicinity will permit it to anchor at all—and passengers and freight are lightered ashore.

I have seen horses hoisted from the deck in their wooden cages and dropped into the sea, where they were liberated. After their first frightened, furious plunges, they headed for the shore, and started out bravely on their long swim. The surf was running high, and for a time it seemed that they could not escape being dashed upon the rocks; but with unerring instinct, they struggled away from one rocky place after another until they reached a strip of smooth sand up which they were borne by the breaking sea, and where they fell for a few moments, exhausted. Then they arose, staggered, threw up their heads and ran as I have never seen horses run—with such wildness, such gladness, such utterance of the joy of freedom in the fling of their legs, in the streaming of mane and tail.

They had been penned in a narrow stall under the forward deck for twelve days; they had been battered by the storms and unable to lie down and rest; they had been plunged from this condition unexpectedly into the ocean and compelled to strike out on a long swim for their lives.

The sudden knowledge of freedom; the smell of sun and air; the very sweet of life itself—all combined to make them almost frantic in the animal expression of their joy.

We put down the powerful glasses with which we had painfully watched every yard of their progress toward the land.

I looked at the pilot. There was a moisture in his eyes, which was not entirely a reflection of that in my own.

It is one hundred and seventy miles from Yakutat to Kayak.

Off this stretch of coast, between Lituya and Cape Suckling, the soundings are moderate and by whalers have long been known as "Fairweather Grounds."

Just before reaching Kayak, Cape Suckling is passed.

The point of this cape is low. It runs up into a considerable hill, which, in turn, sinking to very low land has the appearance of an island. It was named by Cook.

Around this cape lies Comptroller Bay—the bay which should have been named Behring's Bay. It was on the two islands at its entrance that Behring landed in 1741. He named one St. Elias; and to this island Cook, in 1778, gave the name of Kaye, for the excellent reason that the "Reverend Doctor Kaye" gave him two silver two-penny pieces of the date of 1772, which he buried in a bottle on the island, together with the names of his ships and the date of discovery.

Unhappily this immortal island retains the name which Cook lightly bestowed upon it, instead of the name given it by the illustrious Dane. It is now, however, more frequently known as Wingham Island. The settlement of Kayak is upon it. The southern extremity of the larger island retains the name St. Elias for the splendid headland that plunges boldly and challengingly out into the sea. It is a magnificent sight in a storm, when sea-birds are shrieking over it and a powerful surf is breaking upon its base. At all times it is a striking landmark.

I have been to Kayak four times. Landings have always been made by passengers in dories or in tiny launches which come out from the settlement, and which bob up and down like corks.

It requires a cool head to descend a rope-ladder twenty or thirty feet from the deck to a dory that rolls away from the ship with every wave and which may only be entered as it rolls back. There is art in the little kick which one must give each rung against the side of the ship to steady the ladder. At the last comes an awful moment when a woman must hang alone on the last swaying rung and await the return of the dory. If the sea is rough, the ship will probably roll away from the boat. When the sailors, therefore, sing out, "Now! Jump!" she must close her eyes, put her trust in heaven and fore-ordination, and jump.

If she chances to jump just at the right moment; if one sailor catches her just right and another catches *him* just right, she will know by the cheer that arises from hurricane and texas that all is well and she may open her eyes. Under other conditions, other situations arise; but let no woman be deterred by the possibility of the latter from descending a rope-ladder when she has an opportunity. The hair-crinkling moments in an ordinary life are few enough, heaven knows.

There are several business houses and dwellings at Kayak; and an Indian village. The Indian graveyard is very interesting. Tiny houses are built over the graves and surrounded by picket fences. Both are painted white. Through the windows may be seen some of the belongings of the dead. In dishes are different kinds of food and drink, that the deceased may not suffer of hunger or thirst in the bourne to which he may have journeyed. There are implements and weapons for the men; unfinished baskets for the women, with the long strands of warp and woof left ready for the idle hand; for the children, beads and rattles made of bear claws and shells. The houses are on posts a few feet above the graves.

For a number of years Kayak was the base of operation for oil companies. In 1898 the Alaska Development Company staked the country, but later leased their lands to the Alaska Oil and Coal Company—commonly known as the "English" company—for a long term of years, with the privilege of taking up the lease in 1906. This company spent millions of dollars and drilled several wells.

The Alaska Petroleum and Coal Company—known as the Lippy Company—put down two holes, one seventeen hundred feet deep. The cost of drilling is about five thousand dollars a hole of two thousand feet; the rig, laid down, six thousand five hundred dollars.

These wells are situated at Katalla, sixteen miles from Kayak, at the mouth of the Copper River. The oil lands extend from the coast to the Malaspina and Behring glaciers.

Since the recent upspringing of a new town at Katalla, the centre of trade has been transferred from Kayak to this point. Katalla was founded in 1904 by the Alaska Petroleum and Coal Company; but not until the actual commencement of work on the Bruner Railway Company's road, in 1907, from Katalla into the heart of the coal and oil fields, did the place rise to the importance of a northern town.

It has attained a wide fame within a few months on account of the remarkable discoveries of high-grade petroleum and coal in the vicinity.

For many years these two products of Alaska were considered of inferior quality; but it has recently been discovered that they rival the finest of Pennsylvania.

The town has grown as only a new Alaskan, or Puget Sound, town can grow. At night, perhaps, there will be a dozen shacks and as many tents on a town site; the next morning a steamer will anchor in the bay bearing government offices, stores, hotels, saloons, dance-halls, banks, offices for several large companies, electric light plants, gas works, telephones—and before another day dawns, business is in full swing.

For fifteen miles along the Comptroller Bay water front oil wells may be seen, some of the largest oil seepages existing close to the shore. The coal and oil lands of this vicinity, however, are about a hundred miles in length and from twenty to thirty in width.

During the fall and early winter of 1907, Katalla suffered a serious menace to its prosperity, owing to its total lack of a harbor.

The bay is but a mere indentation, and an open roadstead sends its surf to curl upon the unprotected beach. The storms in winter are ceaseless and terrific. Steamers cannot land and anchors will not hold.

As Nome, similarly situated, is cut off from the world for several months by ice, so is Katalla cut off by storms.

Steamer after steamer sails into the roadstead, rolls and tosses in the trough of the sea, lingers regretfully, and sails away, without landing even a passenger, or mail.

In October, 1907, one whole banking outfit, including everything necessary for the opening of a bank, save the cashier,—who was already there,—and the building,—which was waiting,—was taken up on a steamer. Not being able to lighter it ashore, the steamer carried the bank to Cook Inlet.

Upon its return, conditions again made it impossible to enter the bay, and the bank was carried back to Seattle. When the steamer again went north, the bank went, too; when the steamer returned, the bank returned.

In the meantime, other events were shaping themselves in such wise as to render the situation extremely interesting.

A few miles northwest of Katalla, the town of Cordova was established three years ago, with the terminus of the Copper River Railway located there. Mr. M. J. Heney, who had built the White Pass and Yukon Railway, received the contract for the work. The building of wharves in the excellent harbor and the laying out of a town site capable of accommodating twenty thousand people—and one that might have pleased even the fastidious Shelikoff—was energetically begun.

Early in 1907 the Copper River Railway sold its interests to the Northwestern and Copper River Valley Railway, promoted by John Rosene, and financed by the Guggenheims. It was semi-officially announced that the new company would tear up the Cordova tracks and that Katalla would be the terminus of the consolidated line. The announcement precipitated the "boom" at Katalla.

Mr. Heney retired from the new company and spent the summer voyaging down the Yukon.

Immediately upon his return to Seattle in September, he journeyed to New York. In a few days, newspapers devoted columns to the sale of the Rosene interests in the railway, also a large fleet of first-class steamers, and wharves, to the Copper

River and Northwestern Railway Company.

The contract for the immediate building of the road had been secured by Mr. Heney, who had returned to his original surveys. The terminus at once travelled back to Cordova; and the itinerant bank may yet thank its guiding star which prevented it from getting itself landed at Katalla.

Important "strikes" are made constantly in the Tanana country, in the Sushitna, and in the Koyukuk, where pay is found surpassing the best of the Klondike.

The trail from Valdez to Fairbanks may yet be as thickly strewn with eager-eyed stampeders as were the Dyea and Skagway trails a decade ago. Never again, however, in any part of Alaska, can the awful conditions of that time prevail. Steamer, rail, and stage transportation have made travelling in the North luxurious, compared to the horrors endured in the old days.

The Guggenheims have been compelled to carry on a fantastic fight for right of way for the Copper River and Northwestern Railroad. In the summer of 1907, they attempted to lay track at Katalla over the disputed Bruner right of way. The Bruner Company had constructed an immense "go-devil" of railway rails, which, operated by powerful machinery, could be swung back and forth over the disputed point. It was operated by armed men behind fortifications.

The Bruner concern was known as the Alaska-Pacific Transportation and Terminal Company, financed by Pittsburg capital, and proposed building a road to the coal regions, thence to the Copper River. They sought right of way by condemnation proceedings.

The town site of Katalla is owned by the Alaska Petroleum and Coal Company, which had deeded a right of way to the Guggenheims; also, a large tract of land for smelter purposes. At one point it was necessary for the latter to cross the right of way of the Bruner road.

The trouble began in May, when the Bruner workmen dynamited a pile-driver and trestle belonging to the

Guggenheims, who had then approached within one hundred feet of the Bruner right of way.

On July 3 a party of Guggenheim laborers, under the protection of a fire from detachments of armed men, succeeded in laying track over the disputed right of way.

Tony de Pascal daringly led the construction party and received the reward of a thousand dollars offered by the Guggenheims to the man who would successfully lead the attacking forces. Soon afterward, he was shot dead by one of his own men who mistook him for a member of the opposing force. Ten other men were seriously injured by bullets from the Bruner block-houses.

In the autumn of the same year a party of men surveying for the Reynolds Home Railway, from Valdez to the Yukon, met armed resistance in Keystone Canyon from a force of men holding right of way for the Guggenheims. A battle occurred in which one man was killed and three seriously wounded.

The wildest excitement prevailed in fiery Valdez, and probably only the proximity of a United States military post prevented the lynching of the men who did the killing.

Ever since the advent of the Russians, Copper River has been considered one of the bonanzas of Alaska. It was discovered in 1783 by Nagaief, a member of Potap Zaïkoff's party. He ascended it for a short distance and traded with the natives, who called the river Atnah. Rufus Serrebrennikof and his men attempted an exploration, but were killed. General Miles, under Abercrombie, attempted to ascend the river in 1884, with the intention of coming out by the Chilkaht country; but the expedition was a failure. In the following year Lieutenant H. T. Allen successfully ascended the river, crossed the divide to the Tanana, sailed down that stream to the Yukon, explored the Koyukuk, and then proceeded down the Yukon to St. Michael and returned to San Francisco by ocean.

His description of Miles Glacier was the first to be printed. This glacier fronts for a distance of six miles in splendid palisades on Copper River. This and Childs Glacier afford the chief obstacles to navigation on this river, and Mr. A. H. Brooks reports their rapid recession.

Lake Bennett in 1898

The river is regarded as exceedingly dangerous for steamers, but may, with caution, be navigated with small boats. Between the mouth of the Chitina and the head of the broad delta of the Copper River, is the only canyon. It is the famous Wood Canyon, several miles in length and in many places only forty yards wide, with the water roaring through perpendicular stone walls. The Tiekel, Tasnuna, and other streams tributary to this part of the Copper also flow through narrow valleys with precipitous slopes.

The Copper River has its source in the mountains east of its great plateau, whose eastern margin it traverses, and then, passing through the Chugach Mountains, debouches across a wide delta into the North Pacific Ocean between Katalla and Cordova. It rises close to Mount Wrangell, flows northward for forty miles, south and southwest for fifty more, when the Chitina joins it from the east and swells its flood for the remaining one hundred and fifty miles to the coast.

The Copper is a silt-laden, turbulent stream from its source to the sea. Its average fall is about twelve feet to the mile. From the Chitina to its mouth, it is steep-sided and rock-bound; for its entire length, it is weird and impressive.

By land, the distance from Katalla to Cordova is insignificant. It is a distance, however, that cannot as yet be traversed, on account of the delta and other impassable topographic features, which only a railroad can overcome. The distance by water is about one hundred and fifty miles.

In the entrance to Cordova Bay is Hawkins Island, and to the southwest of this island lies Hinchingbroke Island, whose southern extremity, at the entrance to Prince William Sound, was named Cape Hinchingbroke by Cook in 1778. At a point named Snug Corner Bay Cook keeled and mended his ships.

This peerless sound itself—brilliantly blue, greenly islanded, and set round with snow peaks and glaciers, including among the latter the most beautiful one of Alaska, if not the most beautiful of the world, the Columbia—was known as Chugach Gulf—a name to which I hope it may some day return,—until Cook renamed it.

A boat sent out by Cook was pursued by natives in canoes. They seemed afraid to approach the ship; but at a distance sang, stood up in the canoes, extending their arms and holding out white garments of peace. One man stood up, entirely nude, with his arms stretched out like a cross, motionless, for a quarter of an hour.

The following night a few natives came out in the skin-boats of the Eskimos. These boats are still used from this point westward and northward to Nome and up the Yukon as far as the Eskimos have settlements. They are of three kinds. One is a large, open, flat-bottomed boat. It is made of a wooden frame, covered with walrus skin or sealskin, held in place by thongs of the former. This is called an oomiak by the Innuits or Eskimos, and a bidarra by the Russians. It is used by women, or by large parties of men.

A boat for one man is made in the same fashion, but covered completely over, with the exception of one hole in which the occupant sits, and around which is an upright rim. When at sea he wears a walrus-gut coat, completely waterproof, which he ties around the outside of the rim. The coat is securely tied around the wrists, and the hood is drawn tightly around the face; so that no water can possibly enter the boat in the most severe storm. This boat is called a bidarka.

The third, called a kayak, differs from the bidarka only in being longer and having two or three holes.

The walrus-gut coats are called kamelinkas or kamelaykas. They may be purchased in curio stores, and at Seldovia and other places on Cook Inlet. They are now gayly decorated with bits of colored wool and range in price from ten to twenty dollars, according to the amount of work upon them.

There is a difference of opinion regarding the names of the boats. Dall claims that the one-holed boat was called a kayak by the natives, and by the Russians a bidarka; and that the others were simply known as two or three holed bidarkas. The other opinion, which I have given, is that of people living in the vicinity at present.

Each of the men who came out in the bidarkas to visit Cook had a stick about three feet long, the end of which was decorated with large tufts of feathers. Behring's men were received in precisely the same manner at the Shumagin Islands, far to westward, in 1741; their sticks, according to Müller, being decorated with hawks' wings.

These natives were found to be thievish and treacherous, attempting to capture a boat under the ship's very guns and in the face of a hundred men.

Cook then sailed southward and discovered the largest island in the sound, the Sukluk of the natives, which he named Montagu.

Nutchek, or Port Etches, as it was named by Portlock, is just inside the entrance to the sound on the western shore of the island that is now known as Hinchingbroke, but which was formerly called Nutchek.

Here Baranoff, several years later, built the ships that bore his first expedition to Sitka. The Russian trading post was called the Redoubt Constantine and Elena. It was a strong, stockaded fort with two bastions.

There is a salmon cannery at Nutchek, and the furs of the Copper River country were brought here for many years for barter.

Orca is situated about three miles north of Cordova, in Cordova Bay. There is a large salmon cannery at Orca; and the number of sea-birds to be seen in this small bay, filling the air in snowy clouds and covering the precipitous cliffs facing the wharf, is surpassed in only one place on the Alaskan coast—Karluk Bay.

For several years before the founding of Valdez, Orca was used as a port by the argonauts who crossed by way of Valdez Pass to the Copper River mining regions, and by way of the Tanana River to the Yukon.

Prince William Sound is one of the most nobly beautiful bodies of water in Alaska. Its wide blue water-sweeps, its many mountainous, wooded, and snow-peaked islands, the magnificent glaciers which palisade its ice-inlets, and the chain of lofty, snowy mountains that float mistily, like linked pearls, around it through the amethystine clouds, give it a poetic and austere beauty of its own. Every slow turn of the prow brings forth some new delight to the eye. Never does one beautiful snow-dome fade lingeringly from the horizon, ere another pushes into the exquisitely colored atmosphere, in a chaste beauty that fairly thrills the heart of the beholder.

The sound, or gulf, extends winding blue arms in every direction,—into the mainland and into the many islands. It covers an extent of more than twenty-five hundred square miles. The entrance is about fifty miles wide, but is sheltered by countless islands. The largest and richest are Montagu, Hinchingbroke, La Touche, Knight's, and Hawkins. There are many excellent harbors on the shores of the gulf and on the islands, and the Russians built several ships here. In Chalmers Bay Vancouver discovered a remarkable point, which bore stumps of trees cut with an axe, but far below low-water mark at the time of his discovery. He named it Sinking Point.

There is a portage from the head of the gulf to Cook Inlet, which, the earliest Russians learned, had long been used by the natives, who are of the Innuit, or Eskimo, tribe, similar to those of the Inlet, and are called Chugaches. The northern shore of Kenai and the western coast of the Inlet are occupied by Indians of the Athabascan stock.

White Horse, Yukon Territory

Cook found the natives of the gulf of medium size, with square chests and large heads. The complexion of the children and some of the younger women was white; many of the latter having agreeable features and pleasing appearance. They were vivacious, good-natured, and of engaging frankness.

These people, of all ages and both sexes, wore a close robe

reaching to the ankles—sometimes only to the knees—made of the skins of sea-otter, seal, gray fox, raccoon, and pine-marten. These garments were worn with the fur outside. Now and then one was seen made of the down of sea-birds, which had been glued to some other substance. The seams were ornamented with thongs, or tassels, of the same skins.

In rain they wore kamelinkas over the fur robes. Cook's description of a kamelinka as resembling a "gold-beater's leaf" is a very good one.

His understanding of the custom of wearing the labret, however, differs from that of other early navigators. The incision in the lip, he states, was made even in the children at the breast; while La Pérouse and others were of the impression that it was not made until a girl had arrived at a marriageable age.

It appears that the incision in time assumes the shape of real lips, through which the tongue may be thrust.

One of Cook's seamen, seeing for the first time a woman having the incision from which the labret had been removed, fell into a panic of horror and ran to his companions, crying that he "had seen a man with two mouths,"—evidently mistaking the woman for a man. Cook reported that both sexes wore the labret; but this was doubtless an error. When they are clad in the fur garments, which are called parkas, it is difficult to distinguish one sex from the other among the younger people.

I had a rather amusing experience myself at the small native settlement of Anvik on the Yukon. It was midnight, but broad daylight, as we were in the Arctic Circle. The natives were all clad in parkas. Two sitting side by side resembled each other closely. After buying some of their curios, I asked one, indicating the other, "Is she your sister?"

To my confusion, my question was received with a loud burst of laughter, in which a dozen natives, sitting around them, hoarsely and hilariously joined.

They poked the unfortunate object of my curiosity in the ribs, pointed at him derisively, and kept crying—"She! She!" until at last the poor young fellow, not more embarrassed than myself, sprang to his feet and ran away, with laughter and cries of "She! She!" following him.

I have frequently recalled the scene, and feared that the innocent dark-eyed and sweet-smiling youth may have retained the name which was so mirthfully bestowed upon him that summer night.

But since the mistake in sex may be so easily made, I am inclined to the belief that Cook and his men were misled in this particular.

A most remarkable difference of opinion existed between Cook and other early explorers as to the cleanliness of the natives. He found their method of eating decent and cleanly, their persons neat, without grease or dirt, and their wooden dishes in excellent order.

The white-headed eagle was found here, as well as the shag, the great kingfisher of brilliant coloring, the humming-bird, water-fowl, grouse, snipe, and plover. Many other species of water and land fowl have been added to these.

The flora of the islands is brilliant, varied, and luxuriant.

In 1786 John Meares—who is dear to my heart because of his confidence in Juan de Fuca—came to disaster in the Chugach Gulf. Overtaken by winter, he first tried the anchorage at Snug Corner Cove, in his ship, the *Nootka*, but later moved to a more sheltered nook closer to the mainland, in the vicinity of the present native village of Tatitlik.

The ill-provisioned vessel was covered for the winter; spruce beer was brewed, but the men preferred the liquors, which were freely served, and, fresh fish being scarce, scurvy became epidemic. The surgeon was the first to die; but he was followed by many others.

At first, graves were dug under the snow; but soon the survivors were too few and too exhausted for this last service to their mates. The dead were then dropped in fissures of the ice which surrounded their ship.

At last, when the lowest depth of despair had been reached, Captains Portlock and Dixon arrived and furnished relief and assistance.

In 1787-1788 the Chugach Gulf presented a strange appearance to the natives, not yet familiar with the presence of ships. Englishmen under different flags, Russians and Spaniards, were sailing to all parts of the gulf, taking possession in the names of different nations of all the harbors and islands.

In Voskressenski Harbor—now known as Resurrection Bay, where the new railroad town of Seward is situated—the first ship ever built in Alaska was launched by Baranoff, in 1794. It was christened the *Phœnix*, and was followed by many others.

Preparations for ship-building were begun in the winter of 1791. Suitable buildings, storehouses, and quarters for the men were erected. There were no large saws, and planks were hewn out of whole logs. The iron required was collected from wrecks in all parts of the colonies; steel for axes was procured in the same way. Having no tar, Baranoff used a mixture of spruce gum and oil.

Provisions were scarce, and no time was allowed for hunting or fishing. So severe were the hardships endured that no one but Baranoff could have kept up his courage and that of his suffering men, and cheered them on to final success.

The *Phœnix*—which was probably named for an English ship which had visited the Chugach Gulf in 1792—was built of spruce timber, and was seventy-three feet long. It was provided with two decks and three masts. The calking above the water-line was of moss. The sails were composed of fragments of canvas gathered from all parts of the colonies.

On her first voyage to Kadiak, the *Phœnix* encountered a storm which brought disaster to her frail rigging; and instead of sailing proudly into harbor, as Baranoff had hoped, she was ignominiously towed in.

But she was the first vessel built in the colonies to enter that harbor in any fashion, and the Russian joy was great. The event was celebrated by solemn Mass, followed by high eating and higher drinking.

The *Phœnix* was refitted and rerigged and sent out on her triumphal voyage to Okhotsk. There she arrived safely and proudly. She was received with volleys of artillery, the ringing of bells, the celebration of Mass, and great and joyous feasting.

A cabin and deck houses were added, the vessel was painted, and from that time until her loss in the Alaskan Gulf, the *Phœnix* regularly plied the waters of Behring Sea and the North Pacific Ocean between Okhotsk and the Russian colonies in America.

CHAPTER XXII

Ellamar is a small town on Virgin Bay, Prince William Sound, at the entrance to Puerto de Valdes, or Valdez Narrows. It is very prettily situated on a gently rising hill.

It has a population of five or six hundred, and is the home of the Ellamar Mining Company. Here are the headquarters of a group of copper properties known as the Gladdaugh mines.

One of the mines extends under the sea, whose waves wash the buildings. It has been a large and regular shipper for several years. In 1903 forty thousand tons of ore were shipped to the Tacoma smelter, and shipments have steadily increased with every year since.

The mine is practically a solid mass of iron and copper pyrites. It has a width of more than one hundred and twenty-five feet where exposed, and extends along the strike for a known distance of more than three hundred feet.

The vast quantities of gold found in Alaska have, up to the present time, kept the other rich mineral products of the country in the background. Copper is, at last, coming into her own. The year of 1907 brought forth tremendous developments in copper properties. The Guggenheim-Morgan-Rockefeller syndicate has kept experts in every known, or suspected, copper district of the North during the last two years. Cordova, the sea terminus of the new railroad, is in the very heart of one of the richest copper districts. The holdings of this syndicate are already immense and cover every district. The railroad will run to the Yukon, with branches extending into every rich region.

Other heavily financed companies are preparing to rival the Guggenheims, and individual miners will work their claims this year. Experts predict that within a decade Alaska will become one of the greatest copper-producing countries of the world. In the Copper River country alone, north of Valdez, there is more copper, according to expert reports, than Montana or Michigan ever has produced, or ever will produce.

The Ketchikan district is also remarkably rich. At Niblack Anchorage, on Prince of Wales Island, the ore carries five per cent of copper, and the mines are most favorably located on tidewater.

Native copper, associated with gold, has been found on Turnagain Arm, in the country tributary to the Alaska Central Railway.

A half interest in the Bonanza, a copper mine on the western side of La Touche Island, Prince William Sound, was sold last year for more than a million dollars. This mine is not fully developed, but is considered one of the best in Alaska. It has an elevation of two hundred feet. Several tunnels have been driven, and the ore taken out runs high in copper, gold, and silver. One shipment of one thousand two hundred and thirty-five pounds gave net returns of fifty dollars to the ton, after deducting freight to Tacoma, smelting, refining, and an allowance of ninety-five per cent for the silver valuation. A sample taken along one tunnel for sixty feet gave an assay of over nine per cent copper, with one and a quarter ounces of silver.

The Bonanza was purchased in 1900 by Messrs. Beatson and Robertson for seventy-two thousand dollars. There is a good wharf and a tramway line to the mine.

Adjoining the Bonanza on the north is a group of eleven claims owned by Messrs. Esterly, Meenach, and Keyes, which are in course of development. There are many other rich claims on this island, on Knight's, and on others in the sound. Timber is abundant, the water power is excellent, and ore is easily shipped.

There is an Indian village two or three miles from Ellamar. It is the village of Tatitlik, the only one now remaining on the sound, so rapidly are the natives vanishing under the evil influence of civilization. Ten years ago there were nine hundred natives in the various villages on the shores of the sound; while now there are not more than two hundred, at the most generous calculation.

White men prospecting and fishing in the vicinity of the village supply them with liquor. When a sufficient quantity can be purchased, the entire village, men and women, indulges in a prolonged and horrible debauch which frequently lasts for several weeks.

The death rate at Tatitlik is very heavy,—more than a hundred natives having died during 1907.

Passengers have time to visit this village while the steamer loads ore at Ellamar.

The loading of ore, by the way, is a new experience. A steamer on which I was travelling once landed at Ellamar during the night.

We were rudely awakened from our dreams by a sound which Lieutenant Whidbey would have called "most stupendously dreadful." We thought that the whole bottom of the ship must have been knocked off by striking a reef, and we reached the floor simultaneously.

I have no notion how my own eyes looked, but my friend's eyes were as large and expressive as bread-and-butter plates.

"We are going down!" she exclaimed, with tragic brevity.

At that instant the dreadful sound was repeated. We were convinced that the ship was being pounded to pieces under us upon rocks. Without speech we began dressing with that haste that makes fingers become thumbs.

But suddenly a tap came upon our door, and the watchman's voice spoke outside.

"Ladies, we are at Ellamar."

"At Ellamar!"

"Yes. You asked to be called if it wasn't midnight when we landed."

"But what is that *awful* noise, watchman?"

"Oh, we're loading ore," he answered cheerfully, and walked away.

All that night and part of the next day tons upon tons of ore thundered into the hold. We could not sleep, we could not talk; we could only think; and the things we thought shall never be told, nor shall wild horses drag them from us.

We dressed, in desperation, and went up to "the store"; sat upon high stools, ate stale peppermint candy, and listened to "Uncle Josh" telling his parrot story through the phonograph.

Somehow, between the ship and the store, we got ourselves through the night and the early morning hours. After breakfast we found the green and flowery slopes back of the town charming; and a walk of three miles along the shore to the Indian village made us forget the ore for a few hours. But to this day, when I read that an Alaskan ship has brought down hundreds of tons of ore to the Tacoma smelter, my heart goes out silently to the passengers who were on that ship when the ore was loaded.

Seattle Grand Canyon of the Yukon

CHAPTER XXIII

When seen under favorable conditions, the Columbia Glacier is the most beautiful thing in Alaska. I have visited it twice; once at sunset, and again on an all-day excursion from Valdez.

The point on the western side of the entrance to Puerto de Valdés, as it was named by Fidalgo, was named Point Fremantle by Vancouver. Just west of this point and three miles north of the Condé, or Glacier, Island is the nearly square bay upon which the glacier fronts.

Entering this bay from the Puerto de Valdés, one is instantly conscious of the presence of something wonderful and mysterious. Long before it can be seen, this presence is felt, like that of a living thing. Quick, vibrant, thrilling, and inexpressibly sweet, its breath sweeps out to salute the voyager and lure him on; and with every sense alert, he follows, but with no conception of what he is to behold.

One may have seen glaciers upon glaciers, yet not be prepared for the splendor and the magnificence of the one that palisades the northern end of this bay.

The Fremantle Glacier was first seen by Lieutenant Whidbey, to whose cold and unappreciative eyes so many of the most precious things of Alaska were first revealed. He simply described it as "a solid body of compact, elevated ice ... bounded at no great distance by a continuation of the high ridge of snowy mountains."

He heard "thunder-like" noises, and found that they had been produced by the breaking off and headlong plunging into the sea of great bodies of ice.

In such wise was one of the most marvellous things of the world first seen and described.

The glacier has a frontage of about four miles, and its glittering palisades tower upward to a height of from three to four hundred feet. There is a small island, named Heather, in the bay. Poor Whidbey felt the earth shake at a distance of three miles from the falling ice.

In ordinary light, the front of the glacier is beautifully blue. It is a blue that is never seen in anything save a glacier or a floating iceberg—a pale, pale blue that seems to flash out fire with every movement. At sunset, its beauty holds one spellbound. It sweeps down magnificently from the snow peaks which form its fit setting and pushes out into the sea in a solid wall of spired and pinnacled opal which, ever and anon breaking off, flings over it clouds of color which dazzle the eyes. At times there is a display of prismatic colors. Across the front grow, fade and grow again, the most beautiful rainbow shadings. They come and go swiftly and noiselessly, affecting one somewhat like Northern Lights—so still, so brilliant, so mysterious.

There was silence upon our ship as it throbbed in, slowly and cautiously, among the floating icebergs—some of which were of palest green, others of that pale blue I have mentioned, and still others of an enchanting rose color. Even the woman who had, during the whole voyage, taken the finest edge off our enjoyment of every mountain by drawling out, "Oh—how—pretty! George, will you just come here and look at this pretty mountain? It looks good enough to eat"—even this woman was speechless now, for which blessing we gave thanks to God, of which we were not even conscious at the time.

It was still fired as brilliantly upon our departure as upon our entrance into its presence. The June sunset in Alaska draws itself out to midnight; and ever since, I have been tormented with the longing to lie before that glacier one whole June night; to hear its falling columns thunder off the hours, and to watch the changing colors play upon its brilliant front.

Even in the middle of the day a peculiarly soft and rich rose color flashes from it and over it. One who has seen the first snow sifting upon a late rose of the garden may guess what a delicate, enchanting rose color it is.

There are many fine glaciers barricading the inlets and bays in this vicinity; in Port Nell Juan, Applegate Arm, Port Wells, Passage Canal—which leads to the portage to Cook Inlet—and Unakwik Bay; but they are scarcely to be mentioned in the same breath with the Fremantle. The latter has been known as the Columbia since the Harriman expedition in 1899. It has had no rival since the destruction of the Muir.

Either the disagreeable features of the Alaskan climate have been grossly exaggerated, or I have been exceedingly fortunate in the three voyages I have made along the coast to Unalaska, and down the Yukon to Nome. On one voyage I travelled continuously for a month by water, experiencing only three rainy days and three cloudy ones. All the other days were clear and golden, with a blue sky, a sparkling sea, and air that was sweet with sunshine, flowers, and snow. I have never been in Alaska in winter, but I have for three years carefully compared the weather reports of different sections of that country with those of other cold countries; and no intelligent, thoughtful person can do this without arriving at conclusions decidedly favorable to Alaska.

Were Alaska possessed of the same degree of civilization that is enjoyed by St. Petersburg, Chicago, St. Paul, Minneapolis, and New York, we would hear no more of the rigors of the Alaskan climate than we hear of those of the cities mentioned. It is more agreeable than the climate of Montana, Nebraska, or the Dakotas.

With large cities, rich and gay cities; prosperous inhabitants clad in costly furs; luxurious homes, well warmed and brilliantly lighted; railway trains, sleighs, and automobiles for transportation; splendid theatres, libraries, art galleries,—with these and the hundreds of advantages enjoyed by the people of other cold countries, Alaska's winters would hold no terrors.

It is the present loneliness of the winter that appalls. The awful spaces and silences; the limitless snow plains; the endless chains of snow mountains; the silent, frozen rivers; the ice-stayed cataracts; the bitter, moaning sea; the hastily built homes, lacking luxuries, sometimes even comforts; the poverty of congenial companionship; the dearth of intelligent amusements—these be the conditions that make all but the stoutest hearts pause.

But the stout heart, the heart that loves Alaska! Pity him not, though he spend all the winters of his life in its snow-bound fastnesses. *He* is not for pity. Joys are his of which those that pity him know not.

According to a report prepared by Lieutenant-Colonel Glassford, of the United States Signal Corps Service, on February 5, 1906, the temperature was twenty-six degrees above zero in Grand Junction, Colorado, and in Salchia, Alaska; twenty-two degrees in Flagstaff, Arizona, Memphis, Salt Lake, Spokane, and Summit, Alaska; fourteen degrees in Cairo, Illinois, Cincinnati, Little Rock, Pittsburgh, and Della, Alaska; twelve degrees in Santa Fé and in Fort Egbert and Eagle, on the Yukon; ten degrees in Helena, Buffalo, and Workman's, Alaska; zero in Denver, Dodge, Kansas, and Fairbanks and Chena, Alaska; five degrees below in Dubuque, Omaha, and Copper Centre and Matanuska, Alaska; ten degrees below in Huron, Michigan, and in Gokona, Alaska; fifteen degrees below in Bismarck, St. Paul, and in Tanana Crossing, Alaska; twenty degrees below in Fort Brady, Michigan, and in Ketchumstock, Alaska.

White Horse Rapids

Statistics giving the absolute mean minimum temperature in the capital cities of the United States prove that out of the forty-seven cities, thirty-one were as cold or colder than Sitka, and four were colder than Valdez.

On the southern coast of Alaska there are few points where zero is recorded, the average winter weather at Juneau, Sitka, Valdez, and Seward being milder than in Washington, D.C. In the interior, the weather is much colder, but it is the dry, light cold. At Fairbanks, it is true that the thermometer has registered sixty degrees below zero; but it has done the same in the Dakotas and other states, and is unusual. Severely cold weather occurs in Alaska as rarely as in other cold countries, and remains but a few days.

Alaska has unfortunately had the reputation of having an unendurable climate thrust upon her, first by such chill-blooded navigators as Whidbey and Vancouver; and later, by the gold seekers who rushed, frenziedly, into the unsettled wastes, with no preparation for the intense cold which at times prevails.

Almost every winter in Wyoming, Nebraska, Montana, and the Dakotas, children of the prairies and their teachers freeze to death going to or from school, and it is accepted as a matter of course. In Alaska, where hundreds of men traverse hundreds of miles by dog sleds and snow-shoes, with none of the comforts of more civilized countries and with road houses few and far, if two or three in a winter freeze to death, the tragedy is wired to all parts of the world as another mute testimony to the "tremendously horrible" climate of Alaska.

The intense heat, of which dozens of people perish every summer in New York and other eastern states is unknown in Alaska. Cyclones and cloud-bursts are unchronicled. Fatal epidemics of disease among white people have never yet occurred.

As for the summer climate of Alaska, both along the coast and in the interior, it is possessed of a charm and fascination which cannot be described in words.

"You can just *taste* the Alaska climate," said an old Klondiker, on a White Pass and Yukon train. We were standing between cars, clinging to the brakes—sooty-eyed, worn-out with joy as we neared White Horse, but standing and looking still, unwilling to lose one moment of that beautiful trip.

"It tastes different every hundred miles," he went on, with that beam in his eye which means love of Alaska in the heart. "You begun to taste it in Grenville Channel. It tasted different in Skagway, and there's a big change when you get to White Horse. I golly! at White Horse, you'll think you never tasted anything like it; but it don't hold a candle there to the way it tastes going down the Yukon. If you happen to get into the Ar'tic Circle, say, about two in the morning, you dress yourself and hike out on deck, an' I darn! you can taste more'n climate. You can taste the Ar'tic Circle itself! Say, can you guess what it tastes like?"

I could not guess what the Arctic Circle tasted like, and frankly confessed it.

"Well, say, weepin' Sinew! It tastes like icicles made out of them durn little blue flowers you call voylets. I picked some out from under the snow once, an' eat 'em. There was moisture froze all over 'em—so I know how they taste; and that's the way the Ar'tic Circle tastes, with—well, maybe a little *rum* mixed in, the way they fix things up at the Butler down in Seattle. I darn!... Just you remember, when you get to the Circle, an' say, straight goods, if Cyanide Bill ain't right."

"Talkin' about climate," he resumed, as the train hesitated in passing the Grand Canyon, "there's a well at White Horse that's got the climate of the hull Yukon country in it. It's about two blocks toward the rapids from White Pass Hotel. It stands on a vacant lot about fifty steps from the sidewalk, on your right hand goin' toward the Rapids. Well, I darn! I've traipsed over every country on this earth, an' I never tasted such water. Not anywheres! You see, it's dug right down into solid ice an' the sun just melts out a little water at a time, an' everything nice in Alaska tastes in that water—ice an' snow, an' flowers an' sun—"

"Do you write poetry?" I asked, smiling.

His face lightened.

"No; but say—there's a young fellow in White Horse that does. He's wrote a whole book of it. His name's Robert Service. Say, I'd shoot up anybody that said his poetry wasn't the real thing."

"I'm sure it is," said I, hastily.

"You bet it is. You can hear the Yukon roar, an' the ice break up an' go down the river, standin' up on end in chunks twenty feet high, an' carryin' everything with it; you can wade through miles an' miles of flowers an' gether your hands full of 'em an' think there's a woman somewhere waitin' for you to take 'em to her; you can tromp through tundra an' over rocks till your feet bleed; you can go blind lookin' for gold; you can get kissed by the prettiest girl in a Dawson dance hall, an' then get jilted for some younger fellow; you can hear glaciers grindin' up, an' avylanches tearin' down the mountains; you can starve to death an' freeze to death; you can strike a gold mine an' go home to your fambly a millionnaire an' have 'em like you again; you can drink champagne an' eat sour-dough; you can feel the heart break up inside of you—an' yes, I God! you can go down on your knees an' say your prayers again like your mother showed you how! You can do every one of them damn fool things when you're readin' that Service fellow's poetry. So that's why I'm ready to shoot up anybody that says, or intimates, that his poetry ain't the genuine article."

CHAPTER XXIV

Port Valdez—or the Puerto de Valdés, as it was named by Vancouver after Whidbey's exploration—is a fiord twelve miles long and of a beauty that is simply enchanting.

On a clear day it winds like a pale blue ribbon between colossal mountains of snow, with glaciers streaming down to the water at every turn. The peaks rise, one after another, sheer from the water, pearl-white from summit to base.

It has been my happiness and my good fortune always to sail this fiord on a clear day. The water has been as smooth as satin, with a faint silvery tinge, as of frost, shimmering over its blue.

At the end, Port Valdez widens into a bay, and upon the bay, in the shadow of her mountains, and shaded by her trees, is Valdez.

Valdez! The mere mention of the name is sufficient to send visions of loveliness glimmering through the memory. Through a soft blur of rose-lavender mist shine houses, glacier, log-cabins, and the tossing green of trees; the wild, white glacial torrents pouring down around the town; and the pearly peaks linked upon the sky.

Valdez was founded in 1898. During the early rush to the Klondike, one of the routes taken was directly over the glacier. In 1898 about three thousand people landed at the upper end of Port Valdez, followed the glacier, crossed over the summit of the Chugach Mountains, and thence down a fork of the Copper River. The route was dangerous, and attended by many hardships and real suffering.

At first hundreds of tents whitened the level plain at the foot of the glacier; then, one by one, cabins were built, stocks were brought in for trading purposes, saloons and dance halls sprang up in a night,—and Valdez was.

In this year Captain Abercrombie, of the United States Army, crossed the glacier with his entire party of men and horses and reached the Tanana. In the following year, surveys were made under his direction for a military wagon trail over the Chugach Mountains from Valdez to the Tanana, and during the following three years this trail was constructed.

It has proved to be of the greatest possible benefit, not only to the vast country tributary to Valdez, but to the various Yukon districts, and to Nome. After many experiments, it has been chosen by the government as the winter route for the distribution of mail to the interior of Alaska and to Nome. Steamers make connection with a regular line of stages and sleighs. There are frequent and comfortable road houses, and the danger of accident is not nearly so great as it is in travelling by railway in the eastern states.

The Valdez military trail follows Lowe River and Keystone Canyon. Through the canyon the trail is only wide enough for pack trains, and travel is by the frozen river.

The Signal Corps of the Army has constructed many hundreds of miles of telegraph lines since the beginning of the present decade. Nome, the Yukon, Tanana, and Copper River valleys are all connected with Valdez and with Dawson by telegraph. Nome has outside connection by wireless, and all the coast towns are in communication with Seattle by cable.

The climate of Valdez is delightful in summer. In winter it is ten degrees colder than at Sitka, with good sleighing. The annual precipitation is fifty per cent less than along the southeastern coast. Snow falls from November to April.

The long winter nights are not disagreeable. The moon and the stars are larger and more brilliant in Alaska than can be imagined by one who has not seen them, and, with the changeful colors of the Aurora playing upon the snow, turn the northern world into Fairyland.

Valdez has a population of about twenty-five hundred people. It is four hundred and fifty miles north of Sitka, and eighteen hundred miles from Seattle. It is said to be the most northern port in the world that is open to navigation the entire year.

There are two good piers to deep water, besides one at the new town site, an electric light plant and telephone system, two newspapers, a hospital, creditable churches of five or six denominations, a graded school, private club-rooms, a library, a brewery, several hotels and restaurants, public halls, a court-house, several merchandise stores carrying stocks of from fifty to one hundred thousand dollars, a tin and sheet metal factory, saw-mills,—and almost every business, industry, and profession is well represented. There are saloons without end, and dance halls; a saloon in Alaska that excludes women is not known, but good order prevails and disturbances are rare.

The homes are, for the most part, small,—building being excessively high,—but pretty, comfortable, and frequently artistic. There are flower-gardens everywhere. There is no log-cabin so humble that its bit of garden-spot is not a blaze of vivid color. Every window has its box of bloom. La France roses were in bloom in July in the garden of ex-Governor Leedy, of Kansas, whose home is now in Valdez.

The civilization of the town is of the highest. The whole world might go to Alaska and learn a lesson in genuine, simple, refined hospitality—for its key-note is kindness of heart.

The visitor soon learns that he must be chary of his admiration of one of the curios on his host's wall, lest he be begged to accept it.

The Tillicum Club is known in all parts of Alaska. It has a very comfortable club-house, where all visitors of note to the town are entertained. The club occasionally has what its own self calls a "dry night," when ladies are entertained with cards and music. (The adjective does not apply to the entertainment.)

The dogs of Valdez are interesting. They are large, and of every color known to dogdom, the malamutes predominating. They are all "heroes of the trail," and are respected and treated as "good fellows." They lie by twos and threes clear across the narrow board sidewalks; and unless one understands the language of the trail, it is easier to walk around them or to jump over them than it is to persuade them to move. A string of oaths, followed by "*Mush!*" all delivered like the crack of a whip, brings quick results. The dogs hasten to the pier, on a long, wolflike lope, when the whistle of a steamer is heard, and offer the hospitality of the town to the stranger, with waving tails and saluting tongues.

It is a heavy expense to feed these dogs in Alaska, yet few men are known to be so mean as to grudge this expense to dogs who have faithfully served them, frequently saving their lives, on the trail.

The situation of Valdez is absolutely unique. The dauntlessness of a city that would boldly found itself upon a glacier has proved too much for even the glacier, and it is rapidly withdrawing, as if to make room for its intrepid rival in interest. Yet it still is so close that, from the water, it appears as though one might reach out and touch it. The wide blue bay sparkles in front, and snow peaks surround it.

Beautiful, oh, most beautiful, are those peaks at dawn, at sunset, at midnight, at noon. The summer nights in Valdez are never dark; and I have often stood at midnight and watched the amethyst lights on the mountains darken to violet, purple, black,—while the peaks themselves stood white and still, softly outlined against the sky.

But in winter, when mountains, glacier, city, trees, lie white and sparkling beneath the large and brilliant stars, and the sea alone is dark—to stand then and see the great golden moon rising slowly, vibrating, pushing, oh, so silently, so beautifully, above the clear line of snow into the dark blue sky—that is worth ten years of living.

"Why do you not go out to 'the states,' as so many other ladies do in winter?" I asked a grave-eyed young wife on my first visit, not knowing that she belonged to the great Alaskan order of "Stout Hearts and Strong Hearts"—the only order in Alaska that is for women and men.

She looked at me and smiled. Her eyes went to the mountains, and they grew almost as wistful and sweet as the eyes of a young mother watching her sleeping child. Then they came back to me, grave and kind.

"Oh," said she, "how can I tell you why? You have never seen the moon come over those mountains in winter, nor the winter stars shining above the sea."

That was all. She could not put it into words more clearly than that; but he that runs may read.

The site of Valdez is as level as a parade ground to the bases of the near mountains, which rise in sheer, bold sweeps. A line of alders, willows, cotton woods, and balms follows the glacial stream that flows down to the sea on each side of the town.

The glacier behind the town—now called a "dead" glacier—once discharged bergs directly into the sea. The soil upon which the town is built is all glacial deposit. Flowers spring up and bloom in a day. Vegetables thrive and are crisp and delicious—particularly lettuce.

Society is gay in Valdez, as in most Alaskan towns. Fort Liscum is situated across the bay, so near that the distance between is travelled in fifteen minutes by launch. Dances, receptions, card-parties, and dinners, at Valdez and at the fort, occur several times each week, and the social line is drawn as rigidly here as in larger communities.

There is always a dance in Valdez on "steamer night." The officers and their wives come over from the fort; the officers of the ship are invited, as are any passengers who may bear letters of introduction or who may be introduced by the captain of the ship. A large and brightly lighted ballroom, beautiful women, handsomely and fashionably gowned, good music, and a genuine spirit of hospitality make these functions brilliant.

The women of Alaska dress more expensively than in "the states." Paris gowns, the most costly furs, and dazzling jewels are everywhere seen in the larger towns.

All travellers in Alaska unite in enthusiastic praise of its unique and generous hospitality. From the time of Baranoff's lavish, and frequently embarrassing, banquets to the refined entertainments of to-day, northern hospitality has been a proverb.

"Petnatchit copla" is still the open sesame.

CHAPTER XXV

The trip over "the trail" from Valdez to the Tanana country is one of the most fascinating in Alaska.

At seven o'clock of a July morning five horses stood at our hotel door. Two gentlemen of Valdez had volunteered to act as escort to the three ladies in our party for a trip over the trail.

I examined with suspicion the red-bay horse that had been assigned to me.

"Is he gentle?" I asked of one of the gentlemen.

"Oh, I don't know. You can't take any one's word about a horse in Alaska. They call regular buckers 'gentle' up here. The only way to find out is to try them."

This was encouraging.

"Do you mean to tell me," said one of the other ladies, "that you don't know whether these horses have ever been ridden by women?"

"No, I do not know."

She sat down on the steps.

"Then there's no trail for me. I don't know how to ride nor to manage a horse."

After many moments of persuasion, we got her upon a mild-eyed horse, saddled with a cross-saddle. The other lady and myself had chosen side-saddles, despite the assurance of almost every man in Valdez that we could not get over the trail sitting a horse sidewise, without accident.

"Your skirt'll catch in the brush and pull you off," said one, cheerfully.

"Your feet'll hit against the rocks in the canyon," said another.

"You can't balance as even on a horse's back, sideways, and if you don't balance even along the precipice in the canyon, your horse'll go over," said a third.

"Your horse is sure to roll over once or twice in the glacier streams, and you can save yourself if you're riding astride," said a fourth.

"You're certain to get into quicksand somewhere on the trip, and if all your weight is on one side of your horse, you'll pull him down and he'll fall on top of you," said a fifth.

In the face of all these cheerful horrors, our escort said:—

"Ride any way you please. If a woman can keep her head, she will pull through everything in Alaska. Besides, we are not going along for nothing!"

So we chose side-saddles, that having been our manner of riding since childhood.

We had waited three weeks for the glacial flood at the eastern side of the town to subside, and could wait no longer. It was roaring within ten steps of the back door of our hotel; and in two minutes after mounting, before our feet were fairly settled in the stirrups, we had ridden down the sloping bank into the boiling, white waters.

One of the gentlemen rode ahead as guide. I watched his big horse go down in the flood—down, down; the water rose to its knees, to its rider's feet, to *his* knees—

He turned his head and called cheerfully, "Come on!" and we went on—one at a time, as still as the dead, save for the splashing and snorting of our horses. I felt the water, icy cold, rising high, higher; it almost washed my foot from the red-slippered stirrup; then I felt it mounting higher, my skirts floated out on the flood, and then fell, limp, about me. My glance kept flying from my horse's head to our guide, and back again. He was tall, and his horse was tall.

"When it reaches *his* waist," was my agonized thought, "it will be over *my* head!"

The other gentleman rode to my side.

"Keep a firm hold of your bridle," said he, gravely, "and watch your horse. If he falls—"

"Falls! *In here!*"

"They do sometimes; one must be prepared. If he falls—of course you can swim?"

"I never swam a stroke in my life; I never even tried!"

"Is it possible?" said he, in astonishment. "Why, we would not have advised you to come at this time if we had known that. We took it for granted that you wouldn't think of going unless you could swim."

"Oh," said I, sarcastically, "do all the women in Valdez swim?"

"No," he answered, gravely, "but then, they don't go over the trail. Well, we can only hope that he will not fall. When he breaks into a swim—"

"*Swim!* Will he do that?"

"Oh, yes, he is liable to swim any minute now."

"What will I do then?" I asked, quite humbly; I could hear tears in my own voice. He must have heard them, too, his voice was so kind as he answered.

"Sit as quietly and as evenly as possible, and lean slightly forward in the saddle; then trust to heaven and give him his head."

"Does he give you any warning?"

"Not the faintest—ah-h!"

Well might he say "ah-h!" for my horse was swimming. Well might we all say "ah-h!" for one wild glance ahead revealed to my glimmering vision that all our horses were swimming.

I never knew before that horses swam so *low down* in the water. I wished when I could see nothing but my horse's ears that I had not been so stubborn about the saddle.

The water itself was different from any water I had ever seen. It did not flow like a river; it boiled, seethed, rushed, whirled; it pushed up into an angry bulk that came down over us like a deluge. I had let go of my reins and, leaning forward in the saddle, was clinging to my horse's mane. The rapidly flowing water gave me the impression that we were being swept down the stream.

The roaring grew louder in my ears; I was so dizzy that I could no longer distinguish any object; there was just a blur of brown and white water, rising, falling, about me; the sole thought that remained was that I was being swept out to sea with my struggling horse.

Suddenly there was a shock which, to my tortured nerves, seemed like a ship striking on a rock. It was some time before I realized that it had been caused by my horse striking bottom. He was walking—staggering, rather, and plunging; his whole neck appeared, then his shoulders; I released his mane mechanically, as I had acted in all things since mounting, and gathered up the reins.

"That was a nasty one, wasn't it?" said my escort, joining me. "I stayed behind to be of service if you required it. We're getting out now, but there are, at least, ten or fifteen as bad on the trail—if not worse."

As if anything *could* be worse!

I chanced to lift my eyes then, and I got a clear view of the ladies ahead of me. Their appearance was of such a nature that I at once looked myself over—and saw myself as others saw me! It was the first and only time that I have ever wished myself at home when I have been travelling in Alaska.

"Cheer up!" called our guide, over his broad shoulder. "The worst is yet to come."

He spoke more truthfully than even he knew. There was one stream after another—and each seemed really worse than the one that went before. From Valdez Glacier the ice, melted by the hot July sun, was pouring out in a dozen streams that spread over the immense flats between the town and the mouth of Lowe River. There were miles and miles of it. Scarcely would we struggle out of one place that had been washed out deep—and how deep, we never knew until we were into it—when we would be compelled to plunge into another.

At last, wet and chilled, after several narrow escapes from whirlpools and quicksand, we reached a level road leading through a cool wood for several miles. From this, of a sudden, we began to climb. So steep was the ascent and so narrow the path—no wider than the horse's feet—that my horse seemed to have a series of movable humps on him, like a camel; and riding sidewise, I could only lie forward and cling desperately to his mane, to avoid a shameful descent over his tail.

Actually, there were steps cut in the hard soil for the horses to climb upon! They pulled themselves up with powerful plunges. On both sides of this narrow path the grass or "feed," as it is called, grew so tall that we could not see one another's heads above it, as we rode; yet it had been growing only six weeks.

Mingling with young alders, fireweed, devil's-club and elderberry—the latter sprayed out in scarlet—it formed a network across our path, through which we could only force our way with closed eyes, blind as Love.

Bad as the ascent was, the sudden descent was worse. The horse's humps all turned the other way, and we turned with them. It was only by constant watchfulness that we kept ourselves from sliding over their heads.

After another ascent, we emerged into the open upon the brow of a cliff. Below us stretched the valley of the Lowe River. Thousands of feet below wound and looped the blue reaches of the river, set here and there with islands of glistening sand or rosy fireweed; while over all trailed the silver mists of morning. One elderberry island was so set with scarlet sprays of berries that from our height no foliage could be seen.

After this came a scented, primeval forest, through which we rode in silence. Its charm was too elusive for speech. Our horses' feet sank into the moss without sound. There was no underbrush; only dim aisles and arcades fashioned from the gray trunks of trees. The pale green foliage floating above us completely shut out the sun. Soft gray, mottled moss dripped from the limbs and branches of the spruce trees in delicate, lacy festoons.

Soon after emerging from this dreamlike wood we reached Camp Comfort, where we paused for lunch.

This is one of the most comfortable road houses in Alaska. It is situated in a low, green valley; the river winds in front, and snow mountains float around it. The air is very sweet.

It is only ten miles from Valdez; but those ten miles are equal to fifty in taxing the endurance.

We found an excellent vegetable garden at Camp Comfort. Pansies and other flowers were as large and fragrant as I have ever seen, the coloring of the pansies being unusually rich. They told us that only two other women had passed over the trail during the summer.

While our lunch was being prepared, we stood about the immense stove in the immense living room and tried to dry our clothing.

White Horse Rapids in Winter

This room was at least thirty feet square. It had a high ceiling and a rough board floor. In one corner was a piano, in another a phonograph. The ceiling was hung with all kinds of trail apparel used by men, including long boots and heavy stockings, guns and other weapons, and other articles that added a picturesque, and even startling, touch to the big room.

In one end was a bench, buckets of water, tin cups hanging on nails, washbowls, and a little wavy mirror swaying on the wall. The gentlemen of our party played the phonograph while we removed the dust and mud which we had gathered on our journey; afterward, *we* played the phonograph.

Then we all stood happily about the stove to "dry out," and listened to our host's stories of the miners who came out from the Tanana country, laden with gold. As many as seventy men, each bearing a fortune, have slept at Camp Comfort on a single night. We slept there ourselves, on our return journey, but our riches were in other things than gold, and there was no need to guard them. Any man or woman may go to Alaska and enrich himself or herself forever, as we did, if he or she have the desire. Not only is there no need to guard our riches, but, on the contrary, we are glad to give freely to whomsoever would have.

Each man, we were told, had his own way of caring for his gold. One leaned a gunnysack full of it outside the house, where it stood all night unguarded, supposed to be a sack of old clothing, from the carelessness with which it was left there. The owner slept calmly in the attic, surrounded by men whose gold made their hard pillows.

They told us, too, of the men who came back, dull-eyed and empty-handed, discouraged and footsore. They slept long and heavily; there was nothing for them to guard.

Every road house has its "talking-machine," with many of the most expensive records. No one can appreciate one of these machines until he goes to Alaska. Its influence is not to be estimated in those far, lonely places, where other music is not.

In a big store "to Westward" we witnessed a scene that would touch any heart. The room was filled with people. There were passengers and officers from the ship, miners, Russian half-breeds, and full-blooded Aleuts. After several records had filled the room with melody, Calvé, herself, sang "The Old Folks At Home." As that voice of golden velvet rose and fell, the unconscious workings of the faces about me spelled out their life tragedies. At last, one big fellow in a blue flannel shirt started for the door. As he reached it, another man caught his sleeve and whispered huskily:—

"Where you goin', Bill?"

"Oh, anywheres," he made answer, roughly, to cover his emotion; "anywheres, so's I can't hear that damn piece,"—and it was not one of the least of Calvé's compliments.

Music in Alaska brings the thought of home; and it is the thought of home that plays upon the heartstrings of the North. The hunger is always there,—hidden, repressed, but waiting,—and at the first touch of music it leaps forth and casts its shadow upon the face. Who knows but that it is this very heart-hunger that puts the universal human look into Alaskan eyes?

After a good lunch at Camp Comfort, we resumed our journey. There was another bit of enchanting forest; then, of a sudden, we were in the famed Keystone Canyon.

Here, the scenery is enthralling. Solid walls of shaded gray stone rise straight from the river to a height of from twelve to fifteen hundred feet. Along one cliff winds the trail, in many places no wider than the horses' feet. One feels that he must only breathe with the land side of him, lest the mere weight of his breath on the other side should topple him over the sheer, dizzy precipice.

It was amusing to see every woman lean toward the rock cliff. Not for all the gold of the Klondike would I have willingly given one look down into the gulf, sinking away, almost under my horse's feet. Somewhere in those purple depths I knew that the river was roaring, white and swollen, between its narrow stone walls.

Now and then, as we turned a sharp, narrow corner, I could not help catching a glimpse of it; for a moment, horse and rider, as we turned, would seem to hang suspended above it with no strip of earth between. There were times, when we were approaching a curve, that there seemed to be nothing ahead of us but a chasm that went sinking dizzily away; no solid place whereon the horse might set his feet. It was like a nightmare in which one hangs half over a precipice, struggling so hard to recover himself that his heart almost bursts with the effort.

Then, while I held my breath and blindly trusted to heaven, the curve would be turned and the path would glimmer once more before my eyes.

But one false step of the horse, one tiniest rock-slide striking his feet, one unexpected sound to startle him—the mere thought of these possibilities made my heart stop beating.

We finally reached a place where the descent was almost perpendicular and the trail painfully narrow. The horses sank to their haunches and slid down, taking gravel and stones down with them. I had been imploring to be permitted to walk; but now, being far in advance of all but one, I did not ask permission. I simply slipped off my horse and left him for the others to bring with them. The gentleman with me was forced to do the same.

We paused for a time to rest and to enjoy the most beautiful waterfall I saw in Alaska—Bridal Veil. It is on the opposite side of the canyon, and has a slow, musical fall of six hundred feet.

When we went on, the other members of our party had not yet come up with us, nor had our horses appeared. In the narrowest of all narrow places I was walking ahead, when, turning a sharp corner, we met a government pack train, face to face.

The bell-horse stood still and looked at me with big eyes, evidently as scared at the sight of a woman as an old prospector who has not seen one for years.

I looked at him with eyes as big as his own. There was only one thing to do. Behind us was a narrow, V-shaped cave in the stone wall, not more than four feet high and three deep. Into this we backed, Grecian-bend wise, and waited.

We waited a very long time. The horse stood still, blowing his breath loudly from steaming nostrils, and contemplated us. I never knew before that a horse could express his opinion of a person so plainly. Around the curve we could hear whips cracking and men swearing; but the horse stood there and kept his suspicious eyes on me.

"I'll stay here till dark," his eyes said, "but you don't get me past a thing like *that*!"

I didn't mind his looking, but his snorting seemed like an insult.

At last a man pushed past the horse. When he saw us backed gracefully up into the Y-shaped cave, he stood as still as the horse. Finding that neither he nor my escort could think of anything to say to relieve the mental and physical strain, I called out graciously:—

"How do you do, sir? Would you like to get by?"

"I'd like it damn well, lady," he replied, with what I felt to be his very politest manner.

"Perhaps," I suggested sweetly, "if I came out and let the horse get a good look at me—"

"Don't you do it, lady. That 'u'd scare him plumb to death!"

I have always been convinced that he did not mean it exactly as it sounded, but I caught the flicker of a smile on my escort's face. It was gone in an instant.

Suddenly the other horses came crowding upon the bell-horse. There was nothing for him to do but to go past me or to go over the precipice. He chose me as the least of the two evils.

"Nice pony, nice boy," I wheedled, as he went sliding and snorting past.

Then we waited for the next horse to come by; but he did not come. Turning my head, I found him fixed in the same place and the same attitude as the first had been; his eyes were as big and they were set as steadily on me.

Well—there were fifty horses in that government pack train.

Every one of the fifty balked at sight of a woman. There were horses of every color—gray, white, black, bay, chestnut, sorrel, and pinto. The sorrel were the stubbornest of all. To this day, I detest the sight of a sorrel horse.

We stood there in that position for a time that seemed like hours; we coaxed each horse as he balked; and at the last were reduced to such misery that we gave thanks to God that there were only fifty of them and that they couldn't kick sidewise as they passed.

I forgot about the men. There were seven men; and as each man turned the bend in the trail, he stood as still as the stillest horse, and for quite as long a time; and naturally I hesitated to say, "Nice boy, nice fellow," to help him by.

There were more glacier streams to cross. These were floored with huge boulders instead of sand and quicksand. The horses stumbled and plunged powerfully. One misstep here would have meant death; the rapids immediately below the crossing would have beaten us to pieces upon the rocks.

Then came more perpendicular climbing; but at last, at five o'clock, with our bodies aching with fatigue, and our senses finally dulled, through sheer surfeit, to the beauty of the journey, we reached "Wortman's" road house.

This is twenty miles from Valdez; and when we were lifted from our horses we could not stand alone, to say nothing of attempting to walk.

But "Wortman's" is the paradise of road houses. In it, and floating over it, is an atmosphere of warmth, comfort and good cheer that is a rest for body and heart. The beds are comfortable and the meals excellent.

But it was the welcome that cheered, the spirit of genuine kind-heartedness.

The road house stands in a large clearing, with barns and other buildings surrounding it. I never saw so many dogs as greeted us, except in Valdez or on the Yukon. They crowded about us, barking and shrieking a welcome. They were all big malamutes.

After a good dinner we went to bed at eight o'clock. The sun was shining brightly, but we darkened our rooms as much as possible, and instantly fell into the sleep of utter exhaustion.

At one o'clock in the morning we were eating breakfast, and half an hour later we were in our saddles and off for the summit of Thompson Pass to see the sun rise. This brought out the humps in the horses' backs again. We went up into the air almost as straight as a telegraph pole. Over heather, ice, flowers, and snow our horses plunged, unspurred.

It was seven miles to the summit. There were no trees nor shrubs,—only grass and moss that gave a velvety look to peaks and slopes that seemed to be floating around us through the silvery mists that were wound over them like turbans. Here and there a hollow was banked with frozen snow.

When we dismounted on the very summit we could hardly step without crushing bluebells and geraniums.

We set the flag of our country on the highest point beside the trail, that every loyal-hearted traveller might salute it and take hope again, if he chanced to be discouraged. Then we sat under its folds and watched the mists change from silver to pearl-gray; from pearl-gray to pink, amethyst, violet, purple,—and back to rose, gold, and flame color.

One peak after another shone out for a moment, only to withdraw. Suddenly, as if with one leap, the sun came over the mountain line; vibrated brilliantly, dazzlingly, flashing long rays like signals to every quickened peak. Then, while we gazed, entranced, other peaks whose presence we had not suspected were brought to life by those searching rays; valleys appeared, filled with purple, brooding shadows; whole slopes blue with bluebells; and, white and hard, the narrow trail that led on to the pitiless land of gold.

We were above the mountain peaks, above the clouds, level with the sun.

Absolute stillness was about us; there was not one faintest sound of nature; no plash of water, nor sough of wind, nor call of a bird. It was so still that it seemed like the beginning of a new world, with the birth of mountains taking place before our reverent eyes, as one after another dawned suddenly and goldenly upon our vision.

Every time we had stopped on the trail we had heard harrowing stories of saddle-horses or pack-horses having missed their footing and gone over the precipice. The horses are so carefully packed, and the packs so securely fastened on—the last cinch being thrown into the "diamond hitch"—that the poor beasts can roll over and over to the bottom of a canyon without disarranging a pack weighing two hundred pounds—a feat which they very frequently perform.

The military trail is, of necessity, poor enough; but it is infinitely superior to all other trails in Alaska, and is a boon to the prospector. It is a well-defined and well-travelled highway. The trees and bushes are cut in places for a width of thirty feet, original bridges span the creeks when it is possible to bridge them at all, and some corduroy has been laid; but in many places the trail is a mere path, not more than two feet wide, shovelled or blasted from the hillside.

In Alaska there were practically no roads at all until the appointment in 1905 of a road commission consisting of Major W. P. Richardson, Captain G. B. Pillsbury, and Lieutenant L. C. Orchard. Since that year eight hundred miles of trails, wagon and sled roads, numerous ferries, and hundreds of bridges have been constructed. The wagon road-beds are all sixteen feet wide, with free side strips of a hundred feet; the sled roads are twelve feet wide; the trails, eight; and the bridges, fourteen. In the interior, laborers on the roads are paid five dollars a day, with board and lodging; they are given better food than any laborers in Alaska, with the possible exception of those employed at the Treadwell mines and on the Cordova Railroad. The average cost of road work in Alaska is about two thousand dollars a mile; two hundred and fifty for sled road, and one hundred for trails. These roads have reduced freight rates one-half and have helped to develop rich regions that had been inaccessible. Their importance in the development of the country is second to that of railroads only.

The scenery from Ptarmigan Drop down the Tsina River to Beaver Dam is magnificent. Huge mountains, saw-toothed and covered with snow, jut diagonally out across the valley, one after another; streams fall, riffling, down the sides of the mountains; and the cloud-effects are especially beautiful.

Tsina River is a narrow, foaming torrent, confined, for the most part, between sheer hills,—although, in places, it spreads out over low, gravelly flats. Beaver Dam huddles into a gloomy gulch at the foot of a vast, overhanging mountain. Its situation is what Whidbey would have called "gloomily magnificent." In 1905 Beaver Dam was a road house which many chose to avoid, if possible.

The Tiekel road house on the Kanata River is pleasantly situated, and is a comfortable place at which to eat and rest.

For its entire length, the military trail climbs and falls and winds through scenery of inspiring beauty. The trail leading off to the east at Tonsina, through the Copper River, Nizina, and Chitina valleys, is even more beautiful.

Vast plains and hillsides of bloom are passed. Some mountainsides are blue with lupine, others rosy with fireweed; acres upon acres are covered with violets, bluebells, wild geranium, anemones, spotted moccasin and other orchids, buttercups, and dozens of others—all large and vivid of color. It has often been said that the flowers of Alaska are not fragrant, but this is not true.

The mountains of the vicinity are glorious. Mount Drum is twelve thousand feet high. Sweeping up splendidly from a level plain, it is more imposing than Mount Wrangell, which is fourteen thousand feet high, and Mount Blackburn, which is sixteen thousand feet.

The view from the summit of Sour-Dough Hill is unsurpassed in the interior of Alaska. Glacial creeks and roaring rivers; wild and fantastic canyons; moving glaciers; gorges of royal purple gloom; green valleys and flowery slopes; the domed and towered Castle Mountains; the lone and majestic peaks pushing up above all others, above the clouds, cascades spraying down sheer precipices; and far to the south the linked peaks of the Coast Range piled magnificently upon the sky, dim and faintly blue in the great distance,—all blend into one grand panorama of unrivalled inland grandeur.

Crossing the Copper River, when it is high and swift, is dangerous,—especially for a "chechaco" of either sex. (A chechaco is one who has not been in Alaska a year.) Packers are often compelled to unpack their horses, putting all their effects into large whipsawed boats. The halters are taken off the horses and the latter are driven into the roaring torrent, followed by the packers in the boats.

The horses apparently make no effort to reach the opposite shore, but use their strength desperately to hold their own in the swift current, fighting against it, with their heads turned pitifully up-stream. Their bodies being turned at a slight angle, the current, pushing violently against them, forces them slowly, but surely, from sand bar to sand bar, and, finally, to the shore.

It frequently requires two hours to get men, horses, and outfit from shore to shore, where they usually arrive dripping wet. Women who make this trip, it is needless to say, suffer still more from the hardship of the crossing than do men.

In riding horses across such streams, they should be started diagonally up-stream toward the first sand bar above. They lean far forward, bracing themselves at every step against the current and choosing their footing carefully. The horses of the trail know all the dangers, and scent them afar—holes, boulders, irresistible currents, and quicksand; they detect them before the most experienced "trailer" even suspects them.

I will not venture even to guess what the other two women in my party did when they crossed dangerous streams; but for myself, I wasted no strength in trying to turn my horse's head up-stream, or down-stream, or in any other direction. When we went down into the foaming water, I gave him his head, clung to his mane, leaned forward in the saddle,—and prayed like anything. I do not believe in childishly asking the Lord to help one so long as one can help one's self; but when one is on the back of a half-swimming, half-floundering horse in the middle of a swollen, treacherous flood, with holes and quicksand on all sides, one is as helpless as he was the day he was born; and it is a good time to pray.

According to the report of Major Abercrombie, who probably knows this part of Alaska more thoroughly than any one else, there are hundreds of thousands of acres in the Copper River Valley alone where almost all kinds of vegetables, as well as barley and rye, will grow in abundance and mature. Considering the travel to the many and fabulously rich mines already discovered in this valley and adjacent ones, and the cost of bringing in grain and supplies, it may be easily seen what splendid opportunities await the small farmer who will select his homestead judiciously, with a view to the accommodation of man and beast, and the cultivation of food for both. The opportunities awaiting such a man are so much more enticing than the inducements of the bleak Dakota prairies or the wind-swept valleys of the Yellowstone as to be beyond comparison.

Major Abercrombie believes that the valleys of the sub-drainage of the Copper River Valley will in future years supply the demands for cereals and vegetables, if not for meats, of the thousands of miners that will be required to extract the vast deposits of metals from the Tonsina, Chitina, Kotsina, Nizina, Chesna, Tanana, and other famous districts.

The vast importance to the whole territory of Alaska, and to the United States, as well, of the building of the Guggenheim railroad from Cordova into this splendid inland empire may be realized after reading Major Abercrombie's report.

We have been accustomed to mineralized zones of from ten to twelve miles in length; in the Wrangell group alone we have a circle eighty miles in diameter, the mineralization of which is simply marvellous; yet, valuable though these concentrates are, they are as valueless commercially as so much sandstone, without the aid of a railroad and reduction works.

If the group of mines at Butte could deflect a great transcontinental trunk-line like the Great Northern, what will this mighty zone, which contains a dozen properties already discovered,—to say nothing of the unfound, undreamed-of ones,—of far greater value as copper propositions than the richest of Montana, do to advance the commercial interests of the Pacific Coast?

The first discovery of gold in the Nizina district was made by Daniel Kain and Clarence Warner. These two prospectors were urged by a crippled Indian to accompany him to inspect a vein of copper on the head waters of a creek that is now known as Dan Creek.

Not being impressed by the copper outlook, the two prospectors returned. They noticed, however, that the gravel of Dan Creek had a look of placer gold.

They were out of provisions, and were in haste to reach their supplies, fifty miles away; but Kain was reluctant to leave the creek unexamined. He went to a small lake and caught sufficient fish for a few days' subsistence; then, with a shovel for his only tool, he took out five ounces of coarse gold in two days.

In this wise was the rich Nizina district discovered. The Nizina River is only one hundred and sixty miles from Valdez. In Rex Gulch as much as eight ounces of gold have been taken out by one man in a single day. The gold is of the finest quality, assaying over eighteen dollars an ounce.

There is an abundance of timber suitable for building houses and for firewood on all the creeks. There is water at all seasons for sluicing, and, if desired, for hydraulic work.

CHAPTER XXVI

The famous Bonanza Copper Mine is on the mountainside high above the Kennicott Valley, and near the Kennicott Glacier—the largest glacier of the Alaskan interior. This glacier does not entirely fill the valley, and one travels close to its precipitous wall of ice, which dwindles from a height of one hundred feet to a low, gravel-darkened moraine. From the summit of Sour-Dough Hill it may be seen for its whole forty-mile length sweeping down from Mounts Wrangell and Regal.

The Bonanza Mine has an elevation of six thousand feet, and was discovered by the merest chance.

The history of this mine from the day of its discovery is one of the most fascinating of Alaska. In the autumn of 1899 a prospecting party was formed at Valdez, known as the "McClellan" party. The ten individuals composing the party were experienced miners and they contributed money, horses, and "caches," as well as experience. The principal cache was known as the "McCarthy Cabin" cache, and was about fifteen miles east of Copper River on the trail to the Nicolai Mine.

The Nicolai had been discovered early in the summer by R. F. McClellan, who was one of the men composing the "McClellan" party, and others. Another important cache of three thousand pounds of provisions was the "Amy" cache, thirty-five miles from Valdez, just over the summit of Thompson Pass.

The agreement was that the McClellan party was to prospect in the interior in 1900 and 1901, all property located to be for their joint benefit.

The members of the party scattered soon after the organization was completed. Clarence Warner, John Sweeney, and Jack Smith remained in Valdez for the winter, all the others going "out to the states."

In March of 1900 Warner and Smith set out for the interior over the snow. There was no government trail then, and the hardships to be endured were as terrific as were those of the old Chilkoot Pass, on the way to the Klondike. The snow was from six to ten feet deep, and their progress was slow and painful. One went ahead on snow-shoes, the other following; when the trail thus made was sufficiently hard, the hand sleds, loaded with provisions and bedding, were drawn over it by ropes around the men's shoulders. From two to three hundred pounds was a heavy burden for each man to drag through the soft snow.

Climbing the summit, and at other steep places, they were compelled to "relay," by leaving the greater portion of their load beside the trail, pulling only a few pounds for a short distance and returning for more. By the most constant and exhaustive labor they were able to make only five or six miles a day.

They replenished their stores at the "Amy" cache, near the summit, and in May reached the "McCarthy Cabin" cache. Here they found that the Indians had broken in and stolen nearly all the supplies.

When they left Valdez, it was with the expectation that McClellan, or some other member of the party, would bring in their horses to the McCarthy cabin, that their supplies might be packed from that point on horseback,—the snow melting in May making it impossible to use sleds, and no man being able to carry more than a few pounds on his back for so long a journey as they expected to make.

However, McClellan had, during the winter, entered into a contract with the Chitina Exploration Company at San Francisco to do a large amount of development work on the Nicolai Mine during the summer of 1900. He returned to Valdez after Warner and Smith had left, bringing twenty horses, a large outfit of tools and supplies, and fifteen men—among them some of the McClellan prospecting party, who had agreed to work for the season for the Chitina Company.

When this party reached the McCarthy cabin, they found

Warner and Smith there. An endless dispute thereupon began as to the amount of provisions the two men had when the Chitina party arrived,—Warner and Smith claiming that they had five hundred pounds, and the Chitina Company claiming that they were entirely "out of grub," to use miner's language.

Warner and Smith demanded that McClellan should give them two horses belonging to the McClellan prospecting party, which he had brought. This matter was finally settled by McClellan's packing in what remained of Smith and Warner's provisions to the Nicolai Mine, a distance of nearly a hundred miles.

McClellan, as superintendent of the Chitina Company, used, with that company's horses, four of the McClellan party's horses during the entire season, sending them to and from Valdez, packing supplies.

In the meantime, upon reaching the Nicolai Mine, on the 1st of July, Warner and Smith, packing supplies on their backs, set out to prospect. The Chitina Company, in the famous and bitterly contested lawsuit which followed, claimed that they were supplied with the Chitina Company's "grub"; while Smith and Warner claimed that their provisions belonged to the McClellan party.

Steamer "White Horse" in Five-Finger Rapids

After a few days' aimless wandering, they reached a point on the east side of Kennicott Glacier, about twenty miles west of the Nicolai Mine. Here they camped at noon, near a small stream that came running down from a great height.

Their camp was about halfway up a mountain which was six thousand feet high. After a miner's lunch of bacon and beans, they were packing up to resume their wanderings, when Warner, chancing to glance upward, discovered a green streak near the top of the mountain. It looked like grass, and at first he gave it no thought; but presently it occurred to him that, as they were camped above timber-line, grass would not be growing at such a height.

They at once decided to investigate the peculiar and mysterious coloring. The mountain was steep, and it was after a slow and painful climb that they reached the top. Jack Smith stooped and picked up a piece of shining metal.

"My God, Clarence," he said fervently, "it's copper."

It was copper; the richest copper, in the greatest quantities, ever found upon the earth. There were hundreds of thousands of tons of it. There was a whole mountain of it. It was so bright and shining that they, at first, thought it was Galena ore; but they soon discovered that it was copper glance,—a copper ore bearing about seventy-five per cent of pure copper.

The Havemeyers, Guggenheims, and other eastern capitalists became interested. Then, when the marvellous richness of the discovery of Jack Smith and Clarence Warner became known, a lawsuit was begun—hinging upon the grub-stake—which was so full of dramatic incidents, attempted bribery, charges of corruption reaching to the United States Senate and the President himself, that the facts would make a long story, vivid with life, action, and fantastic setting—the scene reaching from Alaska to New York, and from New York to Manila.

The lawsuit was at last settled in favor of the discoverers.

On January 14, 1908, Mr. Smith disposed of his interest in a mine which he had located across McCarthy Creek from the Bonanza, for a hundred and fifty thousand dollars. It will be "stocked" and named "The Bonanza Mine Extension." It is said to be as rich as the great Bonanza itself.

CHAPTER XXVII

In the district which comprises the entire coast from the southern boundary of Oregon to the northernmost point of Alaska there are but forty-five lighthouses. Included in this district are the Strait of Juan de Fuca, Washington Sound, the Gulf of Georgia, and all the tidal waters tributary to the sea straits and sounds of this coast. There are also twenty-eight fog signals, operated by steam, hot air, or oil engines; six fog signals operated by clockwork; two gas-lighted buoys in position; nine whistling-buoys and five bell-buoys in position; three hundred and twenty-two other buoys in position; and four tenders, to visit lighthouses and care for buoys.

The above list does not include post lights, the Umatilla Reef Light vessel, and unlighted day beacons.

It is the far, lonely Alaskan coast that is neglected. The wild, stormy, and immense stretch of coast reaching from Chichagoff Island to Point Barrow in the Arctic Ocean has two light and fog signal stations on Unimak Island and two fixed lights on Cape Stephens. A light and fog signal station is to be built at Cape Hinchingbroke, and a light is to be established at Point Romanoff.

No navigator should be censured for disaster on this dark and dangerous coast. The little *Dora*, running regularly from Seward and Valdez to Unalaska, does not pass a light. Her way is wild and stormy in winter, and the coasts she passes are largely uninhabited; yet there is not a flash of light, unless it be from some volcano, to guide her into difficult ports and around the perilous reefs with which the coast abounds.

A prayer for a lighthouse at the entrance to Resurrection Bay was refused by the department, with the advice that the needs of commerce do not require a light at this point, particularly as there are several other points more in need of such aid. The department further advised that it would require a hundred thousand dollars to establish a light and fog signal station at the place designated, instead of the twenty-five thousand dollars asked.

Meanwhile, ships are wrecked and lives and valuable cargoes are lost,—and will be while the Alaskan coast remains unlighted.

Along the intricate, winding, and exceedingly dangerous channels, straits, and narrows of the "inside passage" of southeastern Alaska, there are only seven light and fog signals, and ten lights; but where the sea-coast belongs to Canada there is sufficient light and ample buoyage protection, as all mariners admit.

Is our government's rigid, and in some instances stubborn, economy in this matter a wise one? Is it a humane one? The nervous strain of this voyage on a conscientious and sensitive master of a ship heavily laden with human beings is tremendous. The anxious faces and unrelaxing vigilance of the officers on the bridge when a ship is passing through Taku Open, Wrangell Narrows, or Peril Straits speak plainly and unmistakably of the ceaseless burden of responsibility and anxiety which they bear. The charting of these waters is incomplete as yet, notwithstanding the faithful service which the Geodetic Survey has performed for many years. Many a rock has never been discovered until a ship went down upon it.

Political influence has been known to establish lights, at immense cost, at points where they are practically luxuries, rather than needs; therefore the government should not be censured for cautiousness in this matter.

But it should be, and it is, censured for not investigating carefully the needs of the Alaskan Coast—the "Great Unlighted Way."

Seward is situated almost as beautifully as Valdez. It is only five years old. It is the sea terminal of the Alaska Central Railway, which is building to the Tanana, through a rich country that is now almost unknown. It will pass within ten miles of Mount McKinley, which rises from a level plain to an altitude of nearly twenty-one thousand feet.

This mountain has been known to white men for nearly a century; yet until very recently it did not appear upon any map, and had no official name. More than fifty years ago the Russian fur traders knew it and called it "Bulshaia,"—signifying "high mountain" or "great mountain." The natives called it "Trolika," a name having the same meaning.

Explorers, traders, and prospectors have seen it and commented upon its magnificent height, yet without realizing its importance, until Mr. W. A. Dickey saw it in 1896 and proposed for it the name of McKinley. In 1902 Mr. Alfred Hulse Brooks, of the United States Geological Survey, with two associates and four camp men, made an expedition to the mountain. Mr. Brooks' report of this expedition is exceedingly interesting. He spent the summer of 1906, also, upon the mountain.

The town site of Seward was purchased from the Lowells, a pioneer family, by Major J. E. Ballaine, for four thousand dollars. It has grown very rapidly. Stumps still stand upon the business streets, and silver-barked log-cabins nestle modestly and picturesquely beside imposing buildings. The bank and the railway company have erected handsome homes. Every business and profession is represented. There are good schools and churches, an electric-light plant, two newspapers, a library and hospital, progressive clubs, and all the modern luxuries of western towns.

When Mr. Seward was asked what he considered the most important measure of his political career, he replied, "The purchase of Alaska; but it will take the people a generation to find it out."

Since the loftiest and noblest peak of North America was doomed to be named for a man, it should have borne the name of this dauntless, loyal, and far-seeing friend of Alaska and of all America. Since this was not to be, it was very fitting that a young and ambitious town on the historic Voskressenski Harbor should bear this honored and forever-to-be-remembered name. If Seward and Valdez would but work together, the region extending from Prince William Sound to Cook Inlet would soon become the best known and the most influential of Alaska, as it is, with the addition of the St. Elias Alps, the most sublimely and entrancingly beautiful.

Voskressenski Harbor, or Resurrection Bay, pushes out in purple waves in front of Seward, and snow peaks circle around it, the lower hills being heavily wooded. There is a good wharf and a safe harbor; the bay extends inland eighteen miles, is completely land-locked, and is kept free of ice the entire year, as is the Bay of Valdez and Cook Inlet, by the Japan current.

It is estimated that the Alaska Central Railway will cost, when completed to Fairbanks, at least twenty-five millions of dollars. Several branches will be extended into different and important mining regions.

The road has a general maximum grade of one per cent. The Coast Range is crossed ten miles from Seward, at an elevation of only seven hundred feet. The road follows the shore of Lake Kenai, Turnagain Arm, and Knik Arm on Cook Inlet; then, reaching the Sushitna River, it follows the sloping plains of that valley for a hundred miles, when, crossing the Alaskan Range, it descends into the vast valley at the head of navigation on the Tanana River, in the vicinity of Chena and Fairbanks.

All of the country which this road is expected to traverse when completed is rich in coal, copper, and quartz and placer gold.

There is a large amount of timber suitable for domestic use throughout this part of the country, spruce trees of three and four feet in diameter being common near the coast; inland, the timber is smaller, but of fair quality.

There is much good agricultural land along the line of the road; the soil is rich and the climatic conditions quite as favorable as those of many producing regions of the northern United States and Europe. Grass, known as "red-top," grows in abundance in the valleys and provides food for horses and cattle. It is expected that, so soon as the different railroads connect the great interior valleys with the sea, the government's offer of three hundred and twenty acres to the homesteader will induce many people to settle there. The Alaska Central Railroad is completed for a distance of fifty-three miles,—more than half the distance to the coal-fields north of Cook Inlet.

Arrangements have been made for the building of a large smelter at Seward, to cost three hundred and fifty thousand dollars, in 1908.

Cook Inlet enjoys well-deserved renown for its scenery. Between it and the Chugach Gulf is the great Kenai Peninsula, whose shores are indented by many deep inlets and bays. The most important of these is Resurrection Bay.

Wood is plentiful along the coast of the peninsula. Cataracts, glaciers, snow peaks, green valleys, and lovely lakes abound.

The peninsula is shaped somewhat like a great pear. Turnagain Arm and an inlet of Prince William Sound almost meet at the north; but the portage mentioned on another page prevents it from being an island. It is crowned by the lofty and rugged Kenai Mountains.

Off its southern coast are several clusters of islands—Pye and Chugatz islands, Seal and Chiswell rocks.

In the entrance to Cook Inlet lie Barren Islands, Amatuli Island, and Ushugat Island.

On a small island off the southern point of the peninsula is a lofty promontory, which Cook named Cape Elizabeth because it was sighted on the Princess Elizabeth's birthday. The lofty, two-peaked promontory on the opposite side of the entrance he named Douglas, in honor of his friend, the Canon of Windsor.

Between the capes, the entrance is sixty-five miles wide; but it steadily diminishes until it reaches a width of but a few miles. There is a passage on each side of Barren Islands.

The Inlet receives the waters of several rivers: the Sushitna, Matanuska, Knik, Yentna,—which flows into the Sushitna near its mouth,—Kaknu, and Kassitof.

Lying near the western shore of the inlet, and just inside the entrance, is an island which rises in graceful sweeps on all sides, directly from the water to a smooth, broken-pointed, and beautiful cone. This cone forms the entire island, and there is not the faintest break in its symmetry until the very crest is reached. It is the volcano of St. Augustine.

A chain of active volcanoes extends along the western shore. Of these, Iliamna, the greatest, is twelve thousand sixty-six feet in height, and was named "Miranda, the Admirable" by Spanish navigators, who may usually be relied upon for poetically significant, or soft-sounding, names. It is clad in eternal snow, but smoke-turbans are wound almost constantly about its brow. It was in eruption in 1854, and running lava has been found near the lower crater. There are many hot and sulphurous springs on its sides.

North of Iliamna is Goryalya, or "The Redoubt," which is a lesser "smoker," eleven thousand two hundred and seventy feet high. It was in eruption in 1867, and ashes fell on islands more than a hundred and fifty miles away.

Iliamna Lake is one of the two largest lakes in Alaska. It is from fifty to eighty miles long and from fifteen to twenty-five wide. A pass at a height of about eight hundred feet affords an easy route of communication between the upper end of the lake and a bay of the same name on Cook Inlet, near the volcano, and has long been in use by white, as well as native, hunters and prospectors. The country surrounding the lake is said to abound in large and small game. Lake Clark, to the north, is connected with Lake Iliamna by the Nogheling River. It is longer than Iliamna, but very much narrower. It lies directly west of the Redoubt Volcano.

Iliamna Lake is connected with Behring Sea by Kvichak River, which flows into Bristol Bay. The lake is a natural hatchery of king salmon, and immense canneries are located on Bristol Bay, which lies directly north of the Aliaska Peninsula.

It is comparatively easy for hunters to cross by the chain of lakes and water-ways from Bristol Bay to Cook Inlet—which is known to sportsmen of all countries, both shores offering everything in the way of game. The big brown bear of the inlet is the same as the famous Kadiak; and hunters come from all parts of the world when they can secure permits to kill them. Moose, caribou, mountain sheep, mountain goat, deer, and all kinds of smaller game are also found. There are many trout and salmon streams on the eastern shore of the inlet, and the lagoons and marshes are the haunts of water-fowl.

The voyage up Cook Inlet is one of the most fascinating that may be taken, as a side trip, in Alaska.

Large steamers touch only at Homer and Seldovia, just inside the entrance. There is a good wharf at Homer, but at Seldovia there is another rope-ladder descent and dory landing. There are a post-office, several stores and houses, and a little Greek-Russian church. Scattered over a low bluff at one side of the settlement are the native huts, half hidden in tall reeds and grasses, and a native graveyard.

Seldovia is not the place to buy baskets, as the only ones to be obtained are of very inferior coloring and workmanship.

My Scotch friend was so fearful that some one else might secure a treasure that she seized the first basket in sight at Seldovia, paying five dollars for it. It was not large, and as for its appearance—!

But with one evil mind we all pretended to envy her and to regret that we had not seen it first; so that, for some time, she stepped out over the tundra with quite a proud and high step, swinging her "buy" proudly at her right side, where all might see and admire.

Presently, however, we came to a hut wherein we stumbled upon all kinds of real treasures—old bows and arrows, kamelinkas, bidarkas, virgin charms, and ivory spears. We all gathered these things unto ourselves—all but my Scotch friend.

She stood by, watching us, silent, ruminative.

She had spent all that she cared to spend on curios in one day on the single treasure which she carried in her hand. We observed that presently she carried it less proudly and that her carriage had less of haughtiness in it, as we went across the beach to the dory.

She took the basket down to the engine-room to have it steamed. I do not know what the engineer said to her about her purchase, but when she came back, her face was somewhat flushed. The Scotch are not a demonstrative race, and when she ever after referred to the chief engineer simply as "that engineer down there," I felt that it meant something. She never again mentioned that basket to me; but I have seen it in six different curio stores trying to get itself sold.

At Seldovia connection is made with small steamers running up the inlet to the head of the arm. Hope and Sunrise are the inspiring names of the chief settlements of the arm.

The tides of Cook Inlet are tremendous. There are fearful tide-rips at the entrance and again about halfway up the inlet, where they appeared "frightful" to Cook and his men. The tide enters Turnagain Arm, at the head of the inlet, in a huge bore, which expert canoemen are said to be able to ride successfully, and to thus be carried with great speed and delightful danger on their way.

Cook thought that the inlet was a river, of which the arm was an eastern branch. Therefore, at the entrance of the latter, he exclaimed in disappointment and chagrin, "Turn again!"—and afterward bestowed this name upon the slender water-way.

He modestly left only a blank for the name of the great inlet itself; and after his cruel death at the hands of natives in the Sandwich Islands, Lord Sandwich directed that it be named Cook's River.

The voyage of two hundred miles to the head of the arm by steamer is slow and sufficiently romantic to satisfy the most sentimental. The steamer is compelled to tie up frequently to await the favorable stage of the tide, affording ample opportunity and time for the full enjoyment of the varied attractions of the trip. The numerous waterfalls are among the finest of Alaska.

Even to-day the trip is attended by the gravest dangers and is only attempted by experienced navigators who are familiar with its unique perils. The very entrance is the dread of mariners. The tide-rips that boil and roar around the naked Barren Islands subject ships to graver danger than the fiercest storms on this wild and stormy coast.

The tides of Turnagain Arm rival those of the Bay of Fundy, entering in tremendous bores that advance faster than a horse can run and bearing everything with resistless force before them. After the first roar of the entering tide is heard, there is but a moment in which to make for safety. There is a tide fall in the arm of from twenty to twenty-seven feet.

The first Russian settlement of the inlet was by the establishment of a fort by Shelikoff, near the entrance, named Alexandrovsk. It was followed in 1786 by the establishment of the Lebedef-Lastuchkin Company on the Kussilof River in a settlement and fort named St. George.

Fort Alexandrovsk formed a square with two bastions, and the imperial arms shone over the entrance, which was protected by two guns. The situation, however, was not so advantageous for trading as that of the other company.

In 1791 the Lebedef Company established another fort, the Redoubt St. Nicholas, still farther up the inlet, just below that narrowing known as the "Forelands," at the Kaknu, or Kenai, River. At this place the shores jut out into three steep, cliffy points which were named by Vancouver West, North, and East Forelands.

Here Vancouver found the flood-tide running with such a violent velocity that the best bower cable proved unable to resist it, and broke. The buoy sank by the strength of the current, and both the anchor and the cable were irrecoverably lost.

Cook did not enter Turnagain Arm, but Vancouver learned

from the Russians that neither the arm nor the inlet was a river; that the arm terminated some thirty miles from its mouth; and that from its head the Russians walked about fifteen versts over a mountain and entered an inlet of Prince William Sound,—thereby keeping themselves in communication with their fellow-countrymen at Port Etches and Kaye Island.

Vancouver sent Lieutenant Whidbey and some men to explore the arm; but having entered with the bore and finding no place where he might escape its ebb, he was compelled to return with it, without making as complete an examination as was desired.

The country bordering upon the bays along Turnagain Arm is low, richly wooded, and pleasant, rising with a gradual slope, until the inner point of entrance is reached. Here the shores suddenly rise to bold and towering eminences, perpendicular cliffs, and mountains which to poor Whidbey, as usual, appeared "stupendous"—cleft by "awfully grand" chasms and gullies, down which rushed immense torrents of water.

The tide rises thirty feet with a roaring rush that is really terrifying to hear and see.

At a Russian settlement Whidbey found one large house, fifty by twenty-four feet, occupied by nineteen Russians. One door afforded the only ventilation, and it was usually closed.

Whidbey and his men were hospitably received and were offered a repast of dried fish and native cranberries; but because of the offensive odor of the house, owing to the lack of ventilation and other unmentionable horrors, they were unable to eat. Perceiving this, their host ordered the cranberries taken away and beaten up with train-oil, when they were again placed before the visitors. This last effort of hospitality proved too much for the politeness of the Englishmen, and they rushed out into the cool air for relief.

Indeed, the Russians appeared to live quite as filthily and disgustingly as the natives, and to have fallen into all their cooking, living, and other customs, save those of painting their faces and wearing ornaments in lips, noses, and ears.

The name "inlet," instead of "river," was first applied to this torrential water-way in 1794 by Vancouver, who also bestowed upon Turnagain the designation of "arm."

Vancouver, upon the invitation of the commanding officer who came out to his ships for that purpose, paid the Redoubt St. Nicholas, near the Forelands, a visit. He was saluted by two guns from a kind of balcony, above which the Russian flag floated on top of a house situated upon a cliff.

Captain Dixon, the most pious navigator I have found, with the exception of the Russians, extolled the Supreme Being for having so bountifully provided in Cook Inlet for the needs of the wretched natives who inhabited the region. The fresh fish and game of all kinds, so easily procured, the rich skins with which to clothe their bodies,—inspired him to praise and thanksgiving.

For the magnificent water-way pushing northward, glaciered, cascaded, blue-bayed, and emerald-valed, with unbroken chains of snow peaks and volcanoes on both sides,—up which the voyager sails charmed and fascinated to-day,—he spoke no enthusiastic word of praise. On the contrary, he found the aspect dreary and uncomfortable. Even Whidbey, the Chilly, could not have given way to deeper shudders than did Dixon in Cook Inlet.

The low land and green valleys close to the shore, grown with trees, shrubbery, and tall grasses, he found "not altogether disagreeable," but it was with shock upon shock to his delicate and outraged feelings that he sailed between the mountains covered with eternal snow. Their "prodigious extent and stupendous precipices ... chilled the blood of the beholder." They were "awfully dreadful."

Dixon, as well as Cook, mentions the wearing of the labret by men, but I still cling to the opinion that they could not distinguish a man from a woman, owing to the attire.

Dixon also reported that the natives have a keen sense of smell, which they quicken by the use of snakeroot. One would naturally have supposed that they would have hunted the forests through and through for some herb, or some dark charm of witchcraft, that would have deprived them utterly and forever of this sense, which is so undesirable a possession to the person living or travelling in Alaska.

The climate of Cook Inlet is more agreeable than that of any other part of Alaska. In the low valleys near the shore the soil is well adapted to the growing of fruits, vegetables, and grain, and to the raising of stock and chickens. Good butter and cheese are made, which, with eggs, bring excellent prices. Roses and all but the tenderest flowers thrive, and berries grow large and of delicious flavor, bearing abundantly.

"Awfully dreadful" scenes are not to be found. It is a pleasure to confess, however, that many features, by their beauty, splendor, and sublimity, fill the appreciative beholder with awe and reverence.

The coal deposits of the region surrounding the inlet are now known to be numerous and important. Coal is found in Kachemak Bay, and Port Graham, at Tyonook, and on Matanuska River, about fifty miles inland from the head of the inlet. It is lignitic and bituminous, but semi-anthracite has been found in the Matanuska Valley.

Lignitic coals have a very wide distribution, but have been, as yet, mined only on Admiralty Island, at Homer and Coal Bay in Cook Inlet, at Chignik and Unga, at several points on the Yukon, and on Seward Peninsula.

The new railroad now building from Cordova will open up not only vast copper districts, but the richest and most extensive oil and coal fields in Alaska, as well.

Semi-anthracite coal exists in commercial quantities, so far as yet discovered, only at Comptroller Bay. A fine quality of bituminous coal also exists there, extending inland for twenty-five miles on the northern tributaries of Behring River and about thirty-five miles east of Copper River, covering an area of about one hundred and twenty square miles.

Southwestern Alaska includes the Cook Inlet region, Kodiak

and adjacent islands, Aliaska Peninsula, and the Aleutian Islands. Coal, mostly of a lignitic character, is widely distributed in all these districts. It has also been discovered in different localities in the Sushitna Basin.

All coal used by the United States government's naval vessels on the Pacific is purchased and transported there from the East at enormous expense. Alaska has vast coal deposits of an exceedingly fine quality lying undeveloped in the Aliaskan Peninsula, two hundred miles farther west than Honolulu, and directly on the route of steamers plying from this country to the Orient. (It is not generally known that the smoke of steamers on their way from Puget Sound to Japan may be plainly seen on clear days at Unalaska.)

This coal is in the neighborhood of Portage Bay, where there is a good harbor and a coaling station. It is reported by geological survey experts to be as fine as Pocahontas coal, and even higher in carbon.

Possibly, in time, the United States government may awaken to a realization of the vast fortunes lying hidden in the undeveloped, neglected, and even scorned resources of Alaska,—not to mention the tremendous advantages of being able to coal its war vessels with Pacific Coast coal.

A Yukon Snow Scene near White Horse

During the spring of 1908 the Alaska-coal land situation was discouraging. A great area of rich coal-bearing land had been withdrawn from entry, because of the amazing presumption of the interior department that the removal of prohibitive restrictions upon entrymen would encourage the formation of monopolies in the mining and marketing of coal.

Secretary Garfield at first inclined strongly to the opinion that the Alaska coal lands should be held by the government for leasing purposes, and that there should be a separate reservation for the navy; and he has not entirely abandoned this opinion.

The withdrawal of the coal lands from entry caused the Copper River and Northwestern Railway Company to discontinue all work on the Katalla branch of the road; nor will it resume until the question of title to the coal lands is settled and the lands themselves admitted to entry.

The fear of monopolies, which is making the interior

department uneasy, is said to have arisen from the fact that it has been absolutely necessary for several entrymen in a coal region to associate themselves together and combine their claims, on account of the enormous expense of opening and operating mines in that country. The surveys alone, which, in accordance with an act passed in 1904, must be borne by the entryman, although this burden is not imposed upon entrymen in the states, are so expensive, particularly in the Behring coal-fields near Katalla, that an entryman cannot bear it alone; while the expense of getting provisions and tools from salt-water into the interior is simply prohibitive to most locators, unless they can combine and divide the expense.

These early discoverers and locators acted in good faith. The lands were entered as coal lands; there was no fraud and no attempt at fraud; not one person sought to take up coal land as homestead, nor with scrip, nor in any fraudulent manner.

There was some carelessness in the observance of new rules and regulations, but there was excuse for this in the fact that Alaska is far from Congress and news travels slowly; also, it has been the belief of Alaskans that when a man, after the infinite labor and deprivation necessary to successful prospecting in Alaska, has found anything of value on the public domain, he could appropriate it with the surety that his right thereto would be recognized and respected; and that any slight mistakes that might be made technically would be condoned, provided that they were honest ones and not made with the intent to defraud the government.

The oldest coal mine in Alaska is located just within the entrance to Cook Inlet, on the western shore, at Coal Harbor. There, in the early fifties, the Russians began extensive operations, importing experienced German miners to direct a large force of Muscovite laborers sent from Sitka, and running their machinery by steam.

Shafts were sunk, and a drift run into the vein for a distance of one thousand seven hundred feet. During a period of three years two thousand seven hundred tons of coal were mined, but the result was a loss to the enterprising Russians.

Its extent was practically unlimited, but the quality was found to be too poor for the use of steamers.

It is only within the past three years that the fine quality of much of the coal found in Alaska has been made known by government experts.

It was inconceivable that Congress should hesitate to enact such laws as would help to develop Alaska; yet it was not until late in the spring that bills were passed which greatly relieved the situation and insured the building of the road upon which the future of this district depends.

CHAPTER XXVIII

Cook Inlet is so sheltered and is favored by a climate so agreeable that it was called "Summer-land" by the Russians.

Across Kachemak Bay from Seldovia is Homer—another town of the inlet blessed with a poetic name. When I landed at its wharf, in 1905, it was the saddest, sweetest place in Alaska. It was but the touching phantom of a town.

We reached it at sunset of a June day.

A low, green, narrow spit runs for several miles out into the waters of the inlet, bordered by a gravelly beach. Here is a railroad running eight miles to the Cook Inlet coal-fields, a telephone line, roundhouses, machine-shops, engines and cars, a good wharf, some of the best store buildings and residences in Alaska,—all painted white with soft red roofs, and all deserted!

On this low and lovely spit, fronting the divinely blue sea and the full glory of the sunset, there was only one human being, the postmaster. When the little *Dora* swung lightly into the wharf, this poor lonely soul showed a pitiable and pathetic joy at this fleeting touch of companionship. We all went ashore and shook hands with him and talked to him. Then we returned to our cabins and carried him a share of all our daintiest luxuries.

When, after fifteen or twenty minutes, the *Dora* withdrew slowly into the great Safrano rose of the sunset, leaving him, a lonely, gray figure, on the wharf, the look on his face made us turn away, so that we could not see one another's eyes.

It was like the look of a dog who stands helpless, lonely, and cannot follow.

I have never been able to forget that man. He was so gentle, so simple, so genuinely pleased and grateful—and so lonely!

As I write, Homer is once more a town, instead of a phantom. I no longer picture him alone in those empty, echoing, red-roofed buildings; but one of my most vivid and tormenting memories of Alaska is of a gray figure, with a little pathetic stoop, going up the path from the wharf, in the splendor of that June sunset, with his dog at his side.

The Act of 1902, commonly known as the Alaska Game Law, defines game, fixes open seasons, restricts the number which may be killed, declares certain methods of hunting unlawful, prohibits the sale of hides, skins, or heads at any time, and prohibits export of game animals, or birds—except for scientific purposes, for propagation, or for trophies—under restrictions prescribed by the Department of Agriculture. The law also authorizes the Secretary of Agriculture, when such action shall be necessary, to place further restrictions on killing in certain regions. The importance of this provision is already apparent. Owing to the fact that nearly all persons who go to Alaska to kill big game visit a few easily accessible localities—notably Kadiak Island, the Kenai Peninsula, and the vicinity of Cook Inlet—it has become necessary to protect the game of these localities by special regulations, in order to prevent its speedy destruction.

The object of the act is to protect the game of the territory so far as possible from the mere "killer," but without causing unnecessary hardship. Therefore, Indians, Eskimos,miners, or explorers actually in need of food, are permitted to kill game for their immediate use. The exception in favor of natives, miners, and explorers must be construed strictly. It must not be used merely as a pretext to kill game out of season, for sport or for market, or to supply canneries or settlements; and, under no circumstances, can the hides or heads of animals thus killed be lawfully offered for sale.

Every person who has travelled in Alaska knows that these laws are violated daily. An amusing incident occurred on the *Dora*, on the first morning "to Westward" from Seward. Far be it from me to eat anything that is forbidden; but I had *seen* fried moose steak in Seward. It resembles slices of pure beef tenderloin, fried.

It chanced that at our first breakfast on the *Dora* I found fried beef tenderloin on the bill of fare, and ordered it. Scarcely had I been served when in came the gentleman from Boston, who, through his alert and insatiable curiosity concerning all things Alaskan and his keen desire to experience every possible Alaskan sensation,—all with the greatest naïveté and good humor,—had endeared himself to us all on our long journey together.

"What's that?" asked he, briskly, scenting a new experience on my plate.

"Moose," said I, sweetly.

"Moose—*moose!*" cried he, excitedly, seizing his bill of fare. "I'll have some. Where is it? I don't see it!"

"Hush-h-h," said I, sternly. "It is not on the bill of fare. It is out of season."

"Then how shall I get it?" he cried, anxiously. "I must have some."

"Tell the waiter to bring you the same that he brought me."

When the dear, gentle Japanese, "Charlie," came to serve him, he shamelessly pointed at my plate.

"I'll have some of that," said he, mysteriously.

Charlie bowed, smiled like a seraph, and withdrew, to return presently with a piece of beef tenderloin.

The gentleman from Boston fairly pounced upon it. We all watched him expectantly. His expression changed from anticipation to satisfaction, delight, rapture.

"That's the most delicious thing I ever ate," he burst forth, presently.

"Do you think so?" said I. "Really, I was disappointed. It tastes very much like beefsteak to me."

"Beefsteak!" said he, scornfully. "It tastes no more like beefsteak than pie tastes like cabbage! What a pity to waste it on one who cannot appreciate its delicate wild flavor!"

Months afterward he sent me a marked copy of a Boston newspaper, in which he had written enthusiastically of the "rare, wild flavor, haunting as a poet's dream," of the moose which he had eaten on the *Dora*.

In addition to the animals commonly regarded as game, walrus and brown bear are protected; but existing laws relating to the fur-seal, sea-otter, or other fur-bearing animals are not affected. The act creates no close season for black bear, and contains no prohibition against the sale or shipment of their skins or heads; but those of brown bear may be shipped only in accordance with regulations.

The Act of 1908 amends the former act as follows:—

It is unlawful for any person in Alaska to kill any wild game, animals, or birds, except during the following seasons: north of latitude sixty-two degrees, brown bear may be killed at any time; moose, caribou, sheep, walrus and sea-lions, from August 1 to December 10, inclusive; south of latitude sixty-two degrees, moose, caribou, and mountain sheep, from August 20 to December 31, inclusive; brown bear, from October 1 to July 1, inclusive; deer and mountain goats, from August 1 to February 1, inclusive; grouse, ptarmigan, shore birds, and water fowl, from September 1 to March 1, inclusive.

The Secretary of Agriculture is authorized, whenever he may deem it necessary for the preservation of game animals or birds, to make and publish rules and regulations which shall modify the close seasons established, or to provide different close seasons for different parts of Alaska, or to place further limitations and restrictions on the killing of such animals or birds in any given locality, or to prohibit killing entirely for a period not exceeding two years in such locality.

It is unlawful for any person at any time to kill any females or yearlings of moose, or for any one person to kill in one year more than the number specified of each of the following game animals: Two moose, one walrus or sea-lion, three caribou; sheep, or large brown bear; or to kill or have in his possession in any one day more than twenty-five grouse or ptarmigan, or twenty-five shore birds or water fowl.

The killing of caribou on the Kenai Peninsula is prohibited

until August 20, 1912.

It is unlawful for any non-resident of Alaska to hunt any of the protected game animals, except deer and goats, without first obtaining a hunting license; or to hunt on the Kenai Peninsula without a registered guide, such license not being transferable and valid only during the year of issue. The fee for this license is fifty dollars to citizens of the United States, and one hundred dollars to foreigners; it is accompanied by coupons authorizing the shipment of two moose,—if killed north of sixty-two degrees,—four deer, three caribou, sheep, goats, brown bear, or any part of said animals. A resident of Alaska may ship heads or trophies by obtaining a shipping license for this purpose. A fee of forty dollars permits the shipment of heads or trophies as follows: one moose, if killed north of sixty-two degrees; four deer, two caribou, two sheep, goats, or brown bear. A fee of ten dollars permits the shipment of a single head or trophy of caribou or sheep; and one of five, that of goat, deer, or brown bear. It costs just one hundred and fifty dollars to ship any part of a moose killed south of sixty-two degrees. Furthermore, before any trophy may be shipped from Alaska, the person desiring to make such shipment shall first make and file with the customs office of the port where the shipment is to be made, an affidavit to the effect that he has not violated any of the provisions of this act; that the trophy has been neither bought nor sold, and is not to be shipped for sale, and that he is the owner thereof.

The Governor of Alaska, in issuing a license, requires the applicant to state whether the trophies are to be shipped through the ports of entry of Seattle, Portland, or San Francisco, and he notifies the collector at the given port as to the name of the license holder, and name and address of the consignee.

After reading these rigid laws, I cannot help wondering whether the Secretary of Agriculture ever saw an Alaskan mountain sheep. If he has seen one and should unexpectedly come across some poor wretch smuggling the head of one out of Alaska, he would—unless his heart is as hard as "stun-cancer," as an old lady once said—just turn his eyes in another direction and refuse to see what was not meant for his vision.

The Alaskan sheep does not resemble those of Montana and other sheep countries. It is more delicate and far more beautiful. There is a deerlike grace in the poise of its head, a fine and sensitive outline to nostril and mouth, a tenderness in the great dark eyes, that is at once startled and appealing; while the wide, graceful sweep of the horns is unrivalled.

The head of the moose, as well as of the caribou, is imposing, but coarse and ugly. The antlers of the delicate-headed deer are pretty, but lack the power of the horns of the Alaskan sheep. The Montana sheep's head is almost as coarse as that of the moose. The dainty ears and soft-colored hair of the Alaskan sheep are fawnlike. From the Alaska Central trains near Lake Kenai, the sheep may be seen feeding on the mountain that has been named for them.

Cape Douglas, at the entrance to Cook Inlet, is the admiration of all save the careful navigator who usually at this point meets such distressing winds and tides that he has no time to devote to the contemplation of scenery.

This noble promontory thrusts itself boldly out into the sea for a distance of about three miles, where it sinks sheer for a thousand feet to the pale green surf that breaks everlastingly upon it. It is far more striking and imposing than the more famous Cape Elizabeth on the eastern side of the entrance to the inlet.

CHAPTER XXIX

The heavy forestation of the Northwest Coast ceases finally at the Kenai Peninsula. Kadiak Island is sparsely wooded in sylvan groves, with green slopes and valleys between; but the islands lying beyond are bare of trees. Sometimes a low, shrubby willow growth is seen; but for the most part the thousands of islands are covered in summer with grasses and mosses, which, drenched by frequent mists and rain, are of a brilliant and dazzling green.

The Aleutian Islands drift out, one after another, toward the coast of Asia, like an emerald rosary on the blue breast of Behring Sea. The only tree in the Aleutian Islands is a stunted evergreen growing at the gate of a residence in Unalaska, on the island of the same name.

The prevailing atmospheric color of Alaska is a kind of misty, rosy lavender, enchantingly blended from different shades of violet, rose, silver, azure, gold, and green. The water coloring changes hourly. One passes from a narrow channel whose waters are of the most delicate green into a wider reach of the palest blue; and from this into a gulf of sun-flecked purple.

The summer voyage out among the Aleutian Islands is lovely beyond all description. It is a sweet, dreamlike drifting through a water world of rose and lavender, along the pale green velvety hills of the islands. There are no adjectives that will clearly describe this greenness to one who has not seen it. It is at once so soft and so vivid; it flames out like the dazzling green fire of an emerald, and pales to the lighter green of the chrysophrase.

Marvellous sunset effects are frequently seen on these waters. There was one which we saw in broad gulfs, which gathered in a point on the purple water about nine o'clock. Every color and shade of color burned in this point, like a superb fire opal; and from it were flung rays of different coloring—so far, so close, so mistily brilliant, and so tremulously ethereal, that in shape and fabric it resembled a vast thistle-down blowing before us on the water. Often we sailed directly into it and its fragile color needles were shattered and fell about us; but immediately another formed farther ahead, and trembled and throbbed until it, too, was overtaken and shattered before our eyes.

At other times the sunset sank over us, about us, and upon us, like a cloud of gold and scarlet dust that is scented with coming rain; but of all the different sunset effects that are but memories now, the most unusual was a great mist of brilliant, vivid green just touched with fire, that went marching down the wide straits of Shelikoff late one night in June.

Early on the morning after leaving Cook Inlet, the "early-decker" will find the *Dora* steaming lightly past Afognak Island through the narrow channel separating it from Marmot Island. This was the most silvery, divinely blue stretch of water I saw in Alaska, with the exception of Behring Sea. The morning that we sailed into Marmot Bay was an exceptionally suave one in June; and the color of the water may have been due to the softness of the day.

We had passed Sea Lion Rocks, where hundreds of these animals lie upon the rocky shelves, with lifted, narrow heads, moving nervously from side to side in serpent fashion, and whom a boat's whistle sends plunging headlong into the sea.

The southern point of Marmot Island is the Cape St. Hermogenes of Behring, a name that has been perpetuated to this day. The steamer passes between it and Pillar Point, and at one o'clock of the same day through the winding, islanded harbor of Kadiak.

This settlement is on the island that won the heart of John Burroughs when he visited it with the famous Harriman Expedition—the Island of Kadiak.

I voyaged with a pilot who had accompanied the expedition.

"Those scientists, now," he said, musingly, one day as he paced the bridge, with his hands behind him. "They were a real study for a fellow like me. The genuine big-bugs in that party were the finest gentlemen you ever saw; but the *little*-bugs—say, they put on more dog than a bogus prince! They were always demanding something they couldn't get and acting as if they was afraid somebody might think they didn't amount to anything. An officer on a ship can always tell a gentleman in two minutes—his wants are so few and his tastes so simple. John Burroughs? Oh, say, every man on the ship liked Mr. Burroughs. I don't know as you'd ought to call him a gentleman. You see, gentlemen live on earth, and he was way up above the earth—in the clouds, you know. He'd look right through you with the sweetest eyes, and never see you. But *flowers*—well, Jeff Davis! Mr. Burroughs could see a flower half a mile away! You could talk to him all day, and he wouldn't hear a word you said to him, any more than if he was deef as a post. I thought he was, the longest while. But Jeff Davis! just let a bird sing on shore when we were sailing along close. His deefness wasn't particularly noticeable then!... He'd go ashore and dawdle 'way off from everybody else, and come back with his arms full of flowers."

Mr. Burroughs was charmed with the sylvan beauty of Kadiak Island; its pale blue, cloud-dappled skies and deep blue, islanded seas; its narrow, winding water-ways; its dimpled hills, silvery streams, and wooded dells; its acres upon acres of flowers of every variety, hue and size; its vivid green, grassy, and mossy slopes, crests, and meadows; its delightful air and singing birds.

He was equally charmed with Wood Island, which is only fifteen minutes' row from Kadiak, and spent much time in its melodious dells, turning his back upon both islands with reluctance, and afterward writing of them appreciative words which their people treasure in their hearts and proudly quote to the stranger who reaches those lovely shores.

The name Kadiak was originally Kaniag, the natives calling themselves Kaniagists or Kaniagmuts. The island was discovered in 1763, by Stephen Glottoff.

His reception by the natives was not of a nature to warm the cockles of his heart. They approached in their skin-boats, but his godson, Ivan Glottoff, a young Aleut interpreter, could not make them understand him, and they fled in apparent fear.

Some days later they returned with an Aleutian boy whom they had captured in a conflict with the natives of the Island of Sannakh, and he served as interpreter.

The natives of Kadiak differ greatly from those of the Aleutian Islands, notwithstanding the fact that the islands drift into one another.

The Kadiaks were more intelligent and ambitious, and of much finer appearance, than the Aleutians.

They were of a fiercer and more warlike nature, and refused to meet the friendly advances of Glottoff. The latter, therefore, kept at some distance from the shore, and a watch was set night and day.

Nevertheless, the Kadiaks made an early-morning attack, firing upon the watches with arrows and attempting to set fire to the ship. They fled in the wildest disorder upon the discharge of firearms, scattering in their flight ludicrous ladders, dried moss, and other materials with which they had expected to destroy the ship.

Within four days they made another attack, provided with wooden shields to ward off the musket-balls.

They were again driven to the shore. At the end of three weeks they made a third and last attack, protected by immense breastworks, over which they cast spears and arrows upon the decks.

As these shields appeared to be bullet-proof and the natives continued to advance, Glottoff landed a body of men and made a fierce attack, which had the desired effect. The savages dropped their shields and fled from the neighborhood.

When Von H. J. Holmberg was on the island, he persuaded an old native to dictate a narrative to an interpreter, concerning the arrival of the first ship—which was undoubtedly Glottoff's.

This narrative is of poignant interest, presenting, as it does, so simply and so eloquently, the "other" point of view—that of the first inhabitant of the country, which we so seldom hear. For this reason, and for the charm of its style, I reproduce it in part:—

"I was a boy of nine or ten years, for I was already set to paddle a bidarka, when the first Russian ship, with two masts, appeared near Cape Aleulik. Before that time we had never seen a ship. We had intercourse with the Aglegnutes, of the Aliaska Peninsula, with the Tnaianas of the Kenai Peninsula, and with the Koloshes, of southeastern Alaska. Some wise men even knew something of the Californias; but of white men and their ships we knew nothing.

"The ship looked like a great whale at a distance. We went out to sea in our bidarkas, but we soon found that it was no whale, but another unknown monster of which we were afraid, and the smell of which made us sick."

(In all literature and history and real life, I know of no single touch of unintentional humor so entirely delicious as this: that any odor could make an Alaskan native, of any locality or tribe, sick; and of all things, an odor connected with a white person! It appears that in more ways than one this old native's story is of value.)

"The people on the ship had buttons on their clothes, and at first we thought they must be cuttle-fish." (More unintentional, and almost as delicious, humor!) "But when we saw them put fire into their mouths and blow out smoke we knew that they must be *devils*."

(Did any early navigator ever make a neater criticism of the natives than these innocent ones of the first white visitors to their shores?)

"The ship sailed by ... into Kaniat, or Alitak, Bay, where it anchored. We followed, full of fear, and at the same time curious to see what would become of the strange apparition, but we did not dare to approach the ship.

"Among our people was a brave warrior named Ishinik, who

was so bold that he feared nothing in the world; he undertook to visit the ship, and came back with presents in his hand,—a red shirt, an Aleut hood, and some glass beads." (Glottoff describes this visit, and the gifts bestowed.)

"He said there was nothing to fear; that they only wished to buy sea-otter skins, and to give us glass beads and other riches for them. We did not fully believe this statement. The old and wise people held a council. Some thought the strangers might bring us sickness.

"Our people formerly were at war with the Fox Island people. My father once made a raid on Unalaska and brought back, among other booty, a little girl left by her fleeing people. As a prisoner taken in war, she was our slave, but my father treated her like a daughter, and brought her up with his own children. We called her Plioo, which means ashes, because she was taken from the ashes of her home. On the Russian ship which came from Unalaska were many Aleuts, and among them the father of our slave. He came to my father's house, and when he found that his daughter was not kept like a slave, but was well cared for, he told him confidentially, out of gratitude, that the Russians would take the sea-otter skins without payment, if they could.

"This warning saved my father. The Russians came ashore with the Aleuts, and the latter persuaded our people to trade, saying, 'Why are you afraid of the Russians? Look at us. We live with them, and they do us no harm.'

"Our people, dazzled by the sight of such quantities of goods, left their weapons in the bidarkas and went to the Russians with the sea-otter skins. While they were busy trading, the Aleuts, who carried arms concealed about them, at a signal from the Russians, fell upon our people, killing about thirty and taking away their sea-otter skins. A few men had cautiously watched the result of the first intercourse from a distance—among them my father." (The poor fellow told this proudly, not understanding that he thus confessed a shameful and cowardly act on his father's part.)

"These attempted to escape in their bidarkas, but they were overtaken by the Aleuts and killed. My father alone was saved by the father of his slave, who gave him his bidarka when my father's own had been pierced by arrows and was sinking.

"In this he fled to Akhiok. My father's name was Penashigak. The time of the arrival of this ship was August, as the whales were coming into the bays, and the berries were ripe.

A Home in the Yukon

"The Russians remained for the winter, but could not find sufficient food in Kaniat Bay. They were compelled to leave the ship in charge of a few watchmen and moved into a bay opposite Aiakhtalik Island. Here was a lake full of herrings and a kind of smelt. They lived in tents through the winter. The brave Ishinik, who first dared to visit the ship, was liked by the Russians, and acted as mediator. When the fish decreased in the lake during the winter, the Russians moved about from place to place. Whenever we saw a boat coming, at a distance, we fled to the hills, and when we returned, no dried fish could be found in the houses.

"In the lake near the Russian camp there was a poisonous kind of starfish. We knew it very well, but said nothing about it to the Russians. We never ate them, and even the gulls would not touch them. Many Russians died from eating them. We injured them, also, in other ways. They put up fox-traps, and we removed them for the sake of obtaining the iron material. The Russians left during the following year."

This native's name was Arsenti Aminak. There are several slight discrepancies between his narrative and Glottoff's account, especially as to time. He does not mention the hostile attacks of his people upon the Russians; and these differences puzzle Bancroft and make him sceptical concerning the veracity of the native's account.

It is barely possible, however, that Glottoff imagined these attacks, as an excuse for his own merciless slaughter of the Kadiaks.

As to the discrepancy in time, it must be remembered that Arsenti Aminak was an old man when he related the events which had occurred when he was a young lad of nine or ten. White lads of that age are not possessed of vivid memories; and possibly the little brown lad, just "set to paddle a bidarka," was not more brilliant than his white brothers.

It is wiser to trust the word of the early native than that of the early navigator—with a few illustrious exceptions.

Kadiak is the second in size of Alaskan islands,—Prince of Wales Island in southeastern Alaska being slightly larger,—and no island, unless it be Baranoff, is of more historic interest and charm. It was from this island that Gregory Shelikoff and his capable wife directed the vast and profitable enterprises of the Shelikoff Company, having finally succeeded, in 1784, in making the first permanent Russian settlement in America at Three Saints Bay, on the southeastern coast of this island. Barracks, offices, counting-houses, storehouses, and shops of various kinds were built, and the settlement was guarded against native attack by two armed vessels.

It was here that the first missionary establishment and school of the Northwest Coast of America were located; and here was built the first great warehouse of logs.

Shelikoff's welcome from the fierce Kadiaks, in 1784, was not more cordial than Glottoff's had been. His ships were repeatedly attacked, and it was not until he had fired upon them, causing great loss of life and general consternation among them, that he obtained possession of the harbor.

Shelikoff lost no time in preparing for permanent occupancy of the island. Dwellings and fortifications were erected. His own residence was furnished with all the comforts and luxuries of civilization, which he collected from his ships, for the purpose of inspiring the natives with respect for a superior mode of living. They watched the construction of buildings with great curiosity, and at last volunteered their own services in the work.

Shelikoff personally conducted a school, endeavoring to teach both children and adults the Russian language and arithmetic, as well as religion.

In 1796 Father Juvenal, a young Russian priest who had been sent to the colonies as a missionary, wrote as follows concerning his work:—

"With the help of God, a school was opened to-day at this place, the first since the attempt of the late Mr. Shelikoff to instruct the natives of this neighborhood. Eleven boys and several grown men were in attendance. When I read prayers they seemed very attentive, and were evidently deeply impressed, although they did not understand the language.... When school was closed, I went to the river with my boys, *and with the help of God*" (the italics are mine) "we caught one hundred and three salmon of large size."

The school prospered and was giving entire satisfaction when Baranoff transferred Father Juvenal to Iliamna, on Cook Inlet.

We now come to what has long appealed to me as the most tragic and heart-breaking story of all Alaska—the story of Father Juvenal's betrayal and death at Iliamna.

Of his last Sabbath's work at Three Saints, Father Juvenal wrote:—

"We had a very solemn and impressive service this morning. Mr. Baranoff and officers and sailors from the ship attended, and also a large number of natives. We had fine singing, and a congregation with great outward appearance of devotion. I could not help but marvel at Alexander Alexandreievitch (Baranoff), who stood there and listened, crossing himself and giving the responses at the proper time, and joined in the singing with the same hoarse voice with which he was shouting obscene songs the night before, when I saw him in the midst of a drunken carousal with a woman seated on his lap. I dispensed with services in the afternoon, because the traders were drunk again, and might have disturbed us and disgusted the natives."

Father Juvenal's pupils were removed to Pavlovsk and placed under the care of Father German, who had recently opened a school there.

The priestly missionaries were treated with scant courtesy by Baranoff, and ceaseless and bitter were the complaints they made against him. On the voyage to Iliamna, Father Juvenal complains that he was compelled to sleep in the hold of the brigantine *Catherine*, between bales of goods and piles of dried fish, because the cabin was occupied by Baranoff and his party.

In his foul quarters, by the light of a dismal lantern, he wrote a portion of his famous journal, which has become a most precious human document, unable to sleep on account of the ribald songs and drunken revelry of the cabin.

He claims to have been constantly insulted and humiliated by Baranoff during the brief voyage; and finally, at Pavlovsk, he was told that he must depend upon bidarkas for the remainder of the voyage to the Gulf of Kenai; and after that to the robbers and murderers of the Lebedef Company.

The vicissitudes, insults, and actual suffering of the voyage are vividly set forth in his journal. It was the 16th of July when he left Kadiak and the 3d of September when he finally reached Iliamna—having journeyed by barkentine to Pavlovsk, by bidarka from island to island and to Cook Inlet, and over the mountains on foot.

He was hospitably received by Shakmut, the chief, who took him into his own house and promised to build one especially for him. A boy named Nikita, who had been a hostage with the Russians, acted as interpreter, and was later presented to Father Juvenal.

This young missionary seems to have been more zealous than diplomatic. Immediately upon discovering that the boy had never been baptized, he performed that ceremony, to the astonishment of the natives, who considered it some dark practice of witchcraft.

Juvenal relates with great naïveté that a pretty young woman asked to have the same ceremony performed upon her, that she, too, might live in the same house with the young priest.

The most powerful shock that he received, however, before the one that led to his death, he relates in the following simple language, under date of September 5, two days after his arrival:—

"It will be a relief to get away from the crowded house of the chief, where persons of all ages and sexes mingle without any regard to decency or morals. To my utter astonishment, Shakmut asked me last night to share the couch of one of his wives. He has three or four. I suppose such abomination is the custom of the country, and he intended no insult. God gave me grace to overcome my indignation, and to decline the offer in a friendly and dignified manner. My first duty, when I have somewhat mastered the language, shall be to preach against such wicked practices, but I could not touch upon such subjects through a boy interpreter."

The severe young priest carried out his intentions so zealously that the chief and his friends were offended. He commanded them to put away all their wives but one.

They had marvelled at his celibacy; but they felt, with the rigid justice of the savage, that, if absolutely sincere, he was entitled to their respect.

However, they doubted his sincerity, and plotted to satisfy their curiosity upon this point. A young Iliamna girl was bribed to conceal herself in his room. Awaking in the middle of the night and finding himself in her arms, the young priest was unable to overcome temptation.

In the morning he was overwhelmed with remorse and a sense of his disgrace. He remembered how haughtily he had spurned Shakmut's offer of peculiar hospitality, and how mercilessly he had criticised Baranoff for his immoral carousals. Remembering these things, as well as the ease with which his own downfall had been accomplished, he was overcome with shame.

"What a terrible blow this is to all my recent hopes!" he wrote, in his pathetic account of the affair in his journal. "As soon as I regained my senses, I drove the woman out, but I felt too guilty to be very harsh with her. How can I hold up my head among the people, who, of course, will hear of this affair?... God is my witness that I have set down the truth here in the face of anything that may be said about it hereafter. I have kept myself secluded to-day from everybody. I have not yet the strength to face the world."

When Juvenal did face the small world of Iliamna, it was to be openly ridiculed and insulted by all. Young girls tittered when he went by; his own boys, whom he had taught and baptized, mocked him; a girl put her head into his room when he was engaged in fastening a heavy bar upon his door, and laughed in his face. Shakmut came and insisted that Juvenal should baptize his several wives the following Sunday. This he had been steadily refusing to do, so long as they lived in daily sin; but now, disgraced, broken in spirit, and no longer able to say, "I am holier than thou," he wearily consented.

"I shall not shrink from my duty to make him relinquish all but one wife, however," he wrote, with a last flash of his old spirit, "when the proper time arrives. If I wink at polygamy now, I shall be forever unable to combat it. Perhaps it is only my imagination, but I think I can discover a lack of respect in Nikita's behavior toward me since yesterday.... My disgrace has become public already, and I am laughed at wherever I go, especially by the women. Of course, they do not understand the sin, but rather look upon it as a good joke. It will require great firmness on my part to regain the respect I have lost for myself, as well as on behalf of the Church. I have vowed to burn no fuel in my bedroom during the entire winter, in order to chastise my body—a mild punishment, indeed, compared to the blackness of my sin."

The following day was the Sabbath. It was with a heavy heart that he baptized Katlewah, the brother of the chief, and his family, the three wives of the chief, seven children, and one aged couple.

The same evening he called on the chief and surprised him in a wild carousal with his wives, in which he was jeeringly invited to join.

Forgetting his disgrace and his loss of the right to condemn for sins not so black as his own, the enraged young priest vigorously denounced them, and told the chief that he must marry one of the women according to the rites of the Church and put away the others, or be forever damned. The chief, equally enraged, ordered him out of the house. On his way home he met Katlewah, who reproached him because his religious teachings had not benefited Shakmut, who was as immoral as ever.

The end was now rapidly approaching. On September 29,

less than a month after his arrival, he wrote: "The chief and his brother have both been here this morning and abused me shamefully. Their language I could not understand, but they spat in my face and, what was worse, upon the sacred images on the walls. Katlewah seized my vestments and carried them off, and I was left bleeding from a blow struck by an ivory club. Nikita has washed and bandaged my wounds; but from his anxious manner I can see that I am still in danger. The other boys have run away. My wound pains me so that I can scarcely—"

The rest is silence. Nikita, who escaped with Juvenal's journal and papers and delivered them to the revered and beloved Veniaminoff, relates that the young priest was here fallen upon and stabbed to death by his enemies.

Many different versions of this pathetic tragedy are given. I have chosen Bancroft's because he seems to have gone more deeply and painstakingly into the small details that add the touch of human interest than any other historian.

The vital interest of the story, however, lies in what no one has told, and what, therefore, no one but the romancer can ever tell.

It lies between the written lines; it lies in the imagination of this austere young priest's remorseful suffering for his sin. There is no sign that he realized—too late, as usual—his first sin of intolerant criticism and condemnation of the sins of others. But neither did he spare himself, nor shrink from the terrible results of his downfall, so unexpected in his lofty and almost flaunting virtue. He was ready, and eager, to chastise his flesh to atone for his sin; and probably only one who has spent a winter in Alaska could comprehend fully the hourly suffering that would result from a total renouncement of fuel for the long, dark period of winter.

Veniaminoff was of the opinion that the assassination was caused not so much by his preaching against polygamy as by the fact that the chiefs, having given him their children to educate at Kadiak, repented of their action, and being unable to recover them, turned against him and slew him as a deceiver, in their ignorance. During the fatal attack upon him, it is said, Juvenal never thought of flight or self-defence, but surrendered himself into their hands without resistance, asking only for mercy for his companions.

CHAPTER XXX

In 1792 Baranoff having risen to the command of the Shelikoff-Golikoff Company, decided to transfer the settlement of Three Saints to the northern end of the island, as a more central location for the distribution of supplies. To-day only a few crumbling ruins remain to mark the site of the first Russian settlement in America—an event of such vital historic interest to the United States that a monument should be erected there by this country.

The new settlement was named St. Paul, and was situated on Pavlovsk Bay, the present site of Kadiak. The great warehouse, built of logs, and other ancient buildings still remain.

It was during the year of Father Juvenal's death—1796—that the first Russo-Greek church was erected at St. Paul. It was about this time that the conversion of twelve thousand natives in the colonies was reported by Father Jossaph. This amazing statement could only have been made after one of Baranoff's banquets—to which the astute governor, desiring that a favorable report should be sent to St. Petersburg, doubtless bade the half-starved priest.

For the Russian-American Company the Kadiaks and Aleuts were obliged to hunt and work, at the will of the officers, and to sell all their furs to the company, at prices established by the latter.

Baranoff, for a time after becoming Chief Director, resided in Kodiak. All persons and affairs in the colonies were under his control; his authority was absolute, his decision final, unless appeal was made to the Directory at Irkutsk; and it was almost impossible for an appeal to reach Irkutsk.

To-day in Kodiak, as in Sitka, the old and the new mingle. Some of the old sod-houses remain, and many that were built of logs; but the majority of the dwellings are modern frame structures, painted white and presenting a neat appearance, in striking contrast to many of the settlements of Alaska where natives reside.

The Greek-Russian church shines white and attractive against the green background of the hill. It is surrounded by a white fence and is shaded by trees.

I called at the priest's residence and was hospitably received by his wife, an intelligent, dark-eyed native woman. The interior of the church is interesting, but lacks the charm and rich furnishings of the one at Sitka. There is a chime of bells in the steeple; and both steeple and dome are surmounted by the peculiar Greek-Russian cross which is everywhere seen in Alaska. It has two short transverse bars, crossing the vertical shaft, one above and one below the main transverse bar, the lower always slanting.

The natives of Kodiak are more highly civilized than in other parts of Alaska. The offspring of Russian fathers and native mothers have frequently married into white or half-breed families, and the strain of dark blood in the offspring of these later marriages is difficult to discern.

I travelled on the *Dora* with a woman whose father had been a Russian priest, married to a native woman at Belkoffski. She had been sent to California for a number of years, and returning, a graduate of a normal school, had married a Russian. She had a comfortable, well-furnished home, and her husband appeared extremely fond and proud of her. Her children were as white as any Russian I have ever seen.

A Russian priest must marry once; but if his wife dies, he cannot marry again.

This law fills my soul with an unholy delight. It persuades a man to appreciate his wife's virtues and to condone her faults. Whatever may be her sins in sight of him and heaven, she is the only one, so far as he is concerned. It must be she, or nobody, to the end of his days. She may fill his soul with rage, but he may not even relieve his feelings by killing her.

The result of this unique religious law is that Russian priests are uncommonly kind and indulgent to their wives.

"Yes, yes, yes, yes, yes," said one who was on the *Dora*, in answer to a question, "I have a wife. She lives in Paris, where my daughter is receiving her education. I am going this year to visit them. Yes, yes, yes."

However, with all the petting and indulgence which the Russian priest lavishes upon his wife, if what I heard be true,—that he is permitted neither to cut nor to wash his hair and beard,—God wot she is welcome to him.

The old graveyard on the hill above Kodiak tempts the visitor, and one may loiter among the old, neglected graves with no fear of snakes in the tall, thick grasses.

At first, a woman receives the statement that there are no snakes in Alaska with open suspicion. It has the sound of an Alaskan joke.

When I first heard it, I was unimpressed. We were nearing a fine field of red-top, already waist-high, and I waited for the gentleman from Boston, who believed everything he heard, and imagined far more, to go prancing innocently through the field.

He went—unhesitatingly, joyously; giving praise to God for his blessings—as, he vowed, he loved to ramble through deep grass, yet would rather meet a hippopotamus alone in a mire than a garter-snake five inches long. The field was the snakiest-looking place imaginable, and when he had passed safely through, I began to have faith in the Alaskan snake story.

The climate of Kadiak Island is delightful. The island is so situated that it is fully exposed to the equalizing influences of the Pacific. The mean annual temperature is four degrees lower than at Sitka, and there is twenty per cent less rainfall.

The coast of Alaska is noted for its rainfall and cloudy weather. Its precipitation is to be compared only to that of the coast of British Columbia, Washington, and Oregon; and it will surprise many people to learn that it is exceeded in the latter district.

The heaviest annual rainfall occurs at Nutchek, with a decided drop to Fort Tongass; then, Orca, Juneau, Sitka, and Fort Liscum. Fort Wrangell, Killisnoo, and Kodiak stand next; while Tyonok, Skaguay, and Kenai record only from fifteen to twenty-five inches.

Kadiak Island is a hundred miles long by about forty in width. Its relief is comparatively low—from three to five thousand feet—and it has many broad, open valleys, gently rounded slopes, and wooded dells.

Lisiansky was told that the Kadiak group of islands was once separated from the Aliaska Peninsula by the tiniest ribbon of water. An immense otter, in attempting to swim through this pass, was caught fast and could not extricate itself. Its desperate struggles for freedom widened the pass into the broad sweep of water now known as the Straits of Shelikoff, and pushed the islands out to their present position. This legend strengthens the general belief that the islands were once a part of the peninsula, having been separated therefrom by one of the mighty upheavals, with its attendant depression, which are constantly taking place.

A native myth is that the original inhabitants were descended from a dog. Another legend is to the effect that the daughter of a great chief north of the peninsula married a dog and was banished with her dog-husband and whelps. The dog tried to swim back, but was drowned, his pups then falling upon the old chief and, having torn him to pieces, reigning in his stead.

In 1791 Shelikoff reported the population of Kadiak Island to be fifty thousand, the exaggeration being for the purpose of enhancing the value of his operations. In 1795 the first actual census of Kadiak showed eighteen hundred adult native males, and about the same number of females. To-day there are probably not five hundred.

I have visited Kadiak Island in June and in July. On both occasions the weather was perfect. Clouds that were like broken columns of pearl pushed languorously up through the misty gold of the atmosphere; the long slopes of the hillside were vividly green in the higher lights, but sank to the soft dark of dells and hollows; here and there shone out acres of brilliant bloom.

To one climbing the hill behind the village, island beyond island drifted into view, with blue water-ways winding through velvety labyrinths of green; and, beyond all, the strong, limitless sweep of the ocean. The winds were but the softest zephyrs, touching the face and hair like rose petals, or other delicate, visible things; and, the air was fragrant with things that grow day and night and that fling their splendor forth in one riotous rush of bloom. Shaken through and through their perfume was that thrilling, indescribable sweetness which abides in vast spaces where snow mountains glimmer and the opaline palisades of glaciers shine.

It is a view to quicken the blood, and to inspire an American to give silent thanks to God that this rich and peerlessly beautiful country is ours.

After the transfer, the village of Kodiak was the headquarters of the Alaska Commercial Company and the Western Fur and Trading Company. The former company still maintains stores and warehouses at this point. The house in which the manager resides occupies a commanding site above the bay. It is historic and commodious, and large house-parties are entertained with lavish hospitality by Mr. and Mrs. Goss, visitors gathering there from adjacent islands and settlements.

There are dances, "when the boats are in," in which the civilized native girls join with a kind of repressed joy that reminds one of New England. They dress well and dance gracefully. Their soft, dark glances over their partners' shoulders haunt even a woman dreamily. A century's silently and gently borne wrongs smoulder now and then in the deep eyes of some beautiful, dark-skinned girl.

Kodiak is clean. One can stand on the hills and breathe.

For several years after the transfer a garrison of United States troops was stationed there. Bridges were built across the streams that flow down through the town, and culverts to drain the marshes. Many of these improvements have been carelessly destroyed with the passing of the years, but their early influence remains.

So charming and so idyllic did this island seem to the Russians that it was with extreme reluctance they moved their capital to Sitka when the change was considered necessary.

We were rowed by native boys across the satiny channel to Wood Island, where Reverend C. P. Coe conducts a successful Baptist Orphanage for native children. Mr. Coe was not at home, but we were cordially received by Mrs. Coe and three or four assistants. Wood Island, or Woody, as it was once called, is as lovely as Kadiak; the site for the buildings of the Orphanage being particularly attractive, surrounded as it is by groves and dells.

There was a pale green, springlike freshness folded over the gently rolling hills and hollows that was as entrancing as the first green mist that floats around the leafing alders on Puget Sound in March.

The Orphanage was established in 1893 by the Woman's American Baptist Home Mission Society of Boston, and the first child was entered in that year. Mr. Coe assumed charge of the Orphanage in 1895, and about one hundred and thirty children have been educated and cared for under his administration. They have come from the east as far as Kayak, and from the west as far as Unga. At present there is but one other Baptist Mission field in Alaska—at Copper Centre.

The purpose of the work is to provide a Christian home and training for the destitute and friendless; to collect children, that they may receive an education; and to give industrial training so far as possible.

There were forty-two children in the home at the time of our visit, and there was a full complement of helpers in the work, including a physician.

The regular industrial work consists of all kinds of housework for the girls. Everything that a woman who keeps house should know is taught to these girls. The boys are taught to plough and sow, to cultivate and harvest the crops, to raise vegetables, to care for stock and poultry. Twenty-five acres are under cultivation, and the hardier grains and vegetables are grown with fair success.

Potatoes yield two hundred and fifty bushels to the acre; and barley, forty bushels. Cattle and poultry thrive and are of exceeding value, fresh milk and vegetables being better than medicines for the welfare of the children. Angora goats require but little care and yield excellent fleece each year.

The most valuable features of the work are the religious training; the furnishing of a comfortable home, warm clothing, clean and wholesome food of sufficient quantity, to children who have been rescued from vice and the most repulsive squalor; the atmosphere of industry, cleanliness, kindness, and love; and the medical care furnished to those who may be suffering because of the vices of their ancestors.

This excellent work is supported by offerings from the Baptist Sunday Schools of New England, and by contributions from the society with the yard-long name by which it was established.

We were offered most delicious ginger-cake with nuts in it and big goblets of half milk and half cream; and we were not surprised that the shy, dark-skinned children looked so happy and so well cared for. We saw their schoolrooms, their play rooms, and their bedrooms, with the little clean cots ranged along the walls.

The children were shy, but made friends with us readily; and holding our hands, led the way to the dells where the violets grew. They listened to stories with large-eyed interest, and were, in general, bright, well-mannered, and attractive children.

It was on Wood Island that the famous and mysterious ice-houses of the American-Russian Ice Company, whose headquarters were in San Francisco, were located. Their ruins still stand on the shore, as well as the deserted buildings of the North American Commercial Company, whose headquarters were here for many years—the furs of the Copper River and Kenai regions having been brought here to be shipped to San Francisco.

One and a Half Millions of Klondyke Gold

The operations of the ice company were shrouded in mystery, many claiming that not a pound of ice was ever shipped to the California seaport from Wood Island. Other authorities, however, affirm that at one time large quantities of ice were shipped to the southern port, and that the agent of the company lived on Wood Island in a manner as autocratic and princely as that of Baranoff himself. The whole island was his park and game preserve; and one of the first roads ever built in Alaska was constructed here, comprising the circuit of the island, a distance of about thirteen miles.

There is a Greek-Russian church and mission on the island.

Not far from Wood Island is Spruce.

"Here," says Tikhmenef, "died the last member of the first clerical mission, the monk Herman. During his lifetime Father Herman built near his dwelling a school for the daughters of the natives, and also cultivated potatoes."

Bancroft pokes fun at this obituary. The growing of potatoes, however, at that time in Alaska must have been of far greater value than any ordinary missionary work. Better to cultivate potatoes than to teach a lot of wretched beings to make the sign of the cross and dabble themselves with holy water—and it is said that this is all the average priest taught a hundred years ago, the poor natives not being able to understand the Russian language.

The Kadiak Archipelago consists of Kadiak, Afognak, Tugidak, Sitkinak, Marmot, Wood, Spruce, Chirikoff (named by Vancouver for the explorer who discovered it upon his return journey to Kamchatka), and several smaller ones. They are all similar in appearance, but smaller and less fertile than Kadiak. A small group northwest of Chirikoff is named the Semidi Islands.

There is a persistent legend of a "lost" island in the Pacific, to the southward of Kadiak.

When the Russian missionaries first came to the colonies in America, they found the natives living "as the seals and the otters lived." They were absolutely without moral understanding, and simply followed their own instincts and desires.

These missionaries were sent out in 1794, by command of the Empress Catherine the Second; and by the time of Sir George Simpson's visit in 1842, their influence had begun to show beneficial results. An Aleutian and his daughter who had committed an unnatural crime suddenly found themselves, because of the drawing of new moral lines, ostracized from the society in which they had been accustomed to move unchallenged. They stole away by night in a bidarka, and having paddled steadily to the southward for four days and nights they sighted an island which had never been discovered by white man or dark. They landed and dwelt upon this island for a year.

Upon their return to Kadiak and their favorable report of their lone, beautiful, and sea-surrounded retreat, a vessel was despatched in search of it, but without success.

To this day it is "Lost" Island. Many have looked for it, but in vain. It is the sailor's dream, and is supposed to be rich in treasure. Its streams are yellow with gold, its mountains green with copper glance; ambergris floats on the waters surrounding it; and all the seals and sea-otters that have been frightened out of the north sun themselves, unmolested, upon its rocks and its floating strands of kelp.

One day it will rise out of the blue Pacific before the wondering eyes of some fortunate wanderer—even as the Northwest Passage, for whose sake men have sailed and suffered and failed and died for four hundred years, at last opened an icy avenue before the amazed and unbelieving eyes of the dauntless Amundsen.

CHAPTER XXXI

Leaving Kodiak, the steamer soon reaches Afognak, on the island of the same name. There is no wharf at this settlement, and we were rowed ashore.

We were greatly interested in this place. The previous year we had made a brief voyage to Alaska. On our steamer was an unmarried lady who was going to Afognak as a missionary. She was to be the only white woman on the island, and she had entertained us with stories which she had heard of a very dreadful and wicked saloon-keeper who had lived near her schoolhouse, and whose evil influence had been too powerful for other missionaries to combat.

"But he can't scare me off!" she declared, her eyes shining with religious ardor. "I'll conquer him before he shall conquer me!"

She was short and stout and looked anything but brave, and as we approached the scene of conflict, we felt much curiosity as to the outcome.

She was on the beach when we landed, stouter, shorter, and more energetic than ever in her movements. She remembered us and proudly led the way up the bank to her schoolhouse. It was large, clean, and attractive. The missionary lived in four adjoining rooms, which were comfortable and homelike. We were offered fresh bread and delicious milk.

She talked rapidly and eagerly upon every subject save the one in which we were so interested. At last, I could endure the suspense no longer.

"And how," asked I, "about the wicked saloon-keeper?"

A dull flush mounted to her very glasses. For a full minute there was silence. Then said she, slowly and stiffly:—

"How about *what* wicked saloon-keeper?"

"Why, the one you told us about last year; who had a poor

abused wife and seven children, and who scared the life out of every missionary who came here."

There was another silence.

"Oh," said she then, coldly. "Well, he was rather hard to get along with at first, but his—er—hum—wife died about three months ago, and he has—er—hum" (the words seemed to stick in her throat) "asked me—he—asked me, you know, to" (she giggled suddenly) "*marry* him, you know.

"I don't know as I will, though," she added, hastily, turning very red, as we stood staring at her, absolutely speechless.

The village of Afognak is located at the southwestern end of Litnik Bay. It is divided into two distinct settlements, the most southerly of which has a population of about one hundred and fifty white and half-breed people. A high, grassy bluff, named Graveyard Point, separates this part of the village from that to the northward, which is entirely a native settlement of probably fifty persons.

The population of the Island of Afognak is composed of Kadiaks, Eskimos, Russian half-breeds, and a few white hunters and fishermen. The social conditions are similar to those existing on the eastern shores of Cook Inlet.

When Alaska was under the control of the Russian-American Company, many men grew old and comparatively useless in its service. These employees were too helpless to be thrown upon their own resources, and their condition was reported to the Russian government.

In 1835 an order was issued directing that such Russian employees as had married native women should be located as permanent settlers when they were no longer able to serve the company. The company was compelled to select suitable land, build comfortable dwellings for them, supply agricultural implements, seed, cattle, chickens, and a year's provisions.

These settlers were exempt from taxation and military duty, and the Russians were known as colonial citizens, the half-breeds as colonial settlers. The eastern shores of Cook Inlet, Afognak Island, and Spruce Island were selected for them. The half-breeds now occupying these localities are largely their descendants. They have always lived on a higher plane of civilization than the natives, and among them may be found many skilled craftsmen.

There is no need for the inhabitants of any of these islands to suffer, for here are all natural resources for native existence. All the hardier vegetables thrive and may be stored for winter use; hay may be provided for cattle; the waters are alive with salmon and cod; bear, fox, mink, and sea-otter are still found.

In summer the men may easily earn two hundred dollars working in the adjacent canneries; while the women, assisted by the old men and children, dry the fish, which is then known as ukala. There is a large demand in the North for ukala, for dog food. There are two large stores in Afognak, representing large trading companies, where two cents a pound is paid for all the ukala that can be obtained.

The white men of Afognak are nearly all Scandinavians, married to, or living with, native women. The school-teacher I have already mentioned was the only white woman, and she told us that we were the first white women who had landed on the island during the year she had spent there. Only once had she talked with white women, and that was during a visit to Kodiak.

The town has a sheltered and attractive site on a level green. There is a large Greek-Russian church, not far from the noisy saloon which is presided over by the saloon-keeper who was once bad, but who has now yielded to the missionary's spell.

Karluk River, on the eastern side of Kadiak Island, is the greatest salmon stream in the world. It is sixteen miles long, less than six feet deep, and so narrow at its mouth that a child could toss a pebble from shore to shore. It seems absurd to enter a canoe to cross this stream, so like a little creek is it, across which one might easily leap.

Yet up this tiny water-way millions of salmon struggle every season to the spawning-grounds in Karluk Lake. Before the coming of canners with traps and gill-nets in 1884, it is said that a solid mass of fish might be seen filling this stream from bank to bank, and from its mouth to the lake in the hills.

In 1890 the largest cannery in the world was located in Karluk Bay, but now that distinction belongs to Bristol Bay, north of the Aliaska Peninsula. (Another "largest in the world" is on Puget Sound!)

Karluk Bay is very small; but several canneries are on its shores, and when they are all in operation, the employees are sufficient in number to make one of the largest towns in Alaska. In 1890 three millions of salmon were packed in the several canneries operating in the bay; in 1900 more than two millions in the two canneries then operating; but, on account of the use of traps and gill-nets, the pack has greatly decreased since then, and during some seasons has proved a total failure.

Fifteen years ago two-thirds of the entire Alaskan salmon pack were furnished by the ten canneries of Kadiak Island, and these secured almost their entire supply from Karluk River. Furthermore, at that time, the canners enjoyed their vast monopoly without tax, license, or any government interference.

Immense fortunes have been made—and lost—in the fish industry during the last twenty years.

The superintendents of these canneries always live luxuriously, and entertain like princes—or Baranoff. Their comfortable houses are furnished with all modern luxuries,—elegant furniture, pianos, hot and cold water, electric baths. Perfectly trained, noiseless Chinamen glide around the table, where dinners of ten or twelve delicate courses are served, with a different wine for each course.

Champagne is a part of the hospitality of Alaska. The cheapest is seven dollars and a half a bottle, and Alaskans seldom buy the cheapest of anything.

It was on a soft gray afternoon that the *Dora* entered Karluk Bay between the two picturesque promontories that plunge boldly out into Shelikoff Straits. It seemed as though all the sea-birds of the world must be gathered there. Our entrance set them afloat from their perches on the rocky cliffs. They filled the air, from shore to shore, like a snow-storm. Their poetic flight and shrill, mournful plaining haunt every memory of Karluk Bay.

Now and then they settled for an instant. A cliff would shine out suddenly—a clear, tremulous white; then, as suddenly, there would be nothing but a sheer height of dark stone veined with green before our bewildered gaze. It was as if a silvery, winged cloud drifted up and down the face of the cliffs and then floated out across the bay.

Several old sailing vessels, or "wind-jammers," lay at anchor. They are used for conveying stores and employees from San Francisco. The many buildings of the canneries give Karluk the appearance of a town—in fact, during the summer, it is a town; while in the winter only a few caretakers of the buildings and property remain.

Men of almost every nationality under the sun may be found here, working side by side.

Ceaseless complaints are made of the lawless conditions existing "to Westward." Besides the thousands of men employed in the canneries of the Kadiak and the Aleutian islands, at least ten thousand men work in the canneries of Bristol Bay. They come from China, Japan, the Sandwich Islands, Norway, Sweden, Finland, Porto Rico, the Philippines, Guam, and almost every country that may be named.

"The prevailing color of Alaska may be 'rosy lavender,'" said a gentleman who knows, "but let me tell you that out there you will find conditions that are neither rosy nor lavender."

There is a United States Commissioner and a Deputy United States Marshal in the district, but they are unable to control these men, many of whom are desperate characters. The superintendents of the canneries are there for the purpose of putting up the season's pack as speedily as possible; and, although they are invariably men who deplore crime, they have been known to condone it, to avoid the taking of themselves or their crews hundreds of miles to await the action of some future term of court.

For many years the District of Alaska has been divided for judicial purposes into three divisions: the first comprising the southeastern Alaska district; the second, Nome and the Seward Peninsula; the third, the vast country lying between these two.

In each is organized a full United States district court. The three judges who preside over these courts receive the salary of five thousand dollars a year,—which, considering the high character of the services required, and the cost of living in Alaska, is niggardly. So much power is placed in the hands of these judges that they are freely called czars by the people of Alaska.

The people of the third district complained bitterly that their court facilities were entirely inadequate. Several murders were committed, and the accused awaited trial for many months. Witnesses were detained from their homes and lawful pursuits. Delays were so vexatious that many crimes remained unpunished, important witnesses rebelling against being held in custody for a whole year before they had an opportunity to testify—the judge of the third district being kept busy along the Yukon and at Fairbanks.

As a partial remedy for some of these abuses of government, Governor Brady, in his report for the year 1904, suggested the creation of a fourth judicial district, to be furnished with a sea-going vessel, which should be under the custody of the marshal and at the command of the court. It was recommended that this vessel be equipped with small arms, a Gatling gun, and ammunition. All the islands which lie along the thousands of miles of shore-line of Kenai and Aliaska peninsulas, Cook Inlet, the Kadiak, Shumagin, and Aleutian chains, and Bristol Bay might be visited in season, and a wholesome respect for law and order be enforced.

The burning question in Alaska has been for many years the

one of home government. As early as 1869 an impassioned plea was made in Sitka that Alaska should be given territorial rights. Yet even the bill for one delegate to Congress was defeated as late as the winter of 1905—whereupon fiery Valdez instantly sent its famous message of secession.

Governor Brady criticised the appointment of United States commissioners by the judges, claiming that there is really no appeal from a commissioner's court to a district court, for the reason that the judge usually appoints some particular protégé and feels bound to sustain his decisions. The governor stated plainly in his report that the most remunerative offices are filled by persons who are peculiarly related, socially or politically, to the judges; that the attorneys and their clients understood this and considered an appeal useless. Governor Brady also declared the fee system, as practised in these commissioners' courts, to be an abomination. Unless there is trouble, the officer cannot live; and the inference is that he, therefore, welcomes trouble.

Whatever of truth there may have been in these pungent criticisms, President Roosevelt endorsed many of the governor's recommendations in his message to Congress; and several have been adopted. During the past two years Alaska has made rapid strides toward self-government, and important reforms have been instituted.

The territory now has a delegate to Congress. Upon the subject of home government the people are widely and bitterly divided. Those having large interests in Alaska are, as a rule, opposed to home government, claiming that it is the politicians and those owning nothing upon which taxes could be levied, who are agitating the subject. These claim that the few who have ventured heavily to develop Alaska would be compelled to bear the entire burden of a heavy taxation, for the benefit of the professional politician, the carpet-bagger, and the impecunious loafer who is "just waiting for something to turn up."

On the other hand, those favoring territorial government claim that it is opposed only by the large corporations which "have been bleeding Alaska for years."

The jurisdiction of the United States commissioners in Alaska is far greater than is that of other court commissioners. They can sit as committing magistrates; as justices of the peace, can try civil cases where the amount involved is one thousand dollars or less; can try criminal cases and sentence to one year's imprisonment; they are clothed with full authority as probate judges; they may act as coroners, notaries, and recorders of precincts.

The third district, presided over by Judge Reid, whose residence is at Fairbanks, is five hundred miles wide by nine hundred miles long. It extends from the North Pacific Ocean to the Arctic Ocean, and from the international boundary on the east to the Koyukuk. The chief means of transportation within this district are steamers along the coast and on the Yukon, and over trails by dog teams.

It is small wonder that a man hesitates long before suing for his rights in Alaska. The expense and hardship of even reaching the nearest seat of justice are unimaginable. One man travelled nine hundred miles to reach Rampart to attend court. The federal court issues all licenses, franchises, and charters, and collects all occupation taxes. Every village or mining settlement of two or three hundred men has a commissioner, whose sway in his small sphere is as absolute as that of Baranoff was.

CHAPTER XXXII

We found only one white woman at Karluk, the wife of the manager of the cannery, a refined and accomplished lady.

Her home was in San Francisco, but she spent the summer months with her husband at Karluk.

We were taken ashore in a boat and were most hospitably received in her comfortable home.

About two o 'clock in the afternoon we boarded a barge and were towed by a very small, but exceedingly noisy, launch up the Karluk River to the hatcheries, which are maintained by the Alaska Packers Association.

It was one of those soft, cloudy afternoons when the coloring is all in pearl and violet tones, and the air was sweet with rain that did not fall. The little make-believe river is very narrow, and so shallow that we were constantly in danger of running aground. We tacked from one side of the stream to the other, as the great steamers do on the Yukon.

On this little pearly voyage, a man who accompanied us told a story which clings to the memory.

"Talk about your big world," said he. "You think it 'u'd be easy to hide yourself up in this God-forgotten place, don't you? Just let me tell you a story. A man come up here a few years ago and went to work. He never did much talkin'. If you ast him a question about hisself or where he come from, he shut up like a steel trap with a rat in it. He was a nice-lookin' man, too, an' he had an education an' kind of nice clean ways with him. He built a little cabin, an' he didn't go 'out' in winter, like the rest of us. He stayed here at Karluk an' looked after things.

"Well, after one-two year a good-lookin' young woman come up here—an' jiminy-cricket! He fell in love with her like greased lightnin' an' married her in no time. I God, but that man was happy. He acted like a plumb fool over that woman. After while they had a baby—an' then he acted like two plumb fools in one. I ain't got any wife an' babies myself an' I God! it ust to make me feel queer in my throat.

"Well, one summer the superintendent's wife brought up a woman to keep house for her. She was a white, sad-faced-lookin' woman, an' when she had a little time to rest she ust to climb up on the hill an' set there alone, watchin' the sea-gulls. I've seen her set there two hours of a Sunday without movin'. Maybe she'd be settin' there now if I hadn't gone and put my foot clean in it, as usual.

"I got kind of sorry for her, an' you may shoot me dead for a fool, but one day I ast her why she didn't walk around the bay an' set a spell with the other woman.

"'I don't care much for women,' she says, never changin' countenance, but just starin' out across the bay.

"'She's got a reel nice, kind husband,' says I, tryin' to work on her feelin's.

"'I don't like husbands,' says she, as short as lard pie-crust.

"'She's got an awful nice little baby,' says I, for if you keep on long enough, you can always get a woman.

"She turns then an' looks at me.

"'It's a girl,' says I, 'an' Lord, the way it nestles up into your neck an' loves you!'

"Her lips opened an' shut, but she didn't say a word; but if you'd look 'way down into a well an' see a fire burnin' in the water, it 'u'd look like her eyes did then.

"'Its father acts like a plumb fool over it an' its mother,' says I. 'The sun raises over there, an' sets over here—but *he* thinks it raises an' sets in that woman an' baby.'

"'The woman must be pretty,' says she, suddenly, an' I never heard a woman speak so bitter.

"'She is,' says I; 'she's got—'

"'Don't tell me what she's got,' snaps she, gettin' up off the ground, kind o' stiff-like. 'I've made up my mind to go see her, an' maybe I'd back out if you told me what she's like. Maybe you'd tell me she had red wavy hair an' blue eyes an' a baby mouth an' smiled like an angel—an' then devils couldn't drag me to look at her.'

"Say, I nearly fell dead, then, for that just described the woman; but I'm no loon, so I just kept still.

"'What's their name?' says she, as we walked along.

"'Davis,' says I; an' mercy to heaven! I didn't know I was tellin' a lie.

"All of a sudden she laughed out loud—the awfullest laugh. It sounded as harrable mo'rnful as a sea-gull just before a storm.

"'*Husband!*' she flings out, jeerin'; '*I* had a husband once. I worshipped the ground he trod on. *I* thought the sun raised an' set in *him*. He carried me on two chips for a while, but I didn't have any children, an' I took to worryin' over it, an' lost my looks an' my disposition. It goes deep with some women, an' it went deep with me. Men don't seem to understand some things. Instid of sympathizin' with me, he took to complainin' an' findin' fault an' finally stayin' away from home.

"'There's no use talkin' about what I suffered for a year; I never told anybody this much before—an' it wa'n't anything to what I've suffered ever since. But one day I stumbled on a letter he had wrote to a woman he called Ruth. He talked about her red wavy hair an' blue eyes an' baby mouth an' the way she smiled like an angel. They were goin' to run away together. He told her he'd heard of a place at the end of the earth where a man could make a lot of money, an' he'd go there an' get settled an' then send for her, if she was willin' to live away from everybody, just for him. He said they'd never see a human soul that knew them.'

"She stopped talkin' all at once, an' we walked along. I was scared plumb to death. I didn't know the woman's name, for he always called her 'dearie,' but the baby's name was Ruth.

"'You've got to feelin' bad now,' says I, 'an' maybe we'd best not go on.'

"'I'm goin' on,' says she.

"After a while she says, in a different voice, kind of hard, 'I put that letter back an' never said a word. I wouldn't turn my hand over to keep a man. I never saw the woman; but I know how she looks. I've gone over it every night of my life since. I know the shape of every feature. I never let on, to him or anybody else. It's the only thing I've thanked God for, since I read that letter—helpin' me to keep up an' never let on. It's the only thing I've prayed for since that day. It wa'n't very long—about a month. He just up an' disappeared. People talked about me awful because I didn't cry, an' take on, an' hunt him.

"'I took what little money he left me an' went away. I got the notion that he'd gone to South America, so I set out to get as far in the other direction as possible. I got to San Francisco, an' then the chance fell to me to come up here. It sounded like the North Pole to me, so I come. I'm awful glad I come. Them sea-gulls is the only pleasure I've had—since; an' it's been four year. That's all.'

"Well, sir, when we got up close to the cabin, I got to shiverin' so's I couldn't brace up an' go in with her. It didn't seem possible it *could* be the same man, but then, such darn queer things do happen in Alaska! Anyhow, I'd got cold feet. I remembered that the cannery the man worked in was shut down, so's he'd likely be at home.

"'I'll go back now,' I mumbles, 'an' leave you womenfolks to get acquainted.'

"I fooled along slow, an' when I'd got nearly to the settlement I heard her comin'. I turned an' waited—an' I God! she won't be any ash-whiter when she's in her coffin. She was steppin' in all directions, like a blind woman; her arms hung down stiff at her sides; her fingers were locked around her thumbs as if they'd never loose; an' some nights, even now, I can't sleep for thinkin' how her eyes looked. I guess if you'd gag a dog, so's he couldn't cry, an' then cut him up *slow*, inch by inch, his eyes 'u'd look like her'n did then. At sight of me her face worked, an' I thought she was goin' to cry; but all at once she burst out into the awfullest laughin' you ever heard outside of a lunatic asylum.

"'Lord God Almighty!' she cries out—'where's his mercy at, the Bible talks about? You'd think he might have a little mercy on an ugly woman who never had any children, wouldn't you—especially when there's women in the world with wavy red hair an' blue eyes—women that smile like angels an' have little baby girls! Oh, Lord, what a joke on me!'

"Well, she went on laughin' till my blood turned cold, but she never told me one word of what happened to her. She went back to California on the first boat that went, but it was two weeks. I saw her several times; an' at sight of me she'd burst out into that same laughin' an' cry out, 'My Lord, what a joke! Did you ever see its beat for a joke?' but she wouldn't answer a thing I ast her. The last time I ever see her, she was leanin' over the ship's side. She looked like a dead woman, but when she see me she waved her hand and burst out laughin'.

A Famous Team of Huskies

"'Do you hear them sea-gulls?' she cries out. 'All they can scream is *Kar*-luk! *Kar*-luk! *Kar*-luk! You can hear'm say it just as plain. *Kar*-luk! I'll hear 'em when I lay in my grave! Oh, my Lord, what a joke!'"

CHAPTER XXXIII

Our progress up Karluk River in the barge was so leisurely that we seemed to be "drifting upward with the flood" between the low green shores that sloped, covered with flowers, to the water. The clouds were a soft gray, edged with violet, and the air was very sweet.

The hatchery is picturesquely situated.

A tiny rivulet, called Shasta Creek, comes tumbling noisily down from the hills, and its waters are utilized in the various "ponds."

The first and highest pond they enter is called the "settling" pond, which receives, also, in one corner, the clear, bubbling waters of a spring, whose upflow, never ceasing, prevents this corner of the pond from freezing. This pond is deeper than the others, and receives the waters of the creek so lightly that the sediment is not disturbed in the bottom, its function being to permit the sediment carried down from the creek to settle before the waters pass on into the wooden flume, which carries part of the overflow into the hatching-house, or on into the lower ponds, which are used for "ripening" the salmon.

There are about a dozen of these ponds, and they are terraced down the hill with a fall of from four to six feet between them.

They are rectangular in shape and walled with large stones and cement. The walls are overgrown with grasses and mosses; and the waters pouring musically down over them from large wooden troughs suspended horizontally above them, and whose bottoms are pierced by numerous augur-holes, produce the effect of a series of gentle and lovely waterfalls.

It is essential that the fall of the water should be as light and as soft as possible, that the fish may not be disturbed and excited—ripening more quickly and perfectly when kept quiet.

These ponds were filled with salmon. Many of them moved slowly and placidly through the clear waters; others struggled and fought to leap their barriers in a seemingly passionate and supreme desire to reach the highest spawning-ground. There is to me something divine in the desperate struggle of a salmon to reach the natural place for the propagation of its kind—the shallow, running upper waters of the stream it chooses to ascend. It cannot be will-power—it can be only a God-given instinct—that enables it to leap cascades eight feet in height to accomplish its uncontrollable desire. Notwithstanding all commercial reasoning and all human needs, it seems to me to be inhumanly cruel to corral so many millions of salmon every year, to confine them during the ripening period, and to spawn them by hand.

In the natural method of spawning, the female salmon seeks the upper waters of the stream, and works out a trough in the gravelly bed by vigorous movements of her body as she lies on one side. In this trough her eggs are deposited and are then fertilized by the male.

The eggs are then covered with gravel to a depth of several feet, such gravel heaps being known as "redds."

To one who has studied the marvellously beautiful instincts of this most human of fishes, their desperate struggles in the ripening ponds are pathetic in the extreme; and I was glad to observe that even the gentlemen of our party frequently turned away with faces full of the pity of it.

A salmon will struggle until it is but a purple, shapeless mass; it will fling itself upon the rocks; the over-pouring waters will bear it back for many yards; then it will gradually recover itself and come plunging and fighting back to fling itself once more upon the same rocks. Each time that it is washed away it is weaker, more bruised and discolored. Battered, bleeding, with fins broken off and eyes beaten out, it still returns again and again, leaping and flinging itself frenziedly upon the stone walls.

Its very rush through the water is pathetic, as one remembers it; it is accompanied by a loud swish and the waters fly out in foam; but its movements are so swift that only a line of silver—or, alas! frequently one of purple—is visible through the beaded foam.

Some discoloration takes place naturally when the fish has been in fresh water for some time; but much of it is due to bruising. A salmon newly arrived from the sea is called a "clean" salmon, because of its bright and sparkling appearance and excellent condition.

There is a tramway two or three hundred yards in length, along which one may walk and view the various ponds. It is used chiefly to convey stock-fish from the corrals to the upper ripening-ponds.

When ripe fish are to be taken from a pond, the water is lowered to a depth of about a foot and a half; a kind of slatting is then put into the water at one end and slidden gently under the fish, which are examined—the "ripe" ones being placed in a floating car and the "green" ones freed in the pond. A stripping platform attends every pond, and upon this the spawning takes place.

The young fish, from one to two years old, before it has gone to sea, is called by a dozen different names, chief of which are parr and salmon-fry. At the end of ten weeks after hatching, the fry are fed tinned salmon flesh,—"do-overs" furnished by the canneries,—which is thoroughly desiccated and put through a sausage-machine.

When the fry are three or four months old, they are "planted." After being freed they work their way gradually down to salt-water, which pushes up into the lagoon, and finally out into the bay. They return frequently to fresh water and for at least a year work in and out with the tides.

The majority of fry cling to the fresh-water vicinity for two years after hatching, at which time they are about eight inches long. The second spring after hatching they sprout out suddenly in bright and glistening scales, which conceal the dark markings along their sides which are known as parr-marks. They are then called "smolt," and are as adult salmon in all respects save size.

In all rivers smolts pass down to the sea between March and June, weighing only a few ounces. The same fall they return as "grilse," weighing from three to five pounds.

After their first spawning, they return during the winter to the sea; and in the following year reascend the river as adult salmon. Males mature sexually earlier than females.

The time of year when salmon ascend from the sea varies greatly in different rivers, and salmon rivers are denominated as "early" or "late."

The hatchery at Karluk is a model one, and is highly commended by government experts. It was established in the spring of 1896, and stripping was done in August of the same year. The cost of the present plant has been about forty thousand dollars, and its annual expenditure for maintenance, labor, and improvements, from ten to twenty thousand. There is a superintendent and a permanent force of six or eight men, including a cook, with additional help from the canneries when it is required.

There are many buildings connected with the hatchery, and all are kept in perfect order. The first season, it is estimated that two millions of salmon-fry were liberated, with a gradual increase until the present time, when forty millions are turned out in a single season.

The superintendent was taken completely by surprise by our visit, but received us very hospitably and conducted us through all departments with courteous explanations. The shining, white cleanliness and order everywhere manifest would make a German housewife green of envy.

At this point Karluk River widens into a lagoon, in which the corrals are wired and netted off somewhat after the fashion of fish-traps, covering an area of about three acres.

Fish for the hatcheries are called "stock-fish." They are secured by seiners in the lagoon opposite the hatcheries, and are then transferred to the corrals. As soon as a salmon has the appearance of ripening, it is removed by the use of seines to the ripening-ponds.

In the hatching-house are more than sixty troughs, fourteen feet in length, sixteen inches in width, and seven inches in depth. The wood of which they are composed is surfaced redwood. The joints are coated with asphaltum tar, with cotton wadding used as calking material. When the trough is completed, it is given one coat of refined tar and two of asphaltum varnish.

In the Karluk hatchery the troughs never leak, owing to this superior construction; and it is said that the importance of this advantage cannot be overestimated.

Leaks make it impossible for the employees to estimate the amount of water in the troughs; repairs startle the young fry and damage the eggs; and the damp floors cause illness among the employees. The Karluk hatchery is noted for its dryness and cleanliness.

The setting of the hatchery is charming. The hills, treeless, pale green, and velvety, slope gently to the river and the lagoon. Now and then a slight ravine is filled with a shrubby growth of a lighter green. Flowers flame everywhere, and tiny rivulets come singing down to the larger stream.

The greenness of the hills continues around the bay, broken off abruptly on Karluk Head, where the soft, veined gray of the stone cliff blends with the green.

The bay opens out into the wide, bold, purple sweep of Shelikoff Strait.

Every body of water has its character—some feature that is peculiarly its own, which impresses itself upon the beholder. The chief characteristic of Shelikoff Strait is its boldness. There is something dauntless, daring, and impassioned in its wide and splendid sweep to the chaste line of snow peaks of the Aleutian Range on the Aliaska Peninsula. It seems to hold a challenge.

I should like to live alone, or almost alone, high on storm-swept Karluk Head, fronting that magnificent scene that can never be twice quite the same. What work one might do there—away from little irritating cares! No neighbors to "drop in" with bits of delicious gossip; no theatres in which to waste the splendid nights; no bridge-luncheons to tempt,—nothing but sunlight glittering down on the pale green hills; the golden atmosphere above the little bay filled with tremulous, winged snow; and miles and miles and miles of purple sea.

CHAPTER XXXIV

"What kind of place is Uyak?" I asked a deck-hand who was a native of Sweden, as we stood out in the bow of the *Dora* one day.

He turned and looked at me and grinned.

"It ees a hal of a blace," he replied, promptly and frankly. "It ees yoost dat t'ing. You vill see."

And I did see. I should, in fact, like to take this frank-spoken gentleman along with me wherever I go, solely to answer people who ask me what kind of place Uyak is—his opinion so perfectly coincides with my own.

There were canneries at Uyak, and mosquitoes, and things to be smelled; but if there be anything there worth seeing, they must first kill the mosquitoes, else it will never be seen.

The air was black with these pests, and the instant we stepped upon the wharf we were black with them, too. Every passenger resembled a windmill in action, as he raced down the wharf toward the cannery, hoping to find relief there; and as he went his nostrils were assailed by an odor that is surpassed in only one place on earth—*Belkoffski!*—and it comes later.

The hope of relief in the canneries proved to be a vain one. The unfortunate Chinamen and natives were covered with mosquitoes as they worked; their faces and arms were swollen; their eyes were fierce with suffering. They did not laugh at our frantic attempts to rid ourselves of the winged pests—as we laughed at one another. There was nothing funny in the situation to those poor wretches. It was a tragedy. They stared at us with desperate eyes which asked:—

"Why don't you go away if you are suffering? You are free to leave. What have you to complain of? *We* must stay."

We went out and tried to walk a little way along the hill; but the mosquitoes mounted in clouds from the wild-rose thickets.

At the end of fifteen minutes we fled back to the steamer and locked ourselves in our staterooms. There we sat down and nursed our grievances with camphor and alcohol.

We sailed up Uyak Bay to the mine of the Kodiak Gold Mining Company. This is a free milling mine and had been a developing property for four years. It was then installing a ten-stamp mill, and had twenty thousand tons of ore blocked out, the ore averaging from fifteen to twenty dollars a ton.

This mine is located on the northern side of Kadiak Island, and has good water power and excellent shipping facilities. Fifty thousand dollars were taken out of the beaches in the vicinity in 1904 by placer mining.

Here, in this lovely, lonely bay, one of the most charming women I ever met spends her summers. She is the wife of one of the owners of the mine, and her home is in San Francisco. She finds the summers ideal, and longs for the novelty of a winter at the mine. She has a canoe and spends most of her time on the water. There are no mosquitoes at the mine; the summers are never uncomfortably hot, and it is seldom, indeed, that the mercury falls to zero in the winter.

From Kadiak Island we crossed Shelikoff Straits to Cold Bay, on the Aliaska Peninsula, which we reached at midnight, and which is the only port that could not tempt us ashore. When our dear, dark-eyed Japanese, "Charlie," played a gentle air upon our cabin door with his fingers and murmured apologetically, "Cold Bay," we heard the rain pouring down our windows in sheets, and we ungratefully replied, "Go away, Charlie, and leave us alone."

No rope-ladders and dory landings for us on such a night, at a place with such a name.

The following day was clear, however, and we sailed all day along the peninsula. To the south of us lay the Tugidak, Trinity, Chirikoff, and Semidi islands.

At six in the evening we landed at Chignik, another uninteresting cannery place. From Chignik on "to Westward" the resemblance of the natives to the Japanese became more remarkable. As they stood side by side on the wharves, it was almost impossible to distinguish one from the other. The slight figures, brown skin, softly bright, dark eyes, narrowing at the corners, and amiable expression made the resemblance almost startling.

At Chignik we had an amusing illustration, however, of the ease with which even a white man may grow to resemble a native.

The mail agent on the *Dora* was a great admirer of his knowledge of natives and native customs and language. *Cham-mi* is a favorite salutation with them. Approaching a man who was sitting on a barrel, and who certainly resembled a native in color and dress, the agent pleasantly exclaimed, "*Cham-mi*."

There was no response; the man did not lift his head; a slouch hat partially concealed his face.

"*Cham-mi!*" repeated the agent, advancing a step nearer.

There was still no response, no movement of recognition.

The mail agent grew red.

"He must be deaf as a post," said he. He slapped the man on the shoulder and, stooping, fairly shouted in his ear, "*Cham-mi*, old man!"

Then the man lifted his head and brought to view the unmistakable features of a Norwegian.

"T'hal with you," said he, briefly. "I'm no tamn Eskimo."

The mail agent looked as though the wharf had gone out from under his feet; and never again did we hear him give the native salutation to any one. The Norwegian had been living for a year among the natives; and by the twinkle in his eye as he again lowered his head it was apparent that he appreciated the joke.

At the entrance to Chignik Bay stands Castle Cape, or Tuliiumnit Point. From the southeastern side it really resembles a castle, with turrets, towers, and domes. It is an immense, stony pile jutting boldly out into the sea, whose sparkling blue waves, pearled with foam, break loudly upon its base. In color it is soft gray, richly and evenly streaked with rose. Sea birds circled, screaming, over it and around it. Castle Cape might be the twin sister of "Calico Bluff" on the Yukon.

Popoff and Unga are the principal islands of the Shumagin group, on one of which Behring landed and buried a sailor named Shumagin. They are the centre of famous cod-fishing grounds which extend westward and northward to the Arctic Ocean, eastward to Cook Inlet, and southeastward to the Straits of Juan de Fuca.

There are several settlements on the Island of Unga—Coal Harbor, Sandy Point, Apollo, and Unga. The latter is a pretty village situated on a curving agate beach. It is of some importance as a trading post.

Finding no one to admit us to the Russo-Greek church, we admitted ourselves easily with our stateroom key; but the tawdry cheapness of the interior scarcely repaid us for the visit. The graveyard surrounding the church was more interesting.

There is no wharf at Unga, but there is one at Apollo, about three miles farther up the bay. We were taken up to Apollo in a sail-boat, and it proved to be an exciting sail. It is not sailing unless the rail is awash; but it seemed as though the entire boat were awash that June afternoon in the Bay of Unga. Scarcely had we left the ship when we were struck by a succession of squalls which lasted until our boat reeled, hissing, up to the wharf at Apollo.

Water poured over us in sheets, drenching us. We could not stay on the seats, as the bottom of the boat stood up in the air almost perpendicularly. We therefore stood up with it, our feet on the lower rail with the sea flowing over them, and our shoulders pressed against the gunwale. Had it not been for the broad shoulders of two Englishmen, our boat would surely have gone over.

It all came upon us so suddenly that we had no time to be frightened, and, with all the danger, it was glorious. No whale—no "right" whale, even—could be prouder than we were of the wild splashing and spouting that attended our tipsy race up Unga Bay.

The wharf floated dizzily above us, and we were compelled to climb a high perpendicular ladder to reach it. No woman who minds climbing should go to Alaska. She is called upon at a moment's notice to climb everything, from rope-ladders and perpendicular ladders to volcanoes. A mile's walk up a tramway brought us to the Apollo.

This is a well-known mine, which has been what is called a "paying proposition" for many years. At the time of our visit it was worked out in its main lode, and the owners had been seeking desperately for a new one. It was discovered the following year, and the Apollo is once more a rich producer.

In a large and commodious house two of the owners of the mine lived, their wives being with them for the summer. They were gay and charming women, fond of society, and pining for the fleshpots of San Francisco. The white women living between Kodiak and Dutch Harbor are so few that they may be counted on one hand, and the luxurious furnishings of their homes in these out-of-the-way places are almost startling in their unexpectedness. We spent the afternoon at the mine, and the ladies returned to the *Dora*with us for dinner. The squalls had taken themselves off, and we had a prosaic return in the mine's launch.

"What do we do?" said one of the ladies, in reply to my question. "Oh, we read, walk, write letters, go out on the water, play cards, sew, and do so much fancy work that when we get back to San Francisco we have nothing to do but enjoy ourselves and brag about the good time we have in Alaska. We are all packed now to go camping—"

"*Camping!*" I repeated, too astonished to be polite.

"Yes, camping," replied she, coloring, and speaking somewhat coldly. "We go in the launch to the most beautiful beach about ten miles from Unga. We stay a month. It is a sheltered beach of white sand. The waves lap on it all day long, blue, sparkling, and warm, and we almost live in them. The hills above the beach are simply covered with the big blueberries that grow only in Alaska.

They are somewhat like the black mountain huckleberry, only more delicious. We can them, preserve them, and dry them, and take them back to San Francisco with us. They are the best things I ever ate—with thick cream on them. I had some in the house; I wish I had thought to offer you some."

She wished she had thought to offer me some!

On the *Dora* we were rapidly getting down to bacon and fish,—being about two thousand miles from Seattle, with no ice aboard in this land of ice,—and I am not enthusiastic about either.

And she wished that she had thought to offer me some Alaskan blueberries that are more delicious than mountain huckleberries, and thick cream!

CHAPTER XXXV

I have heard of steamers that have been built and sent out by missionary or church societies to do good in far and lonely places.

The little *Dora* is not one of these, nor is religion her cargo; her hold is filled with other things. Yet blessings be on her for the good she does! Her mission is to carry mail, food, freight, and good cheer to the people of these green islands that go drifting out to Siberia, one by one. She is the one link that connects them with the great world outside; through her they obtain their sole touch of society, of which their appreciation is pitiful.

Our captain was a big, violet-eyed Norwegian, about forty years old. He showed a kindness, a courtesy, and a patience to those lonely people that endeared him to us.

He knew them all by name and greeted them cordially as they stood, smiling and eager, on the wharves. All kinds of commissions had been intrusted to him on his last monthly trip. To one he brought a hat; to another a phonograph; to another a box of fruit; dogs, cats, chairs, flowers, books—there seemed to be nothing that he had not personally selected for the people at the various ports. Even a little seven-year-old half-breed girl had travelled in his care from Valdez to join her father on one of the islands.

Wherever there was a woman, native or half-breed, he took us ashore to make her acquaintance.

"Come along now," he would say, in a tone of command, "and be nice. They don't get a chance to talk to many women. Haven't you got some little womanly thing along with you that you can give them? It'll make them happy for months."

We were eager enough to talk to them, heaven knows, and to give them what we could; but the "little womanly things" that we could spare on a two months' voyage in Alaska were distressingly few. When we had nothing more that we could give, the stern disapproval in the captain's eyes went to our hearts. Box after box of bonbons, figs, salted almonds, preserved ginger, oranges, apples, ribbons, belts, pretty bags—one after one they went, until, like Olive Schreiner's woman, I felt that I had given up everything save the one green leaf in my bosom; and that the time would come when the captain would command me to give that up, too.

There seems to be something in those great lonely spaces that moves the people to kindness, to patience and consideration—to tenderness, even. I never before came close to such *humanness*. It shone out of people in whom one would least expect to find it.

Several times while we were at dinner the chief steward, a gay and handsome youth not more than twenty-one years old, rushed through the dining room, crying:—

"Give me your old magazines—*quick*! There's a whaler's boat alongside."

A stampede to our cabins would follow, and a hasty upgathering of such literature as we could lay our hands upon.

The whaling and cod-fishing schooners cruise these waters for months without a word from the outside until they come close enough to a steamer to send out a boat. The crew of the steamer, discovering the approach of this boat, gather up everything they can throw into it as it flashes for a moment alongside. Frequently the occupants of the boat throw fresh cod aboard, and then there are smiling faces at dinner. It is my opinion, however, that any one who would smile at cod would smile at anything.

The most marvellous voyage ever made in the beautiful and not always peaceful Pacific Ocean was the one upon which the *Dora* started at an instant's notice, and by no will of her master's, on the first day of January, 1906. Blown from the coast down into the Pacific in a freezing storm, she became disabled and drifted helplessly for more than two months.

During that time the weather was the worst ever known by

seafaring men on the coast. The steamship *Santa Ana* and the United States steamship *Rush* were sent in search of the *Dora*, and when both had returned without tidings, hope for her safety was abandoned.

Eighty-one days from the time she had sailed from Valdez, she crawled into the harbor of Seattle, two thousand miles off her course. She carried a crew of seven men and three or four passengers, one of whom was a young Aleutian lad of Unalaska. As the *Dora* was on her outward trip when blown to sea, she was well stocked with provisions which she was carrying to the islanders; but there was no fuel and but a scant supply of water aboard.

The physical and mental sufferings of all were ferocious; and it was but a feeble cheer that arose from the little shipwrecked band when the *Dora* at last crept up beside the Seattle pier. For two months they had expected each day to be their last, and their joy was now too deep for expression.

The welcome they received when they returned to their regular run among the Aleutian Islands is still described by the settlers.

Cloud Effect on the Yukon

The *Dora* reached Kodiak late on a boisterous night; but her whistle was heard, and the whole town was on the wharf when she docked, to welcome the crew and to congratulate them on their safety. Some greeted their old friends hilariously, and others simply pressed their hands in emotion too deep for expression.

So completely are the people of the smaller places on the route cut off from the world, save for the monthly visits of the *Dora*, that they had not heard of her safety. When, after supposing her to be lost for two months, they beheld her steaming into their harbors, the superstitious believed her to be a spectre-ship.

The greatest demonstration was at Unalaska. A schooner had brought the news of her safety to Dutch Harbor; from there a messenger was despatched to Unalaska, two miles away, to carry the glad tidings to the father of the little lad aboard the *Dora*.

The news flashed wildly through the town. People in bed, or sitting by their firesides, were startled by the flinging open of their door and the shouting of a voice from the darkness outside:—

"The *Dora's* safe!"—but before they could reach the door, messenger and voice would be gone—fleeing on through the town.

At last he reached the Jessie Lee Missionary Home, at the end of the street, where a prayer-meeting was in progress. Undaunted, he flung wide the door, burst into the room, shouting, "The *Dora's* safe!"—and was gone. Instantly the meeting broke up, people sprang to their feet, and prayer gave place to a glad thanksgiving service.

When the *Dora* finally reached Unalaska once more, the whole town was in holiday garb. Flags were flying, and every one that could walk was on the wharf. Children, native and white, carried flags which they joyfully waved. Their welcome was enthusiastic and sincere, and the men on the boat were deeply affected.

The *Dora* is not a fine steamship, but she is stanch, seaworthy, and comfortable; and the islanders are as attached to her as though she were a thing of flesh and blood.

No steamer could have a twelve-hundred-mile route more fascinating than the one from Valdez to Unalaska. It is intensely lovely. Behind the gray cliffs of the peninsula float the snow-peaks of the Aleutian Range. Here and there a volcano winds its own dark, fleecy turban round its crest, or flings out a scarlet scarf of flame. There are glaciers sweeping everything before them; bold headlands plunging out into the sea, where they pause with a sheer drop of thousands of feet; and flowery vales and dells. There are countless islands—some of them mere bits of green floating upon the blue.

At times a kind of divine blueness seems to swim over everything. Wherever one turns, the eye is rested and charmed with blue. Sea, shore, islands, atmosphere, and sky—all are blue. A mist of it rests upon the snow mountains and goes drifting down the straits. It is a warm, delicate, luscious blue. It is like the blue of frost-touched grapes when the prisoned wine shines through.

Sand Point, a trading post on Unga Island, is a wild and picturesque place. It impressed me chiefly, however, by the enormous size of its crabs and starfishes, which I saw in great numbers under the wharf. Rocks, timbers, and boards were incrusted with rosy-purple starfishes, some measuring three feet from the tip of one ray to the tip of the ray nearly opposite. Smaller ones were wedged in between the rays of the larger ones, so that frequently a piling from the wharf to the sandy bottom of the bay, which we could plainly see, would seem to be solid starfish.

As for the crabs—they were so large that they were positively startling. They were three and four feet from tip to tip; yet their movements, as they floated in the clear green water, were exceedingly graceful.

Sand Point has a wild, weird, and lonely look. It is just the place for the desperate murder that was committed in the house that stands alone across the bay,—a dull and neglected house with open windows and banging doors.

"Does no one live there?" I asked the storekeeper's wife.

"Live there!" she repeated with a quick shudder. "No one could be hired at any price to live there."

The murdered man had purchased a young Aleutian girl, twelve years old, for ten dollars and some tobacco. When she grew older, he lived with her and called her his wife. He abused her shamefully. A Russian half-breed named Gerassenoff—the name fits the story—fell in love with the girl, loved her to desperation, and tried to persuade her to run away with him.

She dared not, for fear of the brutal white wretch who owned her, body and soul. Gerassenoff, seeing the cruelties and abuse to which she was daily subjected, brooded upon his troubles until he became partially insane. He entered the house when the man was asleep and murdered him—foully, horribly, cold-bloodedly.

Gerassenoff is now serving a life-sentence in the government penitentiary on McNeil's Island; the man he murdered lies in an unmarked grave; the girl—for the story has its touch of awful humor!—the girl married another man within a twelvemonth.

There is a persistent invitation at Sand Point to the swimmer. The temptation to sink down, down, through those translucent depths, and then to rise and float lazily with the jelly-fishes, is almost irresistible. There is a seductive, languorous charm in the slow curve of the waves, as though they reached soft arms and wet lips to caress. There are more beautiful waters along the Alaskan coast, but none in which the very spirit of the swimmer seems so surely to dwell.

CHAPTER XXXVI

Belkoffski! There was something in the name that attracted my attention the first time I heard it; and my interest increased with each mile that brought it nearer. It is situated on the green and sloping shores of Pavloff Bay, which rise gradually to hills of considerable height. Behind it smokes the active volcano, Mount Pavloff, with whose ashes the hills are in places gray, and whose fires frequently light the night with scarlet beauty.

The *Dora* anchored more than a mile from shore, and when the boat was lowered we joyfully made ready to descend. We were surprised that no one would go ashore with us. Important duties claimed the attention of officers and passengers; yet they seemed interested in our preparations.

"Won't you come ashore with us?" we asked.

"No, I thank you," they all replied, as one.

"Have you ever been ashore here?"

"Oh, yes, thank you."

"Isn't it interesting, then?"

"Oh, very interesting, indeed."

"There is something in their manner that I do not like," I whispered to my companion. "What do you suppose is the matter with Belkoffski."

"Smallpox, perhaps," she whispered back.

"I don't care; I'm going."

"So am I."

"What kind of place is Belkoffski?" I asked one of the sailors who rowed us ashore.

He grinned until it seemed that he would never again be able to get his mouth shut.

"Jou vill see vot kind oof a blace it ees," he replied luminously.

"Is it not a nice place, then?"

"Jou vill see."

We did see.

The tide was so low and the shore so rocky that we could not get within a hundred yards of any land. A sailor named "Nelse" volunteered to carry us on his back; and as nothing better presented itself for our consideration, we promptly and joyfully went pick-a-back.

This was my most painful experience in Alaska. My father used to make stirrups of his hands; but as Nelse did not offer, diffidence kept me from requesting this added gallantry of him. It was well that I went first; for after viewing my friend's progress shoreward, had I not already been upon the beach, I should never have landed at Belkoffski.

For many years Belkoffski was the centre of the sea-otter trade. This small animal, which has the most valuable fur in the world, was found only along the rock shores of the Aliaska Peninsula and the Aleutian Islands. The Shumagins and Sannak islands were the richest grounds. Sea-otter, furnishing the court fur of both Russia and China, were in such demand that they have been almost entirely exterminated—as the fur-bearing seal will soon be.

The fur of the sea-otter is extremely beautiful. It is thick and velvety, its rich brown under-fur being remarkable. The general color is a frosted, or silvery, purplish brown.

The sea-otter frequented the stormiest and most dangerous shores, where they were found lying on the rocks, or sometimes floating, asleep, upon fronds of an immense kelp which was called "sea-otter's cabbage." The hunters would patiently lie in hiding for days, awaiting a favorable opportunity to surround their game.

They were killed at first by ivory spears, which were deftly cast by natives. In later years they were captured in nets, clubbed brutally, or shot. They were excessively shy, and the difficulty and danger of securing them increased as their slaughter became more pitiless. Only natives were allowed to kill otter until 1878, when white men married to native women were permitted by the Secretary of the Treasury to consider themselves, and to be considered, natives, so far as hunting privileges were concerned.

The rarest and most valuable of otter are the deep-sea otter, which never go ashore, as do the "rock-hobbers," unless driven there by unusual storms. "Silver-tips"—deep-sea otter having a silvery tinge on the tips of the fur—bring the most fabulous prices.

The hunting of these scarce and precious animals calls for greater bravery, hardship, perilous hazard, and actual suffering than does the chase of any other fur-bearing animal. Pitiful, shameful, and loathsome though the slaughter of seals be, it is not attended by the exposure and the hourly peril which the otter hunter unflinchingly faces.

Sea-otter swim and sleep upon their backs, with their paws held over their eyes, like sleepy puppies, their bodies barely visible and their hind flippers sticking up out of the water.

The young are born sometimes at sea, but usually on kelp-beds; and the mother swims, sleeps, and even suckles her young stretched at full length in the water upon her back. She carries her offspring upon her breast, held in her forearms, and has many humanly maternal ways with it,—fondling it, tossing it into the air and catching it, and even lulling it to sleep with a kind of purring lullaby.

Both the male and female are fond of their young, caring for it with every appearance of tenderness. In making difficult landings, the male "hauls out" first and catches the young, which the mother tosses to him. Sometimes, when a baby is left alone for a few minutes, it is attacked by some water enemy and killed or turned over, when it invariably drowns. The mother, returning and finding it floating, dead, takes it in her arms and makes every attempt possible to bring it to life. Failing, she utters a wild cry of almost human grief and slides down into the sea, leaving it.

The otter hunters used to go out to sea in their bidarkas, with bows, arrows, and harpoons; several would go together, keeping two or three hundred yards apart and proceeding noiselessly. When one discovered an otter, he would hold his paddle straight up in the air, uttering a loud shout. Then all would paddle cautiously about, keeping a close watch for the otter, which cannot remain under water longer than fifteen or twenty minutes. When it came up, the native nearest its breathing place yelled and held up his paddle, startling it under the water again so suddenly that it could not draw a fair breath. In this manner they forced the poor thing to dive again and again, until it was exhausted and floated helplessly upon the water, when it was easily killed. Frequently two or three hours were required to tire an otter.

This picturesque method of hunting has given place to shooting and clubbing the otter to death as he lies asleep on the rocks. As they come ashore during the fiercest weather, the hunter must brave the most violent storms and perilous surfs to reach the otter's retreat in his frail, but beautiful, bidarka. With his gut kamelinka—thin and yellow as the "gold-beater's leaf"—tied tightly around his face, wrists, and the "man-hole" in which he sits or kneels, his bidarka may turn over and over in the sea without drowning him or shipping a drop of water—on his lucky days. But the unlucky day comes; an accident occurs; and a dark-eyed woman watches and waits on the green slopes of Belkoffski for the bidarka that does not come.

There were only women and children in the village of Belkoffski that June day. The men—with the exception of two or three old ones, who are always left, probably as male chaperons, at the village—were away, hunting.

The beach was alive, and very noisy, with little brown lads, half-bare, bright-eyed, and with faces that revealed much intelligence, kindness, and humor.

They clung to us, begging for pennies, which, to our very real regret, we had not thought to take with us. Candy did not go far, and dimes, even if we had been provided with them, would have too rapidly run into dollars.

Long-stemmed violets and dozens of other varieties of wild flowers covered the slopes. One little creek flowed down to the sea between banks that were of the solid blue of violets.

But the village itself! With one of the prettiest natural locations in Alaska; with singing rills and flowery slopes and a volcano burning splendidly behind it; with little clean-looking brown lads playing upon its sands, a Greek-Russian church in its centre, and a resident priest who ought to know that cleanliness is next to godliness—with all these blessings, if blessings they all be, Belkoffski is surely the most unclean place on this fair earth.

The filth, ignorance, and apparent degradation of these villagers were revolting in the extreme. Nauseous odors assailed us. They came out of the doors and windows; they swam out of barns and empty sheds; they oozed up out of the earth; they seemed, even, to sink upon us out of the blue sky. The sweetness and the freshness of green grass and blowing flowers, of dews and mists, of mountain and sea scented winds, are not sufficient to cleanse Belkoffski—the Caliban among towns.

An educated half-breed Aleutian woman, married to a white man, accompanied us ashore. She was on her way to Unalaska, and had been eager to land at Belkoffski, where she was born.

Her father had been a priest of the Greek-Russian church and her mother a native woman. She had told us much of the kind-heartedness and generosity of the villagers. Her heart was full of love and gratitude to them for their tenderness to her when her father, of blessed memory, had died.

"I have never had such friends since," she said. "They would do anything on earth for those in trouble, and give their own daily food, if necessary. I have never seen anything like it since. Education doesn't put *that* into our hearts. Such sympathy, such tenderness, such understanding of grief and trouble!—and the kind of help that helps most."

If this be the real nature of these people, only the right influence is needed to lift them from their degradation. The larger children—the brown-limbed, joyous children down on the beach—looked clean, probably from spending much time in the healing sea.

The people of the islands do not travel much, and our fellow-voyager had not been to Belkoffski since she was a little girl. For many years she had been living among white people, with all the comforts and cleanliness of a white woman. I watched her narrowly as we went from house to house, looking for baskets.

We had told her we desired baskets, and she had offered to find some for us. After we saw the houses and the women, we would have touched a leper as readily as we would have touched one of the baskets that were brought out for our inspection; but politeness kept us from admitting to her our feeling.

As for her own courtesy and restraint, I have never seen them surpassed by any one. Shock upon shock must have been hers as we passed through that village of her childhood and affection. She went into those noisome hovels without the faintest hesitation; she breathed their atmosphere without complaint; she embraced the women without shrinking.

She knew perfectly why we did not buy the baskets; but she received our excuses with every appearance of believing them to be sincere, and she offered us others with utmost dignity and with the manner of serving us, strangers, in a strange land.

If her delicacy was outraged by the scenes she witnessed, there was not the faintest trace of it visible in her manner. She made no excuses for the people, nor for their manner of living, nor for the village. Belkoffski had been her childhood's home, her father's field; its people had befriended her and had given her love and tenderness when she was in need; therefore, both were sacred and beyond criticism.

When we returned to the ship, she could not have failed to hear the jests and frank opinions of Belkoffski which were freely expressed among the passengers; but her grave, dark face gave no sign that she disapproved, or even that she heard.

A government cutter should be sent to Belkoffski with orders to clean it up, and to burn such portions as are past cleansing.

So far as the Russian priest and the people in his charge are concerned, they would be benefited by less religion and more cleanliness.

Dr. Hutton, an army surgeon stationed at Fort Seward on Lynn Canal, and Judge Gunnison, of Juneau, have recently made an appeal to President Roosevelt for relief for diseased and suffering Indians of Alaska.

Tuberculosis and trachoma prevail among the many tribes and are increasing at an alarming rate, owing to the utter lack of sanitation in the villages. Alaskans travelling in the territory are thrown in constant contact with the Indians. They are encountered on steamers and trains, in stores and hotels. Owing to the pure air and the general healthfulness of the northern climate, Alaskans feel no real alarm over the conditions prevailing as yet; but all feel that the time has arrived when the Indians should be cared for.

Everything purchased of an Indian should be at once fumigated—especially furs, blankets, baskets, and every article that has been handled by him or housed in one of his vile shacks.

The United States Grand Jury recently recommended that medical men be sent by the government to attend the disease-stricken creatures, and that a system of inspection and education along sanitary lines—with special stress laid upon domestic sanitation—should be established.

This system should be extended to the last island of the Aleutian Chain, and in the interior down the Yukon to Nome. The fur trade and the canneries depend largely upon the labor of Indians. The former industry could scarcely be made successful without them. The Indians are rapidly becoming a "vanishing race" in the North, as elsewhere. For the vices that are to-day responsible for their unfortunate condition they are indebted to the white men who have kept them supplied with cheap whiskey ever since the advent of the first American traders who taught them, soon after the purchase of Alaska by the United States, to make "hootchenoo" of molasses, flour, dried apples, or rice, and hops. This highly intoxicating and degrading liquor was known also as molasses-rum. During the latter part of the seventies, six thousand five hundred and twenty-four gallons of molasses were delivered at Sitka and Wrangell.

The loss of their help, however, is not so serious—being merely a commercial loss—as the danger to civilized people by coming in contact with these dreaded diseases. An Indian in Alaska whose eyes are not diseased is an exception, while the ravages of consumption are very frequently visible to the most careless observer. Both diseases are aggravated by such conditions as those existing at Belkoffski. A physician should be stationed there for a few years at least, to teach these poor, kind-hearted people what the Russian priest has not taught them—the science of sanitation.

Bishop Rowe reports that if there were no missionaries to protect the Eskimo and Indians from unscrupulous white whiskey-traders, they would survive but a short time. When they can obtain cheap liquors they go on prolonged and licentious debauches, and are unable to provide for their actual physical needs for the long, hard winter. Their condition then becomes pitiable, and many die of hunger and privation. Prosecutions are made entirely by missionaries. One Episcopal missionary post is conducted by two young women, one of whom was formerly a society woman of Los Angeles. The post is more than a thousand miles from Fairbanks, the nearest city, and one hundred and twenty-five miles from the nearest white settler. It is owing to the reports and the prosecutions of missionaries in all parts of Alaska that the outrages formerly practised upon Eskimo women by licentious white traders are on the decrease.

Federal Commissioner of Education Brown advocates a compulsory school law for Alaska. He favors instruction in modern methods of fishing and of curing fish; in the care of all parts of walrus that are merchantable; in the handling of wooden boats, the tanning and preparing of skins, in coal mining and the elements of agriculture.

In 1907 fifty-two native schools were maintained in Alaska, with two thousand five hundred children enrolled. Ten new school buildings have recently been constructed.

The reindeer service has been one of Alaska's grave scandals, but it has greatly improved during the past year.

The Eskimo, or Innuit, inhabit a broad belt of the coast line bordering on Behring Sea and the Arctic Ocean, as well as along the coast "to Westward" from Yakutat; also the lower part of the Yukon.

Lieutenant Emmons, who is one of the highest authorities on the natives of Alaska and their customs, has frequently reported the deplorable condition of the Eskimo, and the prevalence of tuberculosis and other dread diseases among them.

In 1900 an epidemic of measles and *la grippe* devastated the Northwestern Coast. Out of a total population of three thousand natives about the mouth of the Kuskokwim, fully half died, without medical attendance or nursing, within a few months.

The hospitality and generous kindness of the Eskimo to those in need is proverbial. Ever since their subjection by the early Russians—to whom, also, they would doubtless have shown kindness had they not been afraid of them—no shipwrecked mariner has sought their huts in vain. Often the entire crew of an abandoned vessel has been succored, clothed, and kept from starvation during a whole winter—the season when provisions are scarce and the Eskimo themselves scarcely know how to find the means of existence.

Along the islands, the rivers, and lakes, nature has provided them with food and clothing, if they were but educated to make the most of these blessings.

But the vast country bordering the coast between the Kuskokwim and the Yukon, and extending inland a hundred and fifty miles, is low and swampy. This is the dreariest portion of Alaska. Tundra, swamps, and sluggish rivers abound. There is no game, and the natives live on fish and seal. The winters are severe, the climate is cold and excessively moist. Food has often failed, and the old or helpless are called upon to go alone out upon the storm-swept tundra and yield their hard lives—bitter and cheerless at the best—that the young and strong may live. As late as 1901 Lieutenant Emmons reports that this system of unselfish and heart-breaking suicide was practised; and it is probably still in vogue in isolated places when occasion demands.

This district is so poor and unprofitable that the prospector and the trader have so far passed it by; yet, by some means, the white man's worst diseases have been carried in to them.

These people are in dire need of schools, hospitals, medical treatment, and often simple food and clothing.

Farther north, on Seward Peninsula and along the lower Yukon, the natives who have mingled with the miners and traders could easily be taught to be not only self-supporting but of real value to the communities in which they live. They are intelligent, docile, easily directed, and eager to learn. Lieutenant Emmons found that everywhere they asked for schools, that their children, to whom they are most affectionately devoted, may learn to be "smart like the white man."

They are more humble, dependent, and trustful than the Indians, and could easily be influenced. But people do not go to Alaska to educate and care for diseased and loathsome natives, unless they are paid well for the mission. So long as the natives obey the laws of the country, no one has authority over them. No one is interested in them, or has the time to spare in teaching them. The United States government should take care of these people. It should take measures to protect them from the death-dealing whiskey with which they are supplied; to provide them with schools, hospitals, medical care; it should supply them with reindeer and teach them to care for these animals.

Surely the government of the United States asks not to be informed more than once by such authorities as Lieutenant Emmons, Bishop Rowe, Judge Gunnison, Ex-Governor Brady, and Doctor Hutton that these most wretched beings on the outskirts of the world are begging for education, and that they are sorely in need of medical services.

The government schools in the territory of Alaska are supported by a portion of the license moneys levied on the various industries of the country. Alaska has an area of six hundred thousand square miles and an estimated native and half-breed population of twenty-five thousand; and for these people only fifty-two schools and as many poorly paid teachers!

When I have criticised the Russian Church because it has not taught these people cleanliness, I blush—remembering how my own government has failed them in needs as vital. And when I reflect upon the outrages perpetrated upon them by my own fellow-countrymen—who have deprived them largely of their means of livelihood, robbed them, debauched them, ravished their women, and lured away their young girls—when I reflect upon these things, my face burns with shame that I should ever criticise any other people or any other government than my own.

The recent rapid development of Alaska, and the appropriation of the native food-supplies by miners, traders, canners, and settlers, present a problem that must be solved at once. In regard to the Philippines, we were like a child with a new toy; we could not play with them and experiment with them enough; yet for forty years these dark, gentle, uncomplaining people of our most northern and most splendid possession—beautiful, glorious Alaska—have been patiently waiting for all that we should long ago have given them: protection, interest, and the education and training that would have converted them from diseased and wretched beings into decent and useful people.

According to Lieutenant Emmons, the condition of the Copper River Indians is exceptionally miserable; and of all the native people, either coastal or of the interior, they are most needy and in want of immediate assistance. Reduced in number to barely two hundred and fifty souls, scattered in small communities along the river valleys amidst the loftiest mountains of the continent and under the most rigid climatic conditions, their natural living has been taken from them by the white man, without the establishment of any labor market for their self-support in return.

Prior to 1888 they lived in a very primitive state, and were, even then, barely able to maintain themselves on the not over-abundant game life of the valley, together with the salmon coming up the river for spawning purposes. The mining excitement of that year brought several thousand men into the Copper River Valley, on their way to the Yukon and the Klondike.

They swept the country clean of game, burnt over vast districts, and frequently destroyed what they could not use. About the same time the salmon canneries in Prince William Sound, having exhausted the home streams, extended their operations to the Copper River delta, decreasing the Indians' salmon catch, which had always provided them with food for the bitter winters.

"Wolf"

These Indians are simple, kind-hearted, and have ever been friendly and hospitable to the white man. They respect his cache, although their own has not always been respected by him.

At Copper Centre, which is connected by military wagon road with the coast at Valdez, flour sells for twenty-four dollars a hundredweight, and all other provisions and clothing in proportion; so it may be readily understood that the white people of the interior cannot afford to divide their provisions with the starving Indians, else they would soon be in the same condition themselves. Therefore, for these Indians, too,—fortunately few in number,—the government must provide liberally and at once.

CHAPTER XXXVII

At sunset on the day of our landing at Belkoffski we passed the active volcanoes of Pogromni and Shishaldin, on the island of Unimak. For years I had longed to see Shishaldin; and one of my nightly prayers during the voyage had been for a clear and beautiful light in which to see it. Not to pass it in the night, nor in the rain, nor in the fog; not to be too ill to get on deck in some fashion—this had been my prayer.

For days I had trembled at the thought of missing Shishaldin. To long for a thing for years; to think of it by day and to dream of it by night, as though it were a sweetheart; to draw near to it once, and once only in a lifetime—and then, to pass it without one glimpse of its coveted loveliness!—that would be too bitter a fate to be endured.

In a few earnest words, soon after leaving Valdez, I had acquainted the captain with my desire.

It was his watch when I told him. He was pacing in front of the pilot-house. A cigar was set immovably between his lips. He heard me to the end and then, without looking at me, smiled out into the golden distance ahead of us.

"You fix the weather," said he, "and I'll fix the mountain."

I, or some other, had surely "fixed" the weather.

No such trip had ever been known by the oldest member of the crew. Only one rainy night and one sweet half-cloudy afternoon. For the rest, blue and golden days and nights of amethyst.

But would the captain forget? The thought always made my heart pause; yet there was something in the firm lines of his strong, brown face that made it impossible for me to mention it to him again.

But on that evening I was sitting in the dining room which, when the tables were cleared, was a kind of general family living room, when Charlie came to me with his angelic smile.

"The captain, he say you please come on deck right away."

I went up the companionway and stepped out upon the deck; and there in the north, across the blue, mist-softened sea, in the rich splendor of an Aleutian sunset, trembled and glowed the exquisite thing of my desire.

In the absolute perfection of its conical form, its chaste and delicate beauty of outline, and the slender column of smoke pushing up from its finely pointed crest, Shishaldin stands alone. Its height is not great, only nine thousand feet; but in any company of loftier mountains it would shine out with a peerlessness that would set it apart.

The sunset trembled upon the North Pacific Ocean, changing hourly as the evening wore on. Through scarlet and purple and gold, the mountain shone; through lavender, pearl, and rose; growing ever more distant and more dim, but not less beautiful. At last, it could barely be seen, in a flood of rich violet mist, just touched with rose.

So steadily I looked, and with such a longing passion of greeting, rapture, possession, and farewell in my gaze and in my heart, that lo! when its last outline had blurred lingeringly and sweetly into the rose-violet mist, I found that it was painted in all its delicacy of outline and soft splendor of coloring upon my memory. There it burns to-day in all its loveliness as vividly as it burned that night, ere it faded, line by line, across the widening sea. It is mine. I own it as surely as I own the green hill upon which I live, the blue sea that sparkles daily beneath my windows, the gold-brilliant constellations that move nightly above my home, or the song that the meadow-lark sings to his mate in the April dawn.

The sea breaks into surf upon Shishaldin's base, and snow covers the slender cone from summit to sea level, save for a month or two in summer when it melts around the base. Owing to the mists, it is almost impossible to obtain a sharp negative of Shishaldin from the water.

They played with it constantly. They wrapped soft rose-colored scarfs about its crest; they wound girdles of purple and gold and pearl about its middle; they set rayed gold upon it, like a crown. Now and then, for a few seconds at a time, they drew away completely, as if to contemplate its loveliness; and then, as if overcome and compelled by its dazzling brilliance, they flung themselves back upon it impetuously and crushed it for several moments completely from our view.

Large and small, the islands of the Aleutian Archipelago number about one hundred. They drift for nearly fifteen hundred miles from the point of the Aliaska Peninsula toward the Kamchatkan shore; and Attu, the last one, lies within the eastern hemisphere. This chain of islands, reaching as far west as the Komandórski, or Commander, Islands—upon one of which Commander Behring died and was buried—was named, in 1786, the Catherina Archipelago, by Forster, in honor of the liberal and enlightened Empress Catherine the Second, of Russia.

The Aleutian Islands are divided into four groups. The most westerly are Nearer, or Blizni, Islands, of which the famed Attu is the largest; the next group to eastward is known as Rat, or Kreesi, Islands; then, Andreanoffski Islands, named for Andreanoff, who discovered them, and whose largest island is Atka, where it is said the baskets known as the Attu baskets are now woven.

East of this group are the Fox, or Leesi, Islands. This is the largest of the four Aleutian groups, and contains thirty-one islands, including Unimak, which is the largest in the archipelago. Others of importance in this group are Unalaska, formerly spelled Unalashka; Umnak; Akutan; Akhun; Ukamak; and the famed volcano islands of St. John the Theologian, or Joanna Bogoslova, and the Four Craters. Unimak Pass, the best known and most used passage into Behring Sea, is between Unimak and Akhun islands. Akutan Pass is between Akutan and Unalaska islands; Umnak Pass, between Unalaska and Umnak islands. (These *u*'s are pronounced as though spelled *oo*.)

Unalaska and Dutch Harbor are situated on the Island of Unalaska. By the little flower-bordered path leading up and down the green, velvety hills, these two settlements are fully two miles apart; by water, they seem scarcely two hundred yards from one another. The steamer, after landing at Dutch Harbor, draws her prow from the wharf, turns it gently around a green point, and lays it beside the wharf at Unalaska.

The bay is so surrounded by hills that slope softly to the water, that one can scarcely remember which blue water-way leads to the sea. There is a curving white beach, from which the town of Unalaska received its ancient name of Iliuliuk, meaning "the beach that curves." The white-painted, red-roofed buildings follow this beach, and loiter picturesquely back over the green level to the stream that flows around the base of the hills and finds the sea at the Unalaska wharf.

This is one of the safest harbors in the world. It is one great, sparkling sapphire, set deep in solid emerald and pearl. It is entered more beautifully than even the Bay of Sitka. It is completely surrounded by high mountains, peak rising behind peak, and all covered with a thick, green, velvety nap and crowned with eternal pearl.

The entrance way is so winding that these peaks have the appearance of leaning aside to let us slide through, and then drawing together behind us, to keep out the storms; for ships of the heaviest draught find refuge here and lie safely at anchor while tempests rage outside.

Now and then, between two enchantingly green near peaks, a third shines out white, far, glistening mistily—covered with snow from summit to base, but with a dark scarf of its own internal passion twisted about its outwardly serene brow.

The *Kuro Siwo*, or Japan Current, breaks on the western end of the Aleutian Chain; half flows eastward south of the islands, and carries with it the warm, moist atmosphere which is condensed on the snow-peaks and sinks downward in the fine and delicious mist that gives the grass and mosses their vivid, brilliant, perpetual green. The other half passes northward into Behring Sea and drives the ice back into the "Frozen Ocean." Dall was told that the whalers in early spring have seen large icebergs steadily sailing northward through the strait at a knot and a half an hour, against a very stiff breeze from the north. In May the first whalers follow the Kamchatkan Coast northward, as the ice melts on that shore earlier than on ours. The first whaler to pass East Cape secures the spring trade and the best catch of whales.

The color of the *Kuro Siwo* is darker than the waters through which it flows, and its Japanese name signifies "Black Stream." Passing on down the coast, it carries a warm and vivifying moisture as far southwest as Oregon. It gives the Aleutians their balmy climate. The average winter temperature is about thirty degrees above zero; and the summer temperature, from fifty to sixty degrees.

The volcano Makushin is the noted "smoker" of this island, and there is a hot spring, containing sulphur, in the vicinity, from which loud, cannon-like reports are frequently heard. The natives believe that the mountains fought together and that Makushin remained the victor. These reports were probably supposed to be fired at his command, as warnings of his fortified position to any inquisitive peak that might chance to fire a lava interrogation-point at him.

In June, and again in October, of 1778, Cook visited the vicinity, anchoring in Samghanooda Harbor. There he was visited by the commander of the Russian expedition in this region, Gregorovich Ismaïloff. The usual civilities and gifts were exchanged. Cook sent the Russian some liquid gifts which were keenly appreciated, and was in return offered a sea-otter skin of such value that Cook courteously declined it, accepting, instead, some dried fish and several baskets of lily root.

The Russian settlement was at Iliuliuk, which was distant several miles from Samghanooda. Several of the members of Cook's party visited the settlement, notably Corporal Ledyard, who reported that it consisted of a dwelling-house and two storehouses, about thirty Russians, and a number of Kamchatkans and natives who were used as servants by the Russians. They all lived in the same houses, but ate at three different tables.

Cook considered the natives themselves the most gentle and inoffensive people he had ever "met with" in his travels; while as to honesty, "they might serve as a pattern to the most civilized nation upon earth." He was convinced, however, that this disposition had been produced by the severities at first practised upon them by the Russians in an effort to subdue them.

Cook described them as low of stature, but plump and well-formed, dark-eyed, and dark-haired. The women wore a single garment, loose-fitting, of sealskin, reaching below the knee—the parka; the men, the same kind of garment, made of the skin of birds, with the feathers worn against the flesh. Over this garment, the men wore another made of gut, which I have elsewhere described under the name of kamelinka, or kamelayka. All wore "oval-snouted" caps made of wood, dyed in colors and decorated with glass beads.

The women punctured their lips and wore bone labrets. "It is as uncommon, at Oonalashka, to see a man with this ornament as to see a woman without it," he adds.

The chief was seen making his dinner of the raw head of a large halibut. Two of his servants ate the gills, which were cleaned simply "by squeezing out the slime." The chief devoured large pieces of the raw meat with as great satisfaction as though they had been raw oysters.

These natives lived in barabaras. (This word is pronounced with the accent on the second syllable; the correct spelling cannot be vouched for here, because no two authorities spell it in the same way.)

They were usually made by forming shallow circular excavations and erecting over them a framework of driftwood, or whale-ribs, with double walls filled with earth and stones and covered over with sod.

The roofs contained square openings in the centre for the

escape of smoke; and these low earth roofs were used by the natives as family gathering places in pleasant weather. Here they would sit for hours, doing nothing and gazing blankly at nothing.

The entrance was through a square hole in, or near, the roof. It was reached by a ladder, and descent into the interior was made in the same way, or by means of steps cut in a post. A narrow dark tunnel led to the inner room, which was from ten to twenty feet in diameter.

These barabaras were sometimes warmed only by lamps; but usually a fire was built in the centre, directly under the opening in the roof. Mats and skins were placed on shelves, slightly elevated above the floor, around the walls. Many persons of both sexes and all ages lived in these places; frequently several dwellings were connected by tunnels and had one common hole-entrance. The filth of these airless habitations was nauseating.

Their household furniture consisted of bowls, spoons, buckets, cans, baskets, and one or two Russian pots; a knife and a hatchet were the only tools they possessed.

The huts were lighted by lamps made of flat stones which were hollowed on one side to hold oil, in which dry grass was burned. Both men and women warmed their bodies by sitting over these lamps and spreading their garments around them.

The natives used the bidarka here, as elsewhere.

They buried their dead on the summits of hills, raising little hillocks over the graves. Cook saw one grave covered with stones, to which every one passing added a stone, after the manner fancied by Helen Hunt Jackson a hundred years later; and he saw several stone hillocks that had an appearance of great antiquity.

In Unalaska to-day may still be seen several barabaras. They must be very old, because the native habitations of the coast are constructed along the lines of the white man's dwellings at the present time. They add to the general quaint and picturesque appearance of the town, however. Their sod roofs are overgrown with tall grasses, among which wild flowers flame out brightly.

(Unalaska is pronounced Oö-na-las'-ka, the *a*'s having the

sound of *a* in arm. Aleutian is pronounced in five syllables: Ä-le-oo'-shi-an, with the same sound of *a*.)

The island of Unalaska was sighted by Chirikoff on his return to Kamchatka, on the 4th of September, 1741.

The chronicles of the first expeditions of the Russian traders—or promyshleniki, as they were called—are wrapped in mystery. But it is believed that as early as 1744 Emilian Bassof and Andrei Serebrennikof voyaged into the islands and were rewarded by a catch of sixteen hundred sea-otters, two thousand fur-seals, and as many blue foxes.

Stephan Glottoff was the first to trade with the natives of Unalaska, whom he found peaceable and friendly. The next, however, Korovin, attempted to make a settlement upon the island, but met with repulse from the natives, and several of his party were killed.

Glottoff returned to his rescue, and the latter's expedition was the most important of the earlier ones to the islands. On his previous visit he had found the highly prized black foxes on the island of Unalaska, and had carried a number to Kamchatka.

I have related elsewhere the story of the atrocities perpetrated upon the natives of these islands by the early promyshleniki. During the years between 1760 and 1770 the natives were in active revolt against their oppressors; and it was not until the advent of Solovioff the Butcher that they were tortured into the mild state of submission in which they were found by Cook in 1778, and in which they have since dwelt.

Father Veniaminoff made the most careful study of the Aleutians, beginning about 1824. It has been claimed that this noble and devout priest was so good that he perceived good where it did not exist; and his statements concerning his beloved Aleutians are not borne out by the promyshleniki. Considering the character of the latter, I prefer to believe Veniaminoff.

The most influential Aleuts were those who were most successful in hunting, which seemed to be their highest ambition.

The best hunters possessed the greatest number of wives; and they were never stinted in this luxury. Even Veniaminoff, with his rose-colored glasses on, failed to discover virtue or the faintest moral sense among them.

"They incline to sensuality," he put it, politely. "Before the teachings of the Christian religion had enlightened them, this inclination had full sway. The nearest consanguinity, only, puts limits to their passions. Although polygamy was general, nevertheless there were frequently secret orgies, in which all joined.... The bad example and worse teachings of the early Russian settlers increased their tendency to licentiousness."

Child-murder was rare, owing to the belief that it brought misfortune upon the whole village.

Among the half-breeds, the character of the dark mother invariably came out more strongly than that of the Russian father. They learned readily and intelligently, and fulfilled all church duties imposed upon them cheerfully, punctually, and with apparent pleasure.

Under the teaching of Veniaminoff, the Aleuts were easily weaned from their early Pantheism, and from their savage songs and dances, described by the earlier voyagers. They no longer wore their painted masks and hats, although some treasured them in secret.

The successful hunter, in times of famine or scarcity of food, shared with all who were in need. The latter met him when his boat returned, and sat down silently on the shore. This is a sign that they ask for aid; and the hunter supplies them, without receiving, or expecting, either restitution or thanks. This generosity is like that of the people of Belkoffski; it comes from the heart.

The Aleutians were frequently intoxicated; but this condition did not lead to quarrelling or trouble. Murder and attempts at murder were unknown among them.

If an Aleut were injured, or offended, after the introduction of Christianity, he received and bore the insult in silence. They had no oaths or violent epithets in their language; and they would rather commit suicide than to receive a blow. The sting that lies in cruel words they dreaded as keenly.

Veniaminoff found that the Aleuts would steal nothing more than a few leaves of tobacco, a few swallows of brandy, or a little food; and these articles but rarely.

The most striking trait of character displayed by the Aleut was, and still is, his patience. He never complained, even when slowly starving to death. He sat by the shore; and if food were not offered to him, he would not ask. He was never known to sigh, nor to groan, nor to shed tears.

These people were found to be very sensitive, however, and capable of deep emotion, even though it was never revealed in their faces. They were exceedingly fond of, and tender with, their children, and readily interpreted a look of contempt or ridicule, which invariably offended in the highest degree.

The most beautiful thing recorded of the Aleut is that when one has done him a favor or kindness, and has afterward offended him, he does not forget the former favor, but permits it to cancel the offence.

They scorn lying, hypocrisy, and exaggeration; and they never betray a secret. They are so hospitable that they will deny themselves to give to the stranger that is in need. They detest a braggart, but they never dispute—not even when they know that their own opinion is the correct one.

Veniaminoff admitted that the Aleuts who had lived among the Russians were passionately addicted to the use of liquor and tobacco. But even with their drunkenness, their uncleanness, and their immorality, the Aleutian character seems to have possessed so many admirable, and even unusual, traits that, if the training and everyday influences of these people had been of a different nature from what they have been since they lost Veniaminoff, they would have, ere this, been able to overcome their inherited and acquired vices, and to have become useful and desirable citizens.

They were formerly of a revengeful nature, but after coming

under the influence of Veniaminoff, no instance of revenge was discovered by him.

They learned readily, with but little teaching, not only mechanical things, but those, also, which require deep thought—such as chess, at which they became experts.

One became an excellent navigator, and made charts which were followed by other voyagers for many years. Others worked skilfully in ivory, and the dark-eyed women wove their dreams into the most precious basketry of the world.

CHAPTER XXXVIII

We sailed into the lovely bay of Unalaska on the fourth day of July. The entire village, native and white, had gone on a picnic to the hills.

We spent the afternoon loitering about the deserted streets and the green and flowery hills. One could sit contentedly for a week upon the hills,—as the natives used to sit upon the roofs of their barabaras,—doing nothing but looking down upon the idyllic loveliness shimmering in every direction.

In the centre of the town rises the Greek-Russian church, green-roofed and bulbous-domed, adding the final touch of mysticism and poetry to this already enchanting scene.

At sunset the mists gathered, slowly, delicately, beautifully. They moved in softly through the same strait by which we had entered—little rose-colored masses that drifted up to meet the violet-tinted ones from the other end of the bay. In the centre of the water valley they met and mixed together, and, in their new and more marvellous coloring, pushed up about the town and the lower slopes. Out of them lifted and shone the green roof and domes of the church; more brilliantly above them, napped thick and soft as velvet, glowed the hills; and more lustrously against the saffron sky flashed the pearl of the higher peaks.

There was a gay dinner party aboard the Dora that night. Afterward, we all attended a dance. There was only one white woman in the hall besides my friend and myself; and we three were belles! We danced with every man who asked us to dance, to the most wonderful music I have ever heard. One of the musicians played a violin with his hands and a French harp with his mouth, both at the same time—besides making quite as much noise with one foot as he did with both of the instruments together.

There were several good-looking Aleutian girls at the dance. They had pretty, slender figures, would have been considered well dressed in any small village in the states, and danced with exceeding grace and ease.

We went to this dance not without some qualms of various kinds; but we went for the same reason that "Cyanide Bill" told us he had journeyed three times to the shores of the "Frozen Ocean"—"just to see."

Toward midnight a pretty and stylishly gowned young woman came in with an escort and joined in the dancing. As she whirled past us, with diamonds flashing from her hands, ears, and neck, my inquiring Scotch friend asked a gentleman with whom she was dancing, "Who is the pretty dark-eyed lady? We have not seen her before."

She was completely extinguished for some time by his reply, given with the cheerful frankness of the North.

"Oh, that's Nelly, miss. I don't know any other name for her. We just always call her Nelly, miss."

We returned to the steamer, leaving "Nelly" to twinkle on. Our curiosity was entirely satisfied. We went "to see," and we had seen.

Captain Gray might be called "the lord of Unalaska." He is the "great gentleman" of the place. He has for many years managed the affairs of the Alaska Commercial Company, and he has acted as host to almost every traveller who has voyaged to this lovely isle.

After supper, which was served on the steamer at midnight, we were invited to his home "to finish the evening."

"At one o'clock in the morning!" gasped my companion.

"Hours don't count up here," said our captain. "It is broad daylight. Besides, it is the 4th of July. I think we should accept the invitation."

We did accept it, in the same spirit in which it was given, and it was one of the most profitable of evenings. We found a home of comfort and refinement in the farthest outpost of civilization in the North Pacific. The hours were spent pleasantly with good music, singing, and reading; and delicate refreshments were served.

The sun shone upon my friend's scandalized face as we returned to our steamer. It was nearly five o'clock.

"I know it was innocent enough," said she, "but think how it *sounds*!—a dance, with only three white women present—not to mention 'Nelly'!—a midnight supper, and then an invitation to 'finish the evening'! It sounds like one of Edith Wharton's novels."

"It's Alaska," said the captain. "You want local color—and you're getting it. But let me tell you that you have never been safer in your life than you have been to-night."

"Safe!" echoed she. "I'm not talking about the safety of it. It's the *form* of it."

"Form doesn't count, as yet, in the Aleutians," said the captain. "'There's never a law of God or man runs north of *fifty-three*!'"

"There's surely never a *social* law runs north of it," was the scornful reply.

The next morning we went to the great warehouses of the company, to look at old Russian samovars. Captain Gray personally escorted us through their dim, cobwebby, high-raftered spaces. There was one long counter covered with samovars, and we began eagerly to examine and price them.

Dog-team Express, Nome

The cheapest was twenty-five dollars; and the most expensive, more than a hundred.

"But they are all sold," added Captain Gray, gloomily.

"All sold!" we exclaimed, in a breath. "What—*all*? Every one?"

"Yes; every one," he answered mournfully.

"Why, how very odd," said I, "for them all to be sold, and all to be left here."

"Yes," said he, sighing. "The captain of a government cutter bought them for his friends in Boston. He has gone on up into Behring Sea, and will call for them on his return."

Far be it from me to try to buy anything that is not for sale. I thanked him politely for showing them to us; and we went on to another part of the warehouse.

We found nothing else that was already "sold." We bought several holy-lamps, baskets, and other things.

"I'm sorry about the samovars," said I, as I paid Captain Gray.

"So am I," said he. Then he sighed. "There's one, now," said he, after a moment, thoughtfully. "I might—Wait a moment."

He disappeared, and presently returned with a perfect treasure of a samovar,—old, battered, green with age and use. We went into ecstasies over it.

"I'll take it," I said. "How much is it?"

"It was twenty-five dollars," said he, dismally. "It is sold."

"How very peculiar," said my companion, as we went away, "to keep bringing out samovars that are sold."

For two years my thoughts reverted at intervals to those "sold" samovars at Unalaska. Last summer I went down the Yukon. At St. Michael I was entertained at the famous "Cottage" for several days. One day at dinner I asked a gentleman if he knew Captain Gray.

"Of Unalaska?" exclaimed two or three at once. Then they all burst out laughing.

"We all know him," one said. "Everybody knows him."

"But why do you laugh?"

"Oh, because he is so 'slick' at taking in a tourist."

"In what manner?" asked I, stiffly. I remembered that Captain Gray had asked me if I were a tourist.

They all laughed again.

"Oh, *especially* on samovars."

My face burned suddenly.

"On samovars!"

"Yes. You see he gets a tourist into his warehouses and shows him samovar after samovar—fifty or sixty of them—and tells him that every one is sold. He puts on the most mournful look.

"'This one was twenty-five dollars,' he says. 'A captain on a government cutter bought them to take to Boston.' Then the tourist gets wild. He offers five, ten, twenty dollars more to get one of those samovars. He always gets it; because, you see, Gray wants to sell it to him even worse than he wants to buy it. It always works."

We walked over the hills to Dutch Harbor—once called Lincoln Harbor. There is a stretch of blue water to cross, and we were ferried over by a gentleman having much Fourth-of-July in his speech and upon his breath.

His efforts at politeness are remembered joys, while a sober ferryman would have been forgotten long ago. But the sober ferrymen that morning were like the core of the little boy's apple.

It was the most beautiful walk of my life. A hard, narrow, white path climbed and wound and fell over the vivid green hills; it led around lakes that lay in the hollows like still, liquid sapphire, set with the pearl of clouds; it lured through banks of violets and over slopes of trembling bluebells; it sent out tempting by-paths that ended in the fireweed's rosy drifts; but always it led on—narrow, well-trodden, yet oh, so lonely and so still! Birds sang and the sound of the waves came to us—that was all. Once a little brown Aleutian lad came whistling around the curve in the path, stood still, and gazed at us with startled eyes as soft and dark as a gazelle's; but he was the only human being we saw upon the hills that day.

We saw acres that were deep blue with violets. They were large enough to cover silver half-dollars, and their stems were several inches in length. Fireweed grew low, but the blooms were large and of a deep rose color.

Standing still, we counted thirteen varieties of wild flowers within a radius of six feet. There were the snapdragon, wild rose, columbine, buttercup, Solomon's seal, anemone, larkspur, lupine, dandelion, iris, geranium, monk's-hood, and too many others to name, to be found on the hills of Unalaska. There are more than two thousand varieties of wild flowers in Alaska and the Yukon Territory. The blossoms are large and brilliant, and they cover whole hillsides and fill deep hollows with beautiful color. The bluebells and violets are exquisite. The latter are unbelievably large; of a rich blue veined with silver. They poise delicately on stems longer than those of the hot-house flower; so that we could gather and carry armfuls of them.

The site of Dutch Harbor is green and level. Fronting the bay are the large buildings of the North American Commercial

Company, with many small frame cottages scattered around them. All are painted white, with bright red roofs, and the town presents a clean and attractive appearance.

Dutch Harbor is the prose, and Unalaska the poetry, of the island. There is neither a hotel nor a restaurant at either place. It was one o'clock when we reached Dutch Harbor; we had breakfasted early, and we sought, in vain, for some building that might resemble an "eating-house."

We finally went into the big store, and meeting the manager of the company, asked to be directed to the nearest restaurant.

He smiled.

"There isn't any," he said.

"Is there no place where one may get *something* to eat? Bread and milk? We saw cows upon the hills."

"You would not care to go to the native houses," he replied, still smiling. "But come with me."

He led the way along a neat board walk to a residence that would attract attention in any town. It was large and of artistic design.

"It was designed by Molly Garfield," the young man somewhat proudly informed us. "Her husband was connected with the company for several years, and they built and lived in this house."

The house was richly papered and furnished. It was past the luncheon hour, but we were excellently served by a perfectly trained Chinaman.

For more than a hundred years the great commercial companies—beginning with the Shelikoff Company—have dispensed the hospitality of Alaska, and have acted as hosts to the stranger within their gates. The managers are instructed to sell provisions at reasonable prices, and to supply any one who may be in distress and unable to pay for food.

They frequently entertain, as guests of the company they represent, travellers to these lonely places, not because the latter are in need, but merely as a courtesy; and their hospitality is as free and generous—but not as embarrassing—as that of Baranoff.

That night I sat late alone upon the hills, on a tundra slope that was blue with violets. I could not put my hand down without crushing them. The lights moving across Unalaska were as poignantly interesting as the thoughts that come and go across a stranger's face when he does not know that one is observing.

All the lights and shadows of the vanishing Aleutian race seemed to be moving across the hills, the village, the blue bay.

Scarcely a day has passed that I have not gone back across the blue and emerald water-ways that stretch between, to that lovely place and that luminous hour.

Perhaps, I thought, Veniaminoff may have looked down upon this exquisite scene from this same violeted spot—Veniaminoff, the humble, devout, and devoted missionary, whom I should rather have been than any man or woman whose history I know; Veniaminoff, who *lived*—instead of *wrote*—a great, a sublime, poem.

Unalaska's commercial glory has faded. It was once port of entry for all vessels passing in or out of Behring Sea; the ships of the Arctic whaling fleet called here for water, coal, supplies, and mail; during the years that the *modus vivendi* was in force it was headquarters of the United States and the British fleets patrolling Behring Sea, and lines of captured sealers often lay here at anchor.

During the early part of the present decade Unalaska saw its most prosperous times. Thousands of people waited here for transportation to the Klondike, via St. Michael and the Yukon. Many ships were built here, and one still lies rotting upon the ways.

The Greek church is second in size and importance to the one at Sitka only, and the bishop once resided here. There is a Russian parish school, a government day-school, and a Methodist mission, the Jessie Lee Home. The only white women on the island reside at the Home. The bay has frequently presented the appearance of a naval parade, from the number of government and other vessels lying at anchor.

No traveller will weary soon of Unalaska. There are caves and waterfalls to visit, and unnumbered excursions to make to beautiful places among the hills. Especially interesting is Samghanooda, or English, Harbor, where Cook mended his ships; while Makushin Harbor, on the western coast, where Glottoff and his Russians first landed in 1756, is only thirty miles away.

The great volcano itself is easy of ascent, and the view from its crest is one of the memories of a lifetime. Borka, a tiny village at Samghanooda, is as noted for its Dutch-like cleanliness as Belkoffski is for its filth.

The other islands of the Aleutian chain drift on to westward, lonely, unknown—almost, if not entirely, uninhabited. Now and then a small trading settlement is found, which is visited only by Captain Applegate,—the last remaining white deep-sea otter hunter,—and once a year by a government cutter, or the Russian priest from Unalaska, or a shrewd and wandering trader.

These green and unknown islands are the islands of my dreams—and dreams do "come true" sometimes. This voyage out among the Aleutians is the most poetic and enchanting in the world to-day; and I shall never be entirely happy until I have drifted on out to the farthest island of Attu, lying within the eastern hemisphere, and watched those lonely, dark women, with the souls of poets and artists and the patience of angels, weaving *their* dreams into ravishing beauty and sending them out into the world as the farewell messages of a betrayed and vanishing people. As we treat them for their few remaining years, so let us in the end be treated.

Alaska is to-day the centre of the world's volcanic activity, and the mountainous appearances and disappearances that have been recorded in the Aleutian Islands are marvellous and awesome. To these upheavals in the North Pacific and Behring Sea Whidbey's adjectives, "stupendous," "tremendous," and "awfully dreadful," might be appropriately applied.

On July the fourth, 1907, officers of the revenue cutter *McCulloch* discovered the new peak which they named in honor of their vessel. It was in the vicinity of the famous volcano of Joanna Bogoslova, or Saint John the Theologian.

In 1796 the natives of Unalaska and the adjoining islands for many miles were startled by violent reports, like continued cannonading, followed by frightful tremblings of the earth upon which they stood.

A dense volume of smoke, ashes, and gas descended upon them in a kind of cloud, and shut everything from their view. They were thus enveloped and cannonaded for about ten days, when the atmosphere gradually cleared and they observed a bright light shining upon the sea from thirty to forty miles north of Unalaska. The brave ones of the island went forth in bidarkas and discovered that a small island had risen from the sea to a height of one hundred feet and that it was still rising.

This was the main peak of the Bogosloff group, and it continued to grow until 1825, when it reached a height of about three hundred feet and cooled sufficiently for Russians to land upon it for the first time. The heat was still so intense, however, and the danger from running lava so great, that they soon withdrew to their boats.

In the early eighties, after similar disturbances, another peak arose near the first and joined to it by a low isthmus, upon which stood a rock seventy feet in height, which was named Ship-Rock. In 1891 the isthmus sank out of sight in the sea, and a new peak arose.

Since then no important changes have occurred. The peaks themselves remained too hot and dangerous for examination; but the short voyage out from Unalaska has been a favorite one for tourists who were able to land upon the lower rocks and spend a day gathering specimens and studying the sea-lions that doze in polygamous herds in the warmth, and the shrieking murres that nest in the cliffs and cover them like a tremulous gray-white cloud.

Every inch of space on these cliffs seems to be taken by these birds for the creation of life. On every tiniest shelf they perch upright, black-backed and white-bellied, brooding their eggs—although these hot and steamy cliffs are sufficient incubators to bring forth life out of every egg deposited upon them. When the murres are suddenly disturbed, their eggs slip from their hold and plunge down the cliffs, splattering them with the yellow of their broken yolks.

The last week in July, 1907, I passed close to the Bogosloff Islands, which had grown to the importance of four peaks. Three days later a violent earthquake occurred in this vicinity. Once more dense clouds of smoke descended upon Unalaska and the adjoining islands, and ashes poured upon the sea and land, as far north as Nome, covering the decks of passing steamers to a depth of several inches, and affecting sailors so powerfully that they could only stay on deck for a few moments at a time.

On September the first, the captain and men of the whaler *Herman*, passing the Bogosloff group, beheld a sight to observe which I would cheerfully have yielded several years of life. They saw the two-months-old McCulloch peak burn itself down into the sea, with vast columns of steam ascending miles into the air above it, and the waters boiling madly on all sides. It went down, foot by foot, and the men stood spellbound, watching it disappear. For miles around the sea was violently agitated and was mixed with volcanic ash, which also covered the decks, and at intervals steam poured up unexpectedly out of the ocean.

As soon as possible the revenue cutter *Buffalo* went to the wonderful volcanic group, and it was found that their whole appearance was changed.

There were three peaks where four had been; but whereas they had formerly been separate and distinct islands, they were now connected and formed one island.

This island is two and a half miles long. Perry Peak, which arose in 1906, had increased in height; and there was a crater-like depression on its south side, around which the waters were continually throwing off vast clouds of steam and smoke. Captain Pond reported that rocks as large as a house were constantly rolling down from Perry Peak, and that the whole scene was one of wonderful interest. To his surprise, the colony of sea-lions, which must have been frightened away, had returned, and seemed to be enjoying the steamy heat on the rocks of the main and oldest peak of the group.

The disappearance of McCulloch peak was accompanied by earthquake shocks as far to eastward as Sitka. Makushin, the great volcano of Unalaska, and others, smoked violently, and ashes fell over the Aleutian Islands and the mainland. At the same time uncharted rocks began to make their appearance all along the coast, to the grave danger of navigation.

CHAPTER XXXIX

In the heart of Behring Sea, about two hundred miles north of Unalaska, lie two tiny cloud and mist haunted and wind-racked islands which are the great slaughter-grounds of Alaska. Here, for a hundred and twenty years, during the short seal season each year, men have literally waded through the bloody gore of the helpless animals, which they have clubbed to death by thousands that women may be handsomely clothed.

The surviving members of Vitus Behring's ill-starred expedition carried back with them a large number of skins of the valuable sea-otter. From that date—1742—until about 1770 the promyshleniki engaged in such an unresting slaughter of the otter that it was almost exterminated.

In desperation, they turned, then, to the chase of the fur-seal, and for years sought in vain for the rumored breeding-grounds of this pelagic animal. The islands of St. Paul and St. George were finally discovered in 1786, by Gerassim Pribyloff, who heard the seals barking and roaring through the heavy fogs, and, sailing cautiously on, surprised them as they lay in polygamous groups by the million upon the rocky shores.

Pribyloff was the son of a sailor who had accompanied Behring on the *St. Peter.* He modestly named his priceless discovery "Subov," for the captain and part owner of the trading association for which he worked. He himself was not engaged in sealing, but was simply the first mate of the sloop *St. George.* The Russians, however, renamed the islands for their discoverer; and happily the name has endured.

St. George Island is ten miles in length by from two to four in width. It is higher than the larger St. Paul, which lies twenty-seven miles farther north, and rises more abruptly from the water.

The temperature of these islands is not low, rarely falling to zero; but the wind blows at so great velocity that frequently for days at a time the natives can only go from one place to another by crawling upon their hands and knees.

To conserve the sealing industry, after the purchase of Alaska, the exclusive privilege of killing seals on these islands was granted to the Alaska Commercial Company for a period of twenty years. When this lease expired in 1890, a new one was made out for a like period to the North American Commercial Company, which still holds possession. The company has agents on both islands, and the government maintains an agent and his assistant on St. Paul Island, and an assistant on St. George, to enforce the terms of the concession.

When the Russians first took possession of the Pribyloff Islands, they brought several hundred Aleutians and established them upon the islands in sod houses, where they were held under the usual slave-like conditions of this abused people. They were miserably housed and fed, received only the smallest wage,—from which they were compelled to contribute to the support of the church,—and were held, against their wishes, upon these dreary and inhospitable shores.

With the coming of the American companies all was changed. Comfortable, clean habitations of frame were erected for them; their pay was increased from ten to forty cents each for the removal of pelts; schools and hospitals were provided, children being compelled to attend the former; and the sale of intoxicating liquors was prohibited. There are between a hundred and fifty and two hundred natives on the islands at present.

The houses are lined with tar paper, painted white, with red roofs, and furnished with stoves. There are streets and large storehouses, and the village presents an attractive appearance.

As a result of good care, food, and cleanliness, the natives are able to do twice the amount of work accomplished by the same number under the old conditions. They are healthier, happier, and more industrious.

The value of the fur-seal catch from the time of the purchase of Alaska to the early part of the present decade was more than thirty-five millions of dollars. In 1903 the yearly catch, however, had dwindled from two millions at the time of discovery to twenty-two thousands.

Indiscriminate and reckless slaughter, and particularly the pelagic sealing carried on by poachers—it being impossible to distinguish the males from the females at sea—have nearly exterminated the seals. They will soon be as rare as the sea-otter, which vanished for the same shameless reasons. In the government's lease it is provided that not more than one hundred thousand seals shall be taken in a single year; but of recent years the catch has fallen so far short of that number that the annual rental, which was first set at sixty thousand dollars, has had a sliding, diminishing scale until it has finally reached twelve thousand dollars.

Great trouble has been experienced with pelagic sealers. Pelagic sealing means simply following the seals on their way north and killing them in the deep sea before they reach the breeding-grounds. There have been American poachers, but the majority have been Canadians. The United States government at first claimed exclusive rights to the seals, and patrolled the waters of Behring Sea, as inland waters, frequently seizing vessels belonging to other nations.

The matter, after much bitter feeling on both sides, was finally submitted to the "Paris Tribunal," which did not allow our claim to exclusive sealing rights in Behring Sea. It, however, forbade pelagic sealing within a zone of sixty miles of the Pribyloff islands.

These waters are now patrolled by vessels of both nations; but Japanese vessels are frequently transgressors, the Japanese claiming that they are not bound by the regulations of the Paris Tribunal. Both British and American sealers have been known to fly the Japanese flag when engaged in pelagic sealing in forbidden waters. Trouble of a serious nature with Japan may yet arise over this matter.

The habits and the life of the seal are exceedingly interesting. In many ways these graceful creatures are startlingly human-like, particularly in their appealing, reproachful looks when a death-dealing blow is about to be struck. Some, it is true, yield to a violent, fighting rage,—growing more furious as their helplessness is realized,—and at such times the eyes flame with the green and red fire of hate and passion, and resemble the eyes of a human being possessed with rage and terror.

The bull seals have been called "beach-masters," "polygamists," and "harem-lords."

These old bulls, then, are the first to return to the breeding-grounds in the spring. They begin to "haul out" upon the rocks during the first week in May. Each locates upon his chosen "ground," and awaits the arrival of the females, which does not occur until the last of June. While awaiting their arrival, incessant and terrible fighting takes place among the bulls, frequently to the death—so stubbornly and so ferociously does each struggle to retain the place he has selected in which to receive the females of his harem. The older the bull the more successful is he both in love and in war; and woe betide any young and bold bachelor who dares to pause for but an instant and cast tempting glances at a gay and coquettish young favorite under an old bull's protection. There is instant battle—in which the festive bachelor invariably goes down.

When the females arrive, a very orgy of fighting takes place. An old bull swaggers down to the water, receives a graceful and beautiful female, and beguiles her to his harem. If he but turn his back upon her for an instant another bull seizes her and bears her bodily to his harem; the first bull returns, and the fight is on—the female sometimes being torn to pieces between them, because neither will give her up. The bulls do not mind a small matter like that, however, there being so many females; and it is never the desire for a special female that impels to the fray, but the human-like lust to triumph over one who dares to set himself up as a rival.

The old bulls take possession of the lower rocks, and these they hold from all comers, yet fighting, fighting, fighting, till they are frequently but half-alive masses of torn flesh and fur.

The bachelors are at last forced, foot by foot, past the harems to the higher grounds, where they herd alone. As they are supposed to be the only seals killed for their skin, they are forced by the drivers away from the vicinity of the rookeries, to the higher slopes.

These graceful creatures drag themselves on shore with pitiable awkwardness and helplessness. They proceed painfully, with a kind of rolling movement, uttering plaintive sounds that are neither barks nor bleats. They easily become heated to exhaustion, and pause at every opportunity to rest. When they sink down for this purpose, they either separate their hind flippers, or draw them both to one side.

They are driven carefully and are permitted frequent rests, as heating ruins the fur. They usually rest and cool off, after reaching the killing grounds, while the men are eating breakfast. By seven o'clock the butchery begins.

The seals are still brutally clubbed to death. The killers are spattered with blood and bloody tufts of hair; and by-standers are said to have been horribly pelted by eyeballs bursting like bullets from the sockets, at the force of the blows. The killers aim to stun at the first blow; but the poor things are often literally beaten to death. In either event a sharp stabbing-knife is instantly run to its heart, to bleed it. The crimson life-stream gushes forth, there is a violent quivering of the great, jelly-like bulk; then, all is still. It is no longer a living, beautiful, pleading-eyed animal, but only a portion of some dainty gentlewoman's cloak. I have not seen it with my own eyes, but I have heard, in ways which make me refuse to discredit it, that sometimes the skinning is begun before the seal is dead; that sometimes the razor-like knife is run down the belly before it is run to the heart—not in useless cruelty, but because of the great need of haste. The tender, beseeching eyes, touching cries, and unavailing attempts to escape, of the seal that is being clubbed to death, are things to remember for the rest of one's life. Strong men, unused to the horrible sight, flee from it, sick and tortured with the pity of it; and surely no woman who has ever beheld it could be tempted to buy sealskin.

No effort is made to dispose of the dead bodies of the seals. They are left where they are killed, and the stench arising therefrom is not surpassed even in Belkoffski. It nauseates the white inhabitants of the islands, and drifts out to sea for miles to meet and salute the visitor. It is, however, caviar to the native nostril.

CHAPTER XL

Authorities differ as to the proper boundaries of Bristol Bay, but it may be said to be the vast indentation of Behring Sea lying east of a line drawn from Unimak Island to the mouth of the Kuskokwim River; or, possibly, from Scotch Cap to Cape Newenham would be better. The commercial salmon fisheries of this district are on the Ugashik, Egegak, Naknek, Kvichak, Nushagak, and Wood rivers and the sea-waters leading to them.

Nushagak Bay is about fifteen miles long and ten wide. It is exceedingly shallow, and is obstructed by sand-bars and shoals. The Redoubt-Alexandra was established at the mouth of the river in 1834 by Kolmakoff.

The rivers are all large and, with one exception,—Wood River,—drain the western slope of the Aleutian Chain which, beginning on the western shore of Cook Inlet, extends down the Aliaska Peninsula, crowning it with fire and snow.

There are several breaks in the range which afford easy portages from Bristol Bay to the North Pacific. The rivers flowing into Bristol Bay have lake sources and have been remarkably rich spawning-streams for salmon.

The present chain of islands known as the Aleutians is supposed to have once belonged to the peninsula and to have been separated by volcanic disturbances which are so common in the region.

Four Beauties of Cape Prince of Wales with Sled Reindeer of the American Missionary Herd

The interior of the Bristol Bay country has not been explored. It is sparsely populated by Innuit, or Eskimo, who live in primitive fashion in small settlements,—usually on high bluffs near a river. They make a poor living by hunting and fishing. Their food is largely salmon, fresh and dried; game, seal, and walrus are delicacies. The "higher" the food the greater delicacy is it considered. Decayed salmon-heads and the decaying carcass of a whale that has been cast upon the beach, by their own abominable odors summon the natives for miles to a feast. Their food is all cooked with rancid oil.

Their dwellings are more primitive than those of the island natives, for they have clung to the barabaras and other ancient structures that were in use among the Aleutians when the Russians first discovered them. Near these dwellings are the drying-frames—so familiar along the Yukon—from which hang thousands of red-fleshed salmon drying in the sun. Little houses are erected on rude pole scaffoldings, high out of the reach of dogs, for the storing of this fish when it has become "ukala" and for other provisions. These are everywhere known as "caches."

The Innuit's summer home is very different from his winter home. It is erected above ground, of small pole frames, roofed with skins and open in front—somewhat like an Indian tepee. There is no opening in the roof, all cooking being done in the open air in summer.

These natives were once thrifty hunters and trappers of wild animals, from the reindeer down to the beaver and marten, but the cannery life has so debauched them that they have no strength left for this energetic work.

Formerly every Innuit settlement contained a "kashga," or town hall, which was built after the fashion of all winter houses, only larger. There the men gathered to talk and manage the affairs of their small world. It was a kind of "corner grocery" or "back-room" of a village drug store. The men usually slept there, and in the mornings their wives arose, cooked their breakfast, and carried it to them in the kashga, turning their backs while their husbands ate—it being considered exceedingly bad form for a woman to look at a man when he is eating in public, although they think nothing of bathing together. The habits of the people are nauseatingly filthy, and the interiors of their dwellings must be seen to be appreciated.

Near the canneries the natives obtain work during the summer, but soon squander their wages in debauches and are left, when winter arrives, in a starving condition.

The season is very short in Bristol Bay, but the "run" of salmon is enormous. When this district is operating thirteen canneries, it packs each day two hundred and fifty thousand fish. In Nushagak Bay the fish frequently run so heavily that they catch in the propellers of launches and stop the engines.

Bristol Bay has always been a dangerous locality to navigate. It is only by the greatest vigilance and the most careful use of the lead, upon approaching the shore, that disaster can be averted.

Nearly all the canneries in this region are operated by the Alaska Packers Association, which also operates the greater number of canneries in Alaska.

In 1907 the value of food fishes taken from Alaskan waters was nearly ten millions of dollars; in the forty years since the purchase of that country, one hundred millions, although up to 1885 the pack was insignificant. At the present time it exceeds by more than half a million cases the entire pack of British Columbia, Puget Sound, Columbia River, and the Oregon and Washington coasts.

In 1907 forty-four canneries packed salmon in Alaska, and those on Bristol Bay were of the most importance.

The Nushagak River rivals the Karluk as a salmon stream, but not in picturesque beauty. The Nushagak and Wood rivers were both closed during the past season by order of the President, to protect the salmon industry of the future.

Cod is abundant in Behring Sea, Bristol Bay, and south of the Aleutian, Shumagin, and Kadiak islands, covering an area of thirty thousand miles. Halibut is plentiful in all the waters of southeastern Alaska. This stupid-looking fish is wiser than it appears, and declines to swim into the parlor of a net. It is still caught by hook and line, is packed in ice, and sent, by regular steamer, to Seattle—whence it goes in refrigerator cars to the markets of the east.

Herring, black cod, candle-fish, smelt, tom-cod, whitefish, black bass, flounders, clams, crabs, mussels, shrimp, and five species of trout—steelhead, Dolly Varden, cutthroat, rainbow, and lake—are all found in abundance in Alaska.

Cook, entering Bristol Bay in 1778, named it for the Earl of Bristol, with difficulty avoiding its shoals. He saw the shoaled entrance to a river which he called Bristol River, but which must have been the Nushagak. He saw many salmon leaping, and found them in the maws of cod.

The following day, seeing a high promontory, he sent Lieutenant Williamson ashore. Possession of the country in his Majesty's name was taken, and a bottle was left containing the names of Cook's ships and the date of discovery. To the promontory was given the name which it retains of Cape Newenham.

Proceeding up the coast Cook met natives who were of a friendly disposition, but who seemed unfamiliar with the sight of white men and vessels; they were dressed somewhat like Aleutians, wearing, also, skin hoods and wooden bonnets.

The ships were caught in the shoals of Kuskokwim Bay, but Cook does not appear to have discovered this great river, which is the second in size of Alaskan rivers and whose length is nine hundred miles. In the bay the tides have a fifty-foot rise and fall, entering in a tremendous bore. This vicinity formerly furnished exceedingly fine black bear skins.

Cook's surgeon died of consumption and was buried on an island which was named Anderson, in his memory. Upon an island about four leagues in circuit a rude sledge was found, and the name of Sledge Island was bestowed upon it. He entered Norton Sound, but only "suspected" the existence of a mighty river, completely missing the Yukon.

He named the extreme western point of North America, which plunges out into Behring Sea, almost meeting the East Cape of Siberia, Cape Prince of Wales. In the centre of the strait are the two Diomede Islands, between which the boundary line runs, one belonging to Russia, the other to the United States.

Cook sailed up into the Frozen Ocean and named Icy Cape, narrowly missing disaster in the ice pack. There he saw many herds of sea-horses, or walrus, lying upon the ice in companies numbering many hundreds. They huddled over one another like swine, roaring and braying; so that in the night or in a fog they gave warning of the nearness of ice. Some members of the herd kept watch; they aroused those nearest to them and warned them of the approach of enemies. Those, in turn, warned others, and so the word was passed along in a kind of ripple until the entire herd was awake. When fired upon, they tumbled one over another into the sea, in the utmost confusion. The female defends her young to the very last, and at the sacrifice of her own life, if necessary, fighting ferociously.

The walrus does not in the least resemble a horse, and it is difficult to understand whence the name arose. It is somewhat like a seal, only much larger. Those found by Cook in the Arctic were from nine to twelve feet in length and weighed about a thousand pounds. Their tusks have always been valuable, and have greatly increased in value of recent years, as the walrus diminish in number.

Cook named Cape Denbigh and Cape Darby on either side of Norton Bay; and Besborough Island south of Cape Denbigh.

Going ashore, he encountered a family of natives which he and Captain King describe in such wise that no one, having read the description, can ever enter Norton Sound without recalling it. The family consisted of a man, his wife, and a child; and a fourth person who bore the human shape, and that was all, for he was the most horribly, the most pitiably, deformed cripple ever seen, heard of, or imagined. The husband was blind; and all were extremely unpleasant in appearance. The underlips were bored.

These natives would have evidently sold their souls for iron. For four knives made out of old iron hoop, they traded four hundred pounds of fish—and Cook must have lost his conscience overboard with his anchor in Kuskokwim Bay. He recovered the anchor!

He gave the girl-child a few beads, "whereupon the mother burst into tears, then the father, then the cripple, and, at last, the girl herself."

Many different passages, or sentences, have been called "the most pathetic ever written"; but, myself, I confess that I have never been so powerfully or so lastingly moved by any sentence as I was when I first read that one of Cook's. Almost equalling it, however, in pathos is the simple account of Captain King's of his meeting with the same family. He was on shore with a party obtaining wood when these people approached in a canoe. He beckoned to them to land, and the husband and wife came ashore. He gave the woman a knife, saying that he would give her a larger one for some fish. She made signs for him to follow them.

"I had proceeded with them about a mile, when the man, in crossing a stony beach, fell down and cut his foot very much. This made me stop, upon which the woman pointed to the man's eyes, which, I observed, were covered with a thick, white film. He afterward kept close to his wife, who apprised him of the obstacles in his way. The woman had a little child on her back, covered with a hood, and which I took for a bundle until I heard it cry. At about two miles distant we came upon their open skin-boat, which was turned on its side, the convex part toward the wind, and served for their house. I was now made to perform a singular operation upon the man's eyes. First, I was directed to hold my breath; afterward, to breathe on the diseased eyes; and next, to spit on them. The woman then took both my hands and, pressing them to his stomach, held them there while she related some calamitous history of her family, pointing sometimes to her husband, sometimes to a frightful cripple belonging to the family, and sometimes to her child."

Berries, birch, willow, alders, broom, and spruce were found. Beer was brewed of the spruce.

Cook now sailed past that divinely beautiful shore upon which St. Michael's is situated, and named Stuart Island and Cape Stephens, but did not hear the Yukon calling him. He did find shoal water, very much discolored and muddy, and "inferred that a considerable river runs into the sea." If he had only guessed *how* considerable! Passing south, he named Clerk's, Gore's, and Pinnacle Islands, and returned to Unalaska.

CHAPTER XLI

A famous engineering feat was the building of the White Pass and Yukon Railway from Skaguay to White Horse. Work was commenced on this road in May, 1898, and finished in January, 1900.

Its completion opened the interior of Alaska and the Klondike to the world, and brought enduring fame to Mr. M. J. Heney, the builder, and Mr. E. C. Hawkins, the engineer.

In 1897 Mr. Heney went North to look for a pass through the Coast Range. Up to that time travel to the Klondike had been about equally divided between the Dyea, Skaguay, and Jack Dalton trails; the route by way of the Stikine and Hootalinqua rivers; and the one to St. Michael's by ocean steamers and thence up the Yukon by small and, at that time, inferior steamers.

Mr. Heney and his engineers at once grasped the possibilities of the "Skaguay Trail." This pass was first explored and surveyed by Captain Moore, of Mr. Ogilvie's survey of June, 1887, who named it White Pass, for Honorable Thomas White, Canadian Minister of the Interior. It could not have been more appropriately named, even though named for a man, as there is never a day in the warmest weather that snow-peaks are not in view to the traveller over this pass; while from September to June the trains wind through sparkling and unbroken whiteness.

Mr. Heney, coming out to finance the road, faced serious difficulties and discouragements in America. Owing to the enormous cost of this short piece of road, as planned, as well as the daring nature of its conception, the boldest financiers of this country, upon investigation, declined to entertain the proposition.

Mr. Heney was a young man who, up to that time, although possessed of great ability, had made no marked success—his opportunity not having as yet presented itself.

Recovering from his first disappointment, he undauntedly voyaged to England, where some of the most conservative capitalists, moved and convinced by his enthusiasm and his clear descriptions of the northern country and its future, freely financed the railroad whose successful building was to become one of the most brilliant achievements of the century.

They were entirely unacquainted with Mr. Heney, and after this proof of confidence in him and his project, the word "fail" dropped out of the English language, so far as the intrepid young builder was concerned.

"After that," he said, "I *could not* fail."

He returned and work was at once begun. A man big of body, mind, and heart, he was specially fitted for the perilous and daring work. Calm, low-voiced, compelling in repressed power and unswerving courage and will, he was a harder worker than any of his men.

Associated with him was a man equally large and equally gifted. Mr. Hawkins is one of the most famous engineers of this country, if not of any country.

The difficult miles that these two men tramped; the long, long hours of each day that they worked; the hardships that they endured, unflinching; the appalling obstacles that they overcame—are a part of Alaskan history.

The first twenty miles of this road from Skaguay cost two millions of dollars; the average cost to the summit was a hundred thousand dollars a mile, and now and then a single mile cost a hundred and fifty thousand dollars.

The road is built on mountainsides so precipitous that men were suspended from the heights above by ropes, to prevent disaster while cutting grades. At one point a cliff a hundred and twenty feet high, eighty feet deep, and twenty feet in width was blasted entirely away for the road-bed.

Thirty-five hundred men in all were employed in constructing the road, but thirty of whom died, of accident and disease, during the construction. Taking into consideration the perilous nature of the work, the rigors of the winter climate, and the fact that work did not cease during the worst weather, this is a remarkably small proportion.

A force of finer men never built a railroad. Many were prospectors, eager to work their way into the land of gold; others were graduates of eastern colleges; all were self-respecting, energetic men.

Skaguay is a thousand miles from Seattle; and from the latter city and Vancouver, men, supplies, and all materials were shipped. This was not one of the least of the hindrances to a rapid completion of the road. Rich strikes were common occurrences at that time. In one day, after the report of a new discovery in the Atlin country had reached headquarters, fifteen hundred men drew their pay and stampeded for the new gold fields.

But all obstacles to the building of the road were surmounted. Within eighteen months from the date of beginning work it was completed to White Horse, a distance of one hundred and eleven miles, and trains were running regularly.

A legend tells us that an old Indian chief saw the canoe of his son upset in the waves lashed by the terrific winds that blow down between the mountains. The lad was drowned before the helpless father's eyes, and in his sorrow the old chief named the place Shkag-ua, or "Home of the North Wind." It has been abbreviated to Skaguay; and has been even further disfigured by a *w*, in place of the *u*.

Between salt water and the foot of White Pass Trail, two miles up the canyon, in the winter of 1897-1898, ten thousand men were camped. Some were trying to get their outfits packed over the trail; others were impatiently waiting for the completion of the wagon road which George A. Brackett was building. This road was completed almost to the summit when the railroad overtook it and bought its right of way. It is not ten years old; yet it is always called "the *old* Brackett road."

At half-past nine of a July morning our train left Skaguay for White Horse. We traversed the entire length of the town before entering the canyon. There are low, brown flats at the mouth of the river, which spreads over them in shallow streams fringed with alders and cottonwoods.

Above, on both sides, rose the gray, stony cliffs. Here and there were wooded slopes; others were rosy with fireweed that moved softly, like clouds.

We soon passed the ruined bridge of the Brackett road, the water brawling noisily, gray-white, over the stones.

Our train was a long one drawn by four engines. There were a baggage-car, two passenger-cars, and twenty flat and freight cars loaded with boilers, machinery, cattle, chickens, merchandise, and food-stuffs of all kinds.

After crossing Skaguay River the train turns back, climbing rapidly, and Skaguay and Lynn Canal are seen shining in the distance.... We turn again. The river foams between mountains of stone, hundreds of feet below—so far below that the trees growing sparsely along its banks seem as the tiniest shrubs.

The Brackett road winds along the bed of the river, while the old White Pass, or Heartbreak, Trail climbs and falls along the stone and crumbling shale of the opposite mountain—in many places rising to an altitude of several hundred feet, in others sinking to a level with the river.

The Brackett road ends at White Pass City, where, ten years ago, was the largest tent-city in the world; and where now are only the crumbling ruins of a couple of log cabins, silence, and loneliness.

At White Pass City that was, the old Trail of Heartbreak leads up the canyon of the north fork of the Skaguay, directly away from the railroad. The latter makes a loop of many miles and returns to the canyon hundreds of feet above its bed. The scenery is of constantly increasing grandeur. Cascades, snow-peaks, glaciers, and overhanging cliffs of stone make the way one of austere beauty. In two hours and a half we climb leisurely, with frequent stops, from the level of the sea to the summit of the pass; and although skirting peaks from five to eight thousand feet in height, we pass through only one short tunnel.

It is a thrilling experience. The rocking train clings to the leaning wall of solid stone. A gulf of purple ether sinks sheer on the other side—so sheer, so deep, that one dare not look too long or too intently into its depth. Hundreds of feet below, the river roars through its narrow banks, and in many places the train overhangs it. In others, solid rock cliffs jut out boldly over the train.

After passing through the tunnel, the train creeps across the steel cantilever bridge which seems to have been flung, as a spider flings his glistening threads, from cliff to cliff, two hundred and fifteen feet above the river, foaming white over the immense boulders that here barricade its headlong race to the sea.

Beautiful and impressive though this trip is in the green time and the bloom time of the year, it remains for the winter to make it sublime.

The mountains are covered deeply with snow, which drifts to a tremendous depth in canyons and cuts. Through these drifts the powerful rotary snow-plough cleaves a white and glistening tunnel, along which the train slowly makes its way. The fascinating element of momentary peril—of snow-slides burying the train—enters into the winter trip.

Near Clifton one looks down upon an immense block of stone, the size of a house but perfectly flat, beneath which three men were buried by a blast during the building of the road. The stone is covered with grass and flowers and is marked with a white cross.

At the summit, twenty miles from Skaguay, is a red station named White Pass. A monument marks the boundary between the United States and Yukon Territory. The American flag floats on one side, the Canadian on the other. A cone of rocks on the crest of the hill leading away from the sea marks the direction the boundary takes.

The White Pass Railway has an average grade of three per cent, and it ascends with gradual, splendid sweeps around mountainsides and projecting cliffs.

The old trail is frequently called "Dead Horse Trail." Thousands of horses and mules were employed by the stampeders. The poor beasts were overloaded, overworked, and, in many instances, treated with unspeakable cruelty. It was one of the shames of the century, and no humane person can ever remember it without horror.

At one time in 1897 more than five thousand dead horses were counted on the trail. Some had lost their footing and were dashed to death on the rocks below; others had sunken under their cruel burdens in utter exhaustion; others had been shot; and still others had been brutally abandoned and had slowly starved to death.

"What became of the horses," I asked an old stampeder, "when you reached Lake Bennett? Did you sell them?"

"Lord, no, ma'am," returned he, politely; "there wa'n't nothing left of 'em to sell. You see, they was dead."

"But I mean the ones that did not die."

"There wa'n't any of that kind, ma'am."

"Do you mean," I asked, in dismay, "that they all died?—that none survived that awful experience?"

"That's about it, ma'am. When we got to Lake Bennett there wa'n't any more use for horses. Nobody was goin' the other way—and if they had been, the horses that reached Lake Bennett wa'n't fit to stand alone, let alone pack. The ones that wa'n't shot, died of starvation. Yes, ma'am, it made a man's soul sick."

Boundary lines are interesting in all parts of the world; but the one at the summit of the White Pass is of unusual historic interest. Side by side float the flags of America and Canada. They are about twenty yards from the little station, and every passenger left the train and walked to them, solely to experience a big patriotic American, or Canadian, thrill; to strut, glow, and walk back to the train again. Myself, I gave thanks to God, silently and alone, that those two flags were floating side by side there on that mountain, beside the little sapphire lake, instead of at the head of Chilkoot Inlet.

There are Canadian and United States inspectors of customs at the summit; also a railway agent. Their families live there with them, and there is no one else and nothing else, save the little sapphire lake lying in the bare hills.

Its blue waves lipped the porch whereon sat the young, sweet-faced wife of the Canadian inspector, with her baby in its carriage at her side.

This bit of liquid sapphire, scarcely larger than an artificial pond in a park, is really one of the chief sources of the Yukon—which, had these clear waters turned toward Lynn Canal, instead of away from it, might have never been. It seems so marvellous. The merest breath, in the beginning, might have toppled their liquid bulk over into the canyon through which we had so slowly and so enchantingly mounted, and in an hour or two they might have forced their foaming, furious way to the ocean. But some power turned the blue waters to the north and set them singing down through the beautiful chain of lakes—Lindeman, Bennett, Tagish, Marsh, Labarge—winding, widening, past ramparts and mountains, through canyons and plains, to Behring Sea, twenty-three hundred miles from this lonely spot.

This beginning of the Yukon is called the Lewes River. Far away, in the Pelly Mountains, the Pelly River rises and flows down to its confluence with the Lewes at old Fort Selkirk, and the Yukon is born of their union.

The Lewes has many tributaries, the most important of which is the Hootalinqua—or, as the Indians named it, Teslin—having its source in Teslin Lake, near the source of the Stikine River.

After leaving the summit the railway follows the shores of the river and the lakes, and the way is one of loveliness rather than grandeur. The saltish atmosphere is left behind, and the air tings with the sweetness of mountain and lake.

We had eaten an early breakfast, and we did not reach an eating station until we arrived at the head of Lake Bennett at half after one o'clock; and then we were given fifteen minutes in which to eat our lunch and get back to the train.

I do not think I have ever been so hungry in my life—and *fifteen minutes*! The dining room was clean and attractive; two long, narrow tables, or counters, extended the entire length of the room. They were decorated with great bouquets of wild flowers; the sweet air from the lake blew in through open windows and shook the white curtains out into the room.

The tables were provided with good food, all ready to be eaten. There were ham sandwiches made of lean ham. It was not edged with fat and embittered with mustard; it must have been baked, too, because no boiled ham could be so sweet. There were big brown lima beans, also baked, not boiled, and dill-pickles—no insipid pin-moneys, but good, sour, delicious dills! There were salads, home-made bread, "salt-rising" bread and butter, cakes and cookies and fruit—and huckleberry pie. Blueberries, they are called in Alaska, but they are our own mountain huckleberries.

No twelve-course luncheon, with a different wine for each course, could impress itself upon my memory as did that lunch-counter meal. We ate as children eat; with their pure, animal enjoyment and satisfaction. For fifteen minutes we had not a desire in the world save to gratify our appetites with plain, wholesome food. There was no crowding, no selfishness and rudeness,—as there had been in that wild scene on the excursion-boat, where the struggle had been for place rather than for food,—but a polite consideration for one another. And outside the sun shone, the blue waves sparkled and rippled along the shore, and their music came in through the open windows.

Here, in 1897, was a city of tents. Several thousand men and women camped here, waiting for the completion of boats and rafts to convey themselves and their outfits down the lakes and the river to the golden land of their dreams.

Standing between cars, clinging to a rattling brake, I made the acquaintance of Cyanide Bill, and he told me about it.

"Tents!" said he. "Did you say tents? Hunh! Why, lady, tents was as thick here in '97 and '98 as seeds on a strawberry. They was so thick it took a man an hour to find his own. Hunh! You tripped up every other step on a tent-peg. I guess nobody knows anything about tents unless he was mushin' around Lake Bennett in the summer of '97. From five to ten thousand men and women was camped here off an' on. Fresh ones by the hundred come strugglin', sweatin', dyin', in over the trail every day, and every day hundreds got their rafts finished, bundled their things and theirselves on to 'em, and went tearin' and yellin' down the lake, gloatin' over the poor tired-out wretches that just got in. Often as not they come sneakin' back afoot without any raft and without any outfit and worked their way back to the states to get another. Them that went slow, went sure, and got in ahead of the rushers.

"I wisht you could of seen the tent town!—young fellows right out of college flauntin' around as if they knew somethin'; old men, stooped and gray-headed; gamblers, tin horns, cut-throats, and thieves; honest women, workin' their way in with their husbands or sons, their noses bent to the earth, with heavy packs on their backs, like men; and gay, painted dance-hall girls, sailin' past 'em on horseback and dressed to kill and livin' on the fat of the land. I bet more good women went to the bad on this here layout than you could shake a stick at. It seemed to get on to their nerves to struggle along, week after week, packin' like animals, sufferin' like dogs, et up by mosquitoes and gnats, pushed and crowded out by men—and then to see them gay girls go singin' by, livin' on luxuries, men fallin' all over theirselves to wait on 'em, champagne to drink—it sure did get on to their nerves!

Council City and Solomon River Railroad—A Characteristic Landscape of Seward Peninsula

"You see, somehow, up here, in them days, things didn't seem the way they do down below. Nature kind of gets in her work ahead of custom up here. Wrong don't look so terrible different from right to a woman a thousand miles from civilization. When she sees women all around her walkin' on flowers, and her own feet blistered and bleedin' on stones and thorns, she's pretty apt to ask herself whether bein' good and workin' like a horse pays. And up here on the trail in '97 the minute a woman begun to ask herself that question, it was all up with her. The end was in plain sight, like the nose on a man's face. The dance hall on in Dawson answered the question practical.

"Of course, lots of 'em went in straight and stayed straight; and they're the ones that made Dawson and saved Dawson. You get a handful of good women located in a minin'-camp and you can build up a town, and you can't do it before, mounted police or no mounted police."

I had heard these hard truths of the Trail of Heartbreak before;

but having been worded more vaguely, they had not impressed me as they did now, spoken with the plain, honest directness of the old trail days.

"If you want straight facts about '97," the collector had said to me, "I'll introduce you to Cyanide Bill, out there. He was all through here time and again. He will tell you everything you want to know. But be careful what you ask him; he'll answer anything—and he doesn't talk parlor."

"The hardships such women went through," continued Cyanide Bill, "the insults and humiliations they faced and lived down, ought to of set 'em on a pe-*des*-tal when all was said and done and decency had the upper hand. The time come when the other'ns got their come-upin's; when they found out whether it paid to live straight.

"The world'll never see such a rush for gold again," went on Cyanide Bill, after a pause. "I tell you it takes a lot to make any impress on me, I've been toughenin' up in this country so many years; but when I arrives and sees the orgy goin' on along this trail, my heart up and stood still a spell. The strong ones was all a-trompin' the weak ones down. The weak ones went down and out, and the strong ones never looked behind. Men just went crazy. Men that had always been kind-hearted went plumb locoed and 'u'd trample down their best friend, to get ahead of him. They got just like brutes and didn't know their own selves. It's no wonder the best women give up. Did you ever hear the story of Lady Belle?"

I remembered Lady Belle, probably because of the name, but I had never heard the details of her tragic story, and I frankly confessed that I would like to hear them—"parlor" language or "trail," it mattered not.

"Well,"—he half closed his eyes and stared down the blue lake,—"she come along this trail the first of July, the prettiest woman you ever laid eyes on. Her husband was with her. He seemed to be kind to her at first, but the horrors of the trail worked on him, and he went kind of locoed. He took to abusin' her and blamin' her for everything. She worked like a dog and he treated her about like one; but she never lost her beauty nor her sweetness. She had the sweetest smile I ever saw on any human bein's face; and she was the only one that thought about others.

"'Don't crowd!' she used to cry, with that smile of her'n. 'We're all havin' a hard time together.'

"Well, they lost their outfit in White Horse Rapids; her husband cursed her and said it wouldn't of happened if she hadn't been hell-bent to come along; he took to drinkin' and up and left her there at the rapids. He went back to the states, sayin' he didn't ever want to see her again.

"She was left there without an ounce of grub or a cent of money. Yakataga Pete had been workin' along the trail with a big outfit, and had gone on in ahead. He'd fell in love with her before he knew she was married. He went on up into the cricks, and when he come down to Dawson six months later, she was in a dance hall. Dawson was wild about her. They called her Lady Belle because she was always such a lady.

"Yakataga went straight to her and asked her to marry him. She burst out into the most terrible cryin' you ever hear. 'As if I could ever marry anybody!' she cries out; and that's all the answer he ever got. We found out she had a little blind sister down in the states. She had to send money to keep her in a blind school. She danced and acted cheerful; but her face was as white as chalk, and her big dark eyes looked like a fawn's eyes when you've shot it and not quite killed it, so's it can't get away from you, nor die, nor anything; but she was always just as sweet as ever.

"Two months after that she—she—killed herself. Yakataga was up in the cricks. He come down and buried her."

It was told, the simple and tragic tale of Lady Belle, and presently Cyanide Bill went away and left me.

The breeze grew cooler; it crested the waves with silver. Pearly clouds floated slowly overhead and were reflected in the depths below.

The mountains surrounding Lake Bennett are of an unusual

color. It is a soft old-rose in the distance. The color is not caused by light and shade; nor by the sun; nor by flowers. It is the color of the mountains themselves. They are said to be almost solid mountains of iron, which gives them their name of "Iron-Crowned," I believe; but to me they will always be the Rose-colored Mountains. They soften and enrich the sparkling, almost dazzling, blue atmosphere, and give the horizon a look of sunset even at midday. The color reminded me of the dull old-rose of Columbia Glacier.

Lake Bennett dashes its foam-crested blue waves along the pebbly beaches and stone terraces for a distance of twenty-seven miles. At its widest it is not more than two miles, and it narrows in places to less than half a mile. It winds and curves like a river.

The railway runs along the eastern shore of the lake, and mountains slope abruptly from the opposite shore to a height of five thousand feet. The scenery is never monotonous. It charms constantly, and the air keeps the traveller as fresh and sparkling in spirit as champagne.

For many miles a solid road-bed, four or five feet above the water, is hewn out of the base of the mountains; the terrace from the railway to the water is a solid blaze of bloom; white sails, blown full, drift up and down the blue water avenue; cloud-fragments move silently over the nearer rose-colored mountains; while in the distance, in every direction that the eye may turn, the enchanted traveller is saluted by some lonely and beautiful peak of snow. It is an exquisitely lovely lake.

We had passed Lake Lindeman—named by Lieutenant Schwatka for Dr. Lindeman of the Breman Geographical Society—before reaching Bennett.

Lake Lindeman is a clear and lovely lake seven miles long, half a mile wide, and of a good depth for any navigation required here. A mountain stream pours tumultuously into it, adding to its picturesque beauty.

Sea birds haunt these lakes, drift on to the Yukon, and follow the voyager until they meet their silvery fellows coming up from

Behring Sea.

Between Lakes Lindeman and Bennett the river connecting link is only three quarters of a mile long, about thirty yards wide, and only two or three feet deep. It is filled with shoals, rapids, cascades, boulders, and bars; and navigation is rendered so difficult and so dangerous that in the old "raft" days outfits were usually portaged to Lake Bennett.

During the rush to the Klondike a saw-mill was established at the head of Lake Bennett, and lumber for boat building was sold for one hundred dollars a thousand feet.

The air in these lake valleys on a warm day is indescribably soft and balmy. It is scented with pine, balm, cottonwood, and flowers. The lower slopes are covered with fireweed, larkspur, dandelions, monk's-hood, purple asters, marguerites, wild roses, dwarf goldenrod, and many other varieties of wild flowers. The fireweed is of special beauty. Its blooms are larger and of a richer red than along the coast. Blooms covering acres of hillside seem to float like a rosy mist suspended in the atmosphere. The grasses are also very beautiful, some having the rich, changeable tints of a humming-bird.

The short stream a couple of hundred yards in width connecting Lake Bennett with the next lake—a very small, but pretty one which Schwatka named Nares—was called by the natives "the place where the caribou cross," and now bears the name of Caribou Crossing. At certain seasons the caribou were supposed to cross this part of the river in vast herds on their way to different feeding-grounds, the current being very shallow at this point.

There is a small settlement here now, and boats were waiting to carry passengers to the Atlin mining district. The caribou have now found less populous territories in which to range. In the winter of 1907-1908 they ranged in droves of many thousands—some reports said hundreds of thousands—through the hills and valleys of the Stewart, Klondike, and Sixty-Mile rivers, in the Upper Yukon country.

Miners killed them by the hundreds, dressed them, and stored them in the shafts and tunnels of their mines, down in the eternally frozen caverns of the earth—thus supplying themselves with the most delicious meat for a year. The trek of caribou from the Tanana River valley to the head of White River consumed more than ninety days in passing the head of the Forty-Mile valley—at least a thousand a day passing during that period. They covered from one to five miles in width, and trod the snow down as solidly as it is trodden in a city street. A great wolf-pack clung to the flank of the herd. The wolves easily cut out the weak or tired-out caribou and devoured them.

Caribou Crossing is a lonely and desolate cluster of tents and cabins huddling in the sand on the water's edge. Considerable business is transacted here, and many passengers transfer here in summer to Atlin. In winter they leave the train at Log-Cabin, which we passed during the forenoon, and make the journey overland in sleighs.

The voyage from Caribou Crossing to Atlin is by way of a chain of blue lakes, pearled by snow mountains. It is a popular round-trip tourist trip, which may be taken with but little extra expense from Skaguay.

Tagish Lake, as it was named by Dr. Dawson,—the distinguished British explorer and chief director of the natural history and geological survey of the Dominion of Canada,—was also known as Bove Lake. Ten miles from its head it is joined by Taku Arm—Tahk-o Lake, it was called by Schwatka.

The shores of Tagish Lake are terraced beautifully to the water, the terraces rising evenly one above another. They were probably formed by the regular movement of ice in other ages, when the waters in these valleys were deeper and wider. There are some striking points of limestone in this vicinity, their pearl-white shoulders gleaming brilliantly in the sunshine, with sparkling blue waves dashing against them.

Marsh Lake, and another with a name so distasteful that I

will not write it, are further links in the brilliant sapphire water chain by which the courageous voyagers of the Heartbreak days used to drift hopefully, yet fearfully, down to the Klondike. The bed of a lake which was unintentionally drained completely dry by the builders of the railroad is passed just before reaching Grand Canyon.

The train pauses at the canyon and again at White Horse Rapids, to give passengers a glimpse of these famed and dreaded places of navigation of a decade ago.

At six o'clock in the evening of the day we left Skaguay we reached White Horse.

CHAPTER XLII

This is a new, clean, wooden town, the first of any importance in Yukon Territory. It has about fifteen hundred inhabitants, is the terminus of the railroad, and is growing rapidly. The town is on the banks of Lewes River, or, as they call it here, the Yukon.

There is an air of tidiness, order, and thrift about this town which is never found in a frontier town in "the states." There are no old newspapers huddled into gutters, nor blowing up and down the street. Men do not stand on corners with their hands in their pockets, or whittling out toothpicks, and waiting for a railroad to be built or a mine to be discovered. They walk the streets with the manner of men who have work to do and who feel that life is worth while, even on the outposts of civilization.

All passengers, freight, and supplies for the interior now pass through White Horse. The river bank is lined with vast warehouses which, by the time the river opens in June, are piled to the roofs with freight. The shipments of heavy machinery are large. From the river one can see little besides these warehouses, the shipyards to the south, and the hills.

Passing through the depot one is confronted by the largest hotel, the White Pass, directly across the street. To this we walked; and from an upstairs window had a good view of the town. The streets are wide and level; the whole town site is as level as a parade-ground. The buildings are frame and log; merchandise is fair in quality and style, and in price, high. Mounted police strut stiffly and importantly up and down the streets to and from their picturesque log barracks. One unconsciously holds one's chin level and one's shoulders high the instant one enters a Yukon town. It is in the air.

Excellent grounds are provided for all outdoor sports; and in the evening every man one meets has a tennis racket or a golf stick in his hand, and on his face that look of enthusiastic anticipation which is seen only on a British sportsman's face. No American, however enthusiastic or "keen" he may be on outdoor sports, ever quite gets that look.

There was no key to our door. Furthermore, the door would not even close securely, but remained a few hair breadths ajar. There was no bell; but on our way down to dinner, having left some valuables in our room, we reported the matter to a porter whom we met in the hall, and asked him to lock our door.

"It doesn't lock," he replied politely. "It doesn't even latch, and the key is lost."

Observing our amazed faces, he added, smiling:—

"You don't need it, ladies. You will be as safe as you would be at home. We never lock doors in White Horse."

This was my first Yukon shock, but not my last. My faith in mounted police has always been strong, but it went down before that unlocked door.

"Possibly the people of White Horse never take what does not belong to them," I said; "but a hundred strangers came in on that train. Might not *one* be afflicted with kleptomania?"

"He wouldn't steal here," said the boy, confidently. "Nobody ever does."

There seemed to be nothing more to say. We left our door ajar and, with lingering backward glances, went down to the dining room.

Never shall I forget that dinner. It was as bad as our lunch had been good. The room was hot; the table-cloth was far from being immaculate; the waitress was untidy and ill-bred; and there was nothing that we could eat.

Nor were we fastidious. We neither expected, nor desired, luxuries; we asked only well-cooked, clean, wholesome food; but if this is to be obtained in White Horse, we found it not—although we did not cease trying while we were there.

We went out and walked the clean streets and looked into restaurants, and tried to see something good to eat, or at least a clean table-cloth; but in the end we went hungry to bed. We had wine and graham wafers in our bags, and they consoled; but we craved something substantial, notwithstanding our hearty lunch. It was the air—the light, fresh, sparkling air of mountain, river, and lake—that gave us our appetites.

When we had walked until our feet could no longer support us, we returned to the hotel. On the way, we saw a sign announcing ice-cream soda. We went in and asked for some, but the ice-cream was "all out."

"But we have plain soda," said the man, looking so wistful that we at once decided to have some, although we both detested it.

He fizzed it elaborately into two very small glasses and led us back into a little dark room, where were chairs and tables, and he gave us spoons with which to eat our plain soda. "Let me pay," said my friend, airily; and she put ten cents on the table.

The man looked at it and grinned. He did not smile; he grinned. Then he went away and left it lying there.

We tried to drink the soda-water; then we tried to coax it through straws; finally we tried to eat it with spoons—as others about us were doing; but we could not. It looked like soap-bubbles and it tasted like soap-bubbles.

"He didn't see his ten cents," said my friend, gathering it up. "I suppose one pays at the counter out there. I would cheerfully pay him an extra ten if I had not gotten the taste of the abominable stuff in my mouth."

She laid the ten cents on the counter grudgingly.

The man looked at it and grinned again.

"Them things don't go here," said he. "It's fifty cents."

There was a silence. I found my handkerchief and laughed into it, wishing I had taken a second glass.

"Oh, I see," said she, slowly and sweetly, as a half-dollar slid lingering down her fingers to the counter. "For the spoons. They were worth it."

It was two o'clock before we could leave our windows that night. It was not dark, not even dusk. A kind of blue-white light lay over the town and valley, deepening toward the hills. In the air was that delicious quality which charms the senses like perfumes. Only to breathe it in was a drowsy, languorous joy. At White Horse one opens the magic, invisible gate and passes into the enchanted land of Forgetfulness—and the gate swings shut behind one.

Home and friends seem far away. If every soul that one loves were at death's door, one could not get home in time to say farewell—so why not banish care and enjoy each hour as it comes?

This is the same reckless spirit which, greatly intensified, possessed desperate men when they went to the Klondike ten years ago. There was no telegraph, then, and mails were carried in only once or twice a year. Letters were lost. Men did not hear from their wives, and, discouraged and disheartened, decided that the women had died or had forgotten; so they went the way of the country, and it often came to pass that Heartbreak Trail led to the Land of Heartbreak.

In the morning we learned that the boat for Dawson was not yet "in," and, even if it should arrive during the day,—which seemed to be as uncertain as the opening of the river in spring,—would not leave until some time during the night; so at nine o'clock we took the Skaguay train for the Grand Canyon.

One "oldest" resident of White Horse told us that it was only a mile to the canyon; another oldest one, that it was four miles; still another, that it was five; all agreed that we should take the train out and walk back.

"There's a tram," they told us, "an old, abandoned tram, and you can't get lost. You've only to follow the tram. Why, a *goose* couldn't get lost. Norman McCauley built the tram, and outfits were portaged around the canyon and the rapids two seasons; then the railroad come in and the tram went out of business."

We took our bundles of mosquito netting and boarded the train. In summer the travel is all "in," and we were the only passengers. When the White Pass Railway Company was organized, stock was worth ten dollars a share; now it is worth six hundred and fifty dollars, and it is not for sale. Freight rates are five cents a pound, one hundred dollars a ton, or fifty in car-load lots, from Skaguay to White Horse. Passenger rates are supposed to be twenty cents a mile. We paid seventy-five cents to return to the canyon which we passed the previous day. This rate should make the distance four miles, and we barely had time to arrange our mosquito veils, according to the instructions of the conductor, when the train stopped.

We were told that we might not see a mosquito; and again, that we might not be able to see anything else.

We were put off and left standing ankle-deep in sand, on the brink of a precipice, four miles from any human being—in the wilds of Alaska. At that moment the trainmen looked like old and dear friends.

"The path down is right in front of you," the collector called, as the train started. "Don't be afraid of the bears! They will not harm you at this time of the year."

Bears!

We had considered heat, mosquitoes, losing our way, hunger, exhaustion,—everything, it appeared, except bears. We looked at one another.

"I had not thought of bears."

"Nor had I."

We looked down at the bushes growing along the canyon; little heat-worms glimmered in the still atmosphere.

"Perhaps it is an Alaskan joke," I suggested feebly.

We stood for some time trying to decide whether we should make the descent or return to White Horse, when suddenly the matter was decided for us. I was standing on the brink of the sandy precipice, down which a path went, almost perpendicularly, without bend or pause, to the bank of the river several hundred yards below.

The sandy soil upon which I stood suddenly caved and went

down into the path. I went with it. I landed several yards below the brink, gave one cry, and then—by no will of my own—was off for the canyon.

The caving of the brink had started a sand and gravel slide; and I, knee-deep in it, was going down with it—slowly, but oh, most surely. There was no pausing, no looking back. I could hear my companion calling to me to "stop"; to "wait"; to "be careful"—and all her entreaties were the bitterest irony by the time they floated down to me. So long as the slide did not stop, it was useless to tell me to do so; for I was embedded in it halfway to my waist. We kept going, slowly and hesitatingly; but never slowly enough for me to get out.

It was eighty in the shade, and the sand was hot. I was wearing a white waist, a dark blue cheviot skirt, and patent-leather shoes; and my appearance, when I finally reached level ground and cool alder trees, may be imagined. Furthermore, our trunks had been bonded to Dawson, and I had no extra skirts or shoes with me.

My companion, profiting by my misfortune, had armed herself with an alpenstock and was "tacking" down the slope. It was half an hour before she arrived.

I have never forgiven her for the way she laughed.

We soon forgot the bears in the beauty of the scene before us. We even forgot the comedy of my unwilling descent.

The Lewes River gradually narrows from a width of three or four hundred yards to one of about fifty yards at the mouth of the Grand Canyon, which it enters in a great bore.

The walls of the canyon are perpendicular columns and palisades of basalt. They rise without bend to a height of from one to two hundred feet, and then, set thickly with dark and gloomy spruce trees, slope gradually into mountains of considerable height. The canyon is five-eighths of a mile long, and in that interval the water drops thirty feet. Halfway through, it widens abruptly into a round water chamber, or basin, where the waters boil and seethe in dangerous whirlpools and eddies. Then it again narrows, and the waters rush wildly and tumultuously through walls of dark stone, veined with gray and lavender. The current runs fifteen miles an hour, and rafts "shooting" the rapids are hurled violently from side to side, pushed on end, spun round in whirlpools, buried for seconds in boiling foam, and at last are shot through the final narrow avenue like spears from a catapult—only to plunge madly on to the more dangerous White Horse Rapids.

The waves dash to a height of four or five feet and break into vast sheets of spray and foam. Their roar, flung back by the stone walls, may be heard for a long distance; and that of the rapids drifts over the streets of White Horse like distant, continuous thunder, when all else is still.

We found a difficult way by which, with the assistance of alpenstocks and overhanging tree branches, we could slide down to the very water, just above Whirlpool Basin. We stood there long, thinking of the tragedies that had been enacted in that short and lonely stretch; of the lost outfits, the worn and wounded bodies, the spirits sore; of the hearts that had gone through, beating high and strong with hope, and that had returned broken. It is almost as poignantly interesting as the old trail; and not for two generations, at least, will the perils of those days be forgotten.

It was about noon that, remembering our long walk, we turned reluctantly and set out for White Horse.

Somewhere back of the basin we lost our way. We could not find the "tram"; searching for it, we got into a swamp and could not make our way back to the river; and suddenly the mosquitoes were upon us.

The underbrush was so thick that our netting was torn into shreds and left in festoons and tatters upon every bush; yet I still bear in my memory the vision of my friend floating like a tall, blond bride—for my dark-haired Scotch friend was not with me on the Yukon voyage—through the shadows of that swamp before her bridal veil went to pieces.

Her bridal glory was grief. In a few moments we were both

as black as negroes with mosquitoes; for, desperately though we fought, we could not drive them away. The air in the swamp was heavy and still; our progress was unspeakably difficult—through mire and tall, lush grasses which, in any other country on earth, would have been alive with snakes and crawling things.

The pests bit and stung our faces, necks, shoulders, and arms; they even swarmed about our ankles; while, for our hands—they were soon swollen to twice their original size.

We wept; we prayed; we said evil things in the hearing of heaven; we asked God to forgive us our sins, or, at the very least, to punish us for them in some other way; but I, at least, in the heaviest of my afflictions, did not forget to thank Him because there are no snakes in Alaska or the Yukon. It seemed to me, even, in the fervor of my gratitude, that it had all been planned æons ago for our special benefit in this extreme hour.

But I shall spare the reader a further description of our sufferings.

I had always considered the Alaskan mosquito a joke. I did not know that they torture men and beasts to a terrible death. They mount in a black mist from the grass; it is impossible for one to keep one's eyes open. Dogs, bears, and strong men have been known to die of pain and nervous exhaustion under their attacks.

After an hour of torture we forced our way through the network of underbrush back to the river, and soon found a narrow path. There was a slight breeze, and the mosquitoes were not so aggressive. There was still a three-mile walk, along the shore bordering the rapids, before we could rest; and during the last mile each step caused such agony that we almost crawled.

When we removed our shoes, we found them full of blood. Our feet were blistered; the blisters had broken and blistered again.

Teller

But we had seen the Grand Canyon of the Yukon—which Schwatka in an evil hour named Miles, for the distinguished army-general—and White Horse Rapids; and seeing them was worth the blisters and the blood. And we know how far it is from the head of the canyon to White Horse town. No matter what the three "oldest" settlers, the railway folders, Schwatka, and all the others say,—*we know.* It is fifteen miles! Also, among those who scoff at Rex Beach for having the villain in his last novel eaten up by mosquitoes on the Yukon, we are not to be included.

Numerous and valuable copper mines lie within a radius of fifteen miles from White Horse. The more important ones are those of the Pennsylvania syndicate, The B. N. White Company, The Arctic Chief, The Grafter, the Anaconda, and the Best Chance. The Puebla, operated by B. N. White, lies four miles northwest of town. It makes a rich showing of magnetite, carrying copper values averaging four and five per cent, with a small by-product of gold and silver.

In the summer of 1907 this mine had in sight two hundred

and fifty thousand tons of pay ore. The deepest development then obtained had a hundred-foot surface showing three hundred feet in width, and stripped along with the strike of the vein seven hundred feet, showing a solid, unbroken mass of ore. Tunnels and crosscuts driven from the bottom of the shaft showed the body to be the same width and the values the same as the surface outcrop.

The Arctic Chief ranks second in importance; and extensive development work is being carried on at all the mines. The railway is building out into the mining district.

Six-horse stages are run from White Horse to Dawson after the river closes. The distance is four hundred and thirty-five miles; the fare in the early autumn and late spring is a hundred and twenty-five dollars; in winter, when sleighing is good, sixty dollars.

White Horse was first named Closeleigh by the railway company; but the name was not popular. At one place in the rapids the waves curving over rocks somewhat resemble a white horse, with wildly floating mane and tail of foam. This is said to be the origin of the name.

White Horse is only eight years old. The hotel accommodations, if one does not mind a little thing like not being able to eat, are good. The rooms are clean and comfortable and filled with sweet mountain and river air.

At eight o'clock that evening the steamer *Dawson* struggled up the river and landed within fifty yards of the hotel. We immediately went aboard; but it was nine o'clock the next morning before we started, so we had another night in White Horse.

The Yukon steamers are four stories high, with a place for a roof garden. I could do nothing for some time but regard the *Dawson* in silent wonder. It seemed to glide along on the surface of the water, like a smooth, flat stone when it is "skipped."

The lower deck is within a few inches of the water; and high above is the pilot-house, with its lonely-looking captain and pilot;

and high, oh, very high, above them—like a charred monarch of a Puget Sound forest—rises the black smoke-stack, from which issue such vast funnels of smoke and such slow and tremendous breathing.

This breathing is a sound that haunts every memory of the Yukon. It is not easy to describe, it is so slow and so powerful. It is not quite like a cough—unless one could cough *in*instead of *out*; it is more like a sobbing, shivering in-drawing of the breath of some mighty animal. It echoes from point to point, and may be heard for several miles on a still day. Day and night it moves through the upper air, and floats on ahead, often echoing so insistently around some point which the steamer has not turned, that the "cheechaco" is deluded into the belief that another steamer is approaching.

The captains and pilots of the Yukon are the loneliest-looking men! First of all, they are so far away from everybody else; and second, passengers, particularly women, are not permitted to be in the pilot-house, nor on the texas, nor even on the hurricane-deck, of steamers passing through Yukon Territory.

Between White Horse and Lake Lebarge the river is about two hundred yards wide. The water is smooth and deep. It loiters along the shore, but the current is strong and bears the steamer down with a rush, compelling it to zigzag ceaselessly from shore to shore.

Going down the Yukon for the first time, one's heart stands still nearly half the time. The steamer heads straight for one shore, approaches it so closely that its bow is within six inches of it, and then swings powerfully and starts for the opposite shore—its great stern wheel barely clearing the rocky wall.

The serious vexations and real dangers of navigation in this great river, from source to mouth, are the sand and gravel bars. One may go down the Yukon from White Horse to St. Michael in fourteen days; and one may be a month on the way—pausing, by no will of his own, on various sand-bars.

The treacherous current changes hourly. It is seldom found

twice the same. It washes the sand from side to side, or heaps it up in the middle—creating new channels and new dangers. The pilot can only be cautious, untiringly watchful—and lucky. The rest he must leave to heaven.

It is twenty-seven miles from White Horse to Lake Lebarge. Midway, the Tahkeena River flows into the Lewes, running through banks of clay.

Lake Lebarge is thirty-two miles long and three and a half wide. The day was suave. The water was silvery blue, and as smooth as satin; gray, deeply veined cliffs were reflected in the water, whose surface was not disturbed by a ripple or wave; the air was soft; farther down the river were forest fires, and just sufficient haze floated back to give the milky old-rose lights of the opal to the atmosphere. There is one small island in the lake. It was not named; and it received the name—as Vancouver would say—of Fireweed Isle, because it floated like a rosy cloud on the pale blue water.

The Indians called this lake Kluk-tas-si, and Schwatka favored retaining it; but the French name has endured, and it is not bad.

The Lake Lebarge grayling and whitefish are justly famed. Steamers stop at some lone fisherman's landing and take them down to Dawson, where they find ready sale. At Lower Lebarge there is a post-office and a telegraph station. Our steamer paused; two men came out in a boat, delivered a large supply of fish, received a few parcels of mail, and went swinging back across the water.

A dreary log-cabin stood on the bank, labelled "Clark's Place." A woman in a scarlet dress, walking through the reeds beside the beach, made a bit of vivid color. It seemed very, very lonely—with that kind of loneliness that is unendurable.

A quarter of a mile farther, around a bend in the shore, the boat landed at the telegraph station, where the Canadian flag was flying.

The different reaches of the Yukon are called locally by very confusing names. The river rising in Summit Lake on the

White Pass railway is called both Lewes and Yukon; the stretch immediately below Lake Lebarge is called Lewes, Thirty-Mile, and Yukon. When we reach the old Hudson Bay post of Selkirk, however, our perplexities over this matter are at an end. The Pelly River here joins the Lewes, and all agree that the splendid river that now surges on to the sea is the Yukon.

It is daylight all the time, and no one should sleep between White Horse and Dawson. Not an hour of this beautiful voyage on the Upper Yukon should be wasted.

The banks are high and bold, for the most part springing sheer out of the water in columns and pinnacles of solid stone. There are also forestated slopes rising to peaks of snow; and the same kind of clay cliffs that we saw at White Horse, white and shining in the bluish light of morning, but more beautiful still in the mysterious rosy shadows of midnight.

There are some striking columns of red rock along Lake Lebarge, and their reflections in the water at sunset of a still evening are said to be entrancing: "two warm pictures of rosy red in the sinking sun, joined base to base by a thread of silver, at the edge of the other shore."

There are many high hills of soft gray limestone, veined and shaded with the green of spruce; vast slopes, timbered heavily; low valleys and picturesque mouths of rivers.

Five-Finger, or Rink, Rapids is caused by a contraction of the river from its usual width to one of a hundred and fifty yards. Five bulks of stone, rising to a perpendicular height of forty or fifty feet, are stretched across the channel. The steamer seems to touch the stone walls as it rushes through on the boiling rapids.

The Upper Ramparts of the Yukon begin at Fort Selkirk. Here the waters cut through the lower spurs of the mountains, and for a distance of a hundred and fifty miles, reaching to Dawson, the scenery is sublime.

"Quiet Sentinel" is a rocky promontory which, seen in profile, resembles the face and entire figure of a woman. She stands with her head slightly bowed, as if in prayer, with loose draperies flowing in classic lines to her feet, and with a rose held to her lips. One of the greatest singers of the present time might have posed for the "Quiet Sentinel."

Rivers and their valleys are more famed in the northern interior than towns. Teslin, Tahkeena, Teslintoo, Big and Little Salmon, Pelly, Stewart, White, Forty-Mile, Indian, Sixty-Mile, Macmillan, Klotassin, Porcupine, Chandlar, Koyukuk, Unalaklik, Tanana, Mynook,—these be names to conjure with in the North; while those south of the Yukon and tributary to other waters have equal fame.

As for the Klondike, it is the only stream of its size, being but the merest creek and averaging a hundred feet in width, which has given its name to one whole country and to a portion of another country. During the past decade it has not been unusual to hear the name Klondike Country applied to all Alaska and that part of Canada adjacent to the Klondike district. The tiny, gold-bearing creeks, from ten to twenty feet wide, tributary to the Klondike, are known by name and fame in all parts of the world to-day. They are Bonanza, Hunker, Too-Much-Gold, Eldorado, Rock, North Fork, All-Gold, Gold-Bottom, and others of less importance. The Bonanza flows into the Klondike at Dawson, and it is but a half-hour's walk to the dredge at work in this stream.

In 1833 Baron Wrangell directed Michael Tebeňkoff to establish Fort St. Michael's on the small island in Norton Sound to which the name of the fort was given. Three years later it was attacked by natives, but was successfully defended by Kurupanoff, who was in charge.

In 1836 a Russian named Glasunoff entered the delta of the Yukon, ascending the river as far as the mouth of the Anvik River. In 1838 Malakoff extended the exploration as far as Nulato, where he established a Russian post and placed Notarmi in command.

When the garrison returned to St. Michael's on account of

the failure of provisions, the following winter, natives destroyed the fort and all buildings which had been erected. It was rebuilt and again destroyed in 1839. In 1841 it once more arose under Derabin, who remained in command. The following year Lieutenant Zagoskin reached Nulato, ascending to Nowikakat in 1843.

The Russians were therefore established on the lower Yukon several years before the English established themselves upon the upper river.

In 1840 Mr. Robert Campbell was sent by Sir George Simpson to explore the Upper Liard River. Mr. Campbell ascended the river to its head waters, crossed the mountains, and descended the Pelly River to the Lewes, where, eight years later, he established Fort Selkirk.

This famous trading post was short-lived. In 1851 it was attacked by a band of savage Chilkahts and was surrendered, without resistance, by Mr. Campbell, who had but two men with him at the time. They were not molested by the Indians, who plundered and burned the warehouses and forts.

Only the chimneys of the fort were found by Lieutenant Schwatka in 1883. As late as 1890 this point was considered the head of navigation on the Yukon.

In 1847 Fort Yukon was established by Mr. A. H. McMurray, of the Hudson Bay Company. Following McMurray and Campbell, came Joseph Harper, Jack McQuesten, and A. H. Mayo, who established a trading post on the Yukon at Fort Reliance, six miles below the mouth of the Klondike.

In 1860 Robert Kennicott reached Fort Yukon, and in the following spring descended to a point that was for several years known as "the Small Houses"—the most attractive name in the Yukon country. In 1865 an expedition was organized in San Francisco by the Western Union Telegraph Company for the purpose of building a telegraph line from San Francisco to Behring Strait—which was to be crossed by cable to meet the Russian government line at the mouth of the Amoor River. One party, headed by Robert Kennicott, was sent by ocean to the mouth of the Yukon; and another, in charge of Michael Byrnes, up the inside route to the Stikine River. Going from that river to the head waters of the Taku, they followed the chain of lakes and the Hootalinqua River to the Lewes, which they reached on the Tahco Arm of Lake Tagish. At that time it became known that the Atlantic cable had proven to be a success, and the daring and hazardous northern project was abandoned.

As late as the date of this expedition it was not determined positively whether the Kwihkpak was one of the mouths of the Yukon, or a separate river. Upon the recall of the telegraph expedition, the only portion of the great river that had not been explored was the short distance between Lake Tagish and Lake Lebarge.

There have been several claimants for the honor of having been the first white man to cross the divide between Lynn Canal and the head waters of the Yukon. The first was a mythological, nameless Scotchman employed by the Hudson Bay Company, who is supposed to have reached Fort Selkirk in 1864, and to have proceeded alone over the old "grease-trail" of the Chilkahts to Lynn Canal. He fell into the hands of the Indians and was held until ransomed by the captain of the *Labouchere*. Because he had long, flowing locks of red hair, he was supposed to be a kind of white shaman, and his life was spared by the savages. This story is doubted by many authorities.

The honor was claimed, also, by George Holt, who is known to have crossed one of the passes in 1872, and twice in later years. James Wynn, of Juneau, went over in 1879 and returned in 1880.

About this time the Indians seemed to realize that packing over the trail might become more profitable than acting as middlemen between the coast Indians and those of the interior. In 1881 and 1882 small parties of miners, and even one or two travelling alone, crossed unmolested. In 1883 Lieutenant Schwatka had his outfit packed over the Dyea—Taiya, or Dayay, it was then called—Trail; and then, dismissing his packers, built rafts and made his perilous way down the unknown river—portaging, "shooting" the Grand Canyon, White Horse, and Rink Rapids, sticking on sand-bars, almost dying of mosquitoes, and, saddest of all for us who come after him, naming every object that met his eyes with the deplorable taste of Vancouver.

Of a river, called Kut-lah-cook-ah by the Chilkahts, he complacently remarks:—

"I shortened its name and called it after Professor Nourse, of the United States Naval Observatory."

Nourse, Saussure, Perrier, Payer, Bennett, Wheaton, Prejevalsky, Richards, Watson, Nares, Bove, Marsh, McClintock, Miles, Richthofen, Hancock, d'Abbadie, Daly, Nordenskiold, Yon Wilczek; these be the choice namings that he bestowed upon the beautiful objects along the Yukon. It is, perhaps, a cause for thankfulness that he did not rename the Yukon *Schwatka* or *Ridderbjelka*! However, many of his namings have died a natural death.

The name Yukon is said to have first been applied to the river in 1846 by Mr. J. Bell, of the Hudson Bay Company, who went over from the MacKenzie and descended the Porcupine to the great river which the Indians called Yukon. He retained the name, although for some time it was spelled Youkon. For this, may he ever be of blessed memory. I should like to contribute to a monument to perpetuate his name and fame.

To-day Fort Selkirk is of some importance as a trading post and because of the successful farming of the vicinity, and all passing steamers call there. Joseph Harper was located there at the time of George Carmack's brilliant discovery of gold on Bonanza Creek, in August, 1896. Harper and Joseph Ladue, who was settled as a trader at Sixty-Mile, immediately transferred their stocks to the junction of the Yukon, Klondike, and Bonanza, and established the town which they named Dawson, in honor of Dr. George M. Dawson.

In 1887 Mr. William Ogilvie headed a Canadian exploring party into the Yukon. His boats were towed up to Taiya Inlet by the United States naval vessel *Pinta*; and while waiting there for supplies, he, having asked for, and received, authority from Commander Newell, made surveys at the heads of the inlets. It was only through the intercession of the commander, furthermore, that Mr. Ogilvie was permitted by the Chilkahts to proceed over the pass. "I am strongly of the opinion," Mr. Ogilvie says in his report, "that these Indians would have been much more difficult to deal with if they had not known that Commander Newell remained in the inlet to see that I got through in safety."

Miners had been going over the trail for several years, but the Chilkahts were enraged at the British because employees of the Hudson Bay Company had killed some of their tribe.

In the meantime Dr. George M. Dawson, heading another Dominion party, was working along the Stikine River.

Dr. Dawson and Mr. Ogilvie—afterward governor of Yukon territory—made extensive surveys and explorations throughout the Yukon district; their reports upon the country are voluminous, thorough, and of much interest. They were both men of superior attainments, and their influence upon the country and upon the people who rushed into the new mining district was great. To-day the name of ex-Governor Ogilvie is heard more frequently in the Klondike than that of any other person, even though his residence is elsewhere. He served as governor during the reckless and picturesque days when to be a governor meant to be a man in the highest sense of the word.

CHAPTER XLIII

Dawson! It was a name to stir men's blood ten years ago,—a wild, picturesque, lawless mining-camp, whose like had never been known and never will be known again.

All kinds and conditions of men and women were represented. Miners, prospectors, millionnaires, adventurers, wanderers, desperadoes; brave-hearted, earnest women, dissolute dance-hall girls, and, more dangerous still, the quiet, seductive adventuress—they were all there, side by side, tent by tent, cabin by cabin.

Almost daily new discoveries were made and stampedes occurred. Every little creek flowing into the Klondike was found rich in gold. The very names that these creeks received—All-Gold, Too-Much-Gold, Gold-Bottom—turned men's blood to fire. The whole country seemed to have gone mad of excitement and the lust for gold. The white mountain passes grew black with struggling human beings—fighting, falling, rising, fighting on. It was like the blind stampeding of crazed animals upon a plain; nothing could check them save exhaustion or death. When the fever burned out in one and left him low, another sprang to take his place. Dawson, like Skaguay, grew from dozens to hundreds in a day; from hundreds to thousands; tents gave place to cabins; cabins, to substantial frame buildings.

Ah, to have been there in the old days! Who would not have suffered the early hardships, paid the price, and paid it cheerfully, for the sake of seeing the life and being a part of it before it was too late?

Now it is forever too late. The glory of what it once was is all that remains. To-day Dawson is so quiet, so dull, so respectable, that one unconsciously yawns in its face.

But men's eyes still kindle when their memories of old days

are stirred.

"They were great times," they say, looking at one another.

"They could only come once. They were times of blood and gold; of dance and song; of glitter and show—and starvation and death. We worked all day and danced or gambled all night. Our only passions were for women and gold. If we couldn't get the women we wanted, the men that did get 'em fought their way to 'em, inch by inch; if we couldn't dig the gold out of the earth, we got it in some other way.

"All the best buildings were occupied by saloons. Every saloon had a dance-hall in the back of it; not that the girls had to keep to their quarters, either—they had the run of the whole shebang. Every saloon had its gambling rooms, too—unless the tables and games were right out in the open. I tell you, it was tough. You can't begin to understand the situation unless you'd been here. There wasn't a hotel nor a corner where a man could go in and get warm except in a saloon—and with the thermometer fooling in the neighborhood of fifty below, he didn't stand around outside with his hands in his pockets, not to any great extent. Most likely his pockets was naturally froze shut, anyhow, and the only way he could get 'em thawed out was to go into a saloon. *That* thawed 'em quick enough. It not only thawed 'em out; it most gen'rally thawed 'em wide open.

"I tell you, the worst element in a mining-camp is women. They follow a man and console him when he's down on his luck; they follow him through thick and thin; and they get such a hold on him that, when he wants to get back to decent ways and decent women, he just naturally can't do it. Young fellows don't realize it. They don't see it being done; they see it after it is done and can't be undone.

"As soon as the mounted police took holt of Dawson, with Inspector Constantine at the head, there was a sure change. Still, even the mounted-police doctrine does have some drawbacks. I noticed they couldn't make the post-office clerks turn out letters unless you slipped two-three dollars into their outstretched hands. I noticed that."

To-day Dawson is a pretty, clean-streeted town built of log and frame buildings. In the hottest summer the earth never thaws deeper than eighteen inches, and no foundation can be obtained for brick buildings. For the same reason plastering is not advisable, the uneven freezing and thawing proving ruinous to both brick and plaster.

The first objects to greet the visitor's eyes are the large buildings of the great commercial and transportation companies of the North, along the bank of the river. Passing through these one finds one's self upon a busy, but unconventional, thoroughfare. Dawson is built solidly to the hill, extending about a mile along the water-front; and the most attractive part of the town is the village of picturesque log cabins climbing over the lower slopes of the hill. They are not large, but they are all built with the roof extending over a wide front porch. The entire roof of each cabin is covered several inches deep with earth, and at the time of our visit—the first week of August—these roofs were grown with brilliant green grasses and flowers to a height of from twelve to eighteen inches. They were literally covered with the bloom of a dozen or more varieties of wild flowers. Every window had its flaming window-box; every garden, its gay beds; and there were even boxes set on square fence posts and running the entire length of fences themselves, from which vines drooped and trailed and flowers blew. Standing at the river and looking toward the hill, the whole town seemed a mass of bloom sloping up to the green, which, in turn, sloped on up to the blue.

We had heard so much about the exorbitant prices of the Klondike, that we were simply speechless when a very jolly, sandy-haired Scotch gentleman offered to take our two steamer trunks, three heavy suit cases, and two shawl-straps to the hotel which we had blindly chosen, for the sum of two dollars. We had expected to pay five; and when he first asked two and a half, we stood as still as though turned to stone—and all for joy. He, however, evidently mistaking our silence, doubtless felt the prick of the stern conscience of his ancestors, for he hastily added:—

"Well, seeing you're ladies, we'll call it an even two."

We agreed to the price coldly, pretending to consider it an outrage.

"My name is Angus McDonald," said he, with reproach. "When a McDonald says that his price is the lowest in the town, his word may be taken. If you come to Dawson twenty years from now, Angus will be standing here waiting to handle your baggage at the lowest price."

We gave him our keys and he attended to all the customs details for us. We had left Seattle on the evening of the 24th of July; had stopped for several hours at Ketchikan, Wrangell, Metlakahtla, Juneau, Treadwell, and Taku Glacier; a day and a night at Skaguay; two nights and a day at White Horse; had made short pauses at Selkirk and Lower Lebarge—to say nothing of hours spent in "wooding-up," which is a picturesque and sure feature of Yukon voyages; and at noon on the fifth day of August we were settled at the "Kenwood"—the dearest hotel at which it has ever been my good fortune to tarry even for a day. I do not mean the most stylish, nor the most elegant, nor even the most comfortable; nor do I mean the dearest in price; but the dearest to my heart. It is kept in a neat, cheerful, and homelike style by Miss Kinney—who had almost as many malamute puppies, by the way, as she had guests.

When we gave Mr. Angus McDonald our keys, it was not quite decided as to our hotel; but when we learned that we were sufficiently respectable in appearance to be accepted by Miss Kinney, we telephoned for our trunks. Then we forgot all about paying for them, and set out for a walk. When we returned, luncheon was being served; our trunks were in our rooms, but—Mr. Angus McDonald had gone off with our keys! We did not know then what we know now; that Mr. Angus McDonald and his retained keys are a Dawson joke. It seems that whenever one does not pay in advance for the delivery of his trunks, Mr. McDonald drives away with the keys in his pocket, whistling the merriest of Scotch tunes.

The joke has its embarrassments, particularly when one has descended to the Grand Canyon of the Yukon in a sand-slide.

The traveller in Alaska who desires to retain his own self-respect and that of his fellow-man will never criticise a price nor ask to have it reduced. He is expected to contribute liberally to every church he enters, every Indian band he hears play, every charitable institution that may present its merits for his consideration, every purse that may be made up on steamers, whatsoever its object may be. Fees are from fifty cents to five dollars. A waiter on a Yukon steamer threw a quarter back at a man who had innocently slipped it into his hand. Later, I saw him in the centre of a group of angry waiters and cabin-boys to whom he was relating his grievance.

Family of King's Island Eskimos living under Skin Boat, Nome

Since one is constantly changing steamers, and has a waiter, a cabin-boy, a night-boy, and frequently a stewardess to fee on each steamer, this must be counted as one of the regular expenses of the trip.

Other expenses we found to be greatly exaggerated on the "outside." Aside from our amusing experience with soap-bubble soda at White Horse and a bill for eight dollars and fifty cents for the poor pressing of three plain dress skirts and one jacket at Nome, we found nothing to criticise in northern prices.

The best rooms at the "Kenwood" were only two dollars a day, and each meal was one dollar—whether one ate little or whether one ate much. It was always the latter with us; for I have never been so hungry except at Bennett. I am convinced that the climate of the Yukon will cure every disease and every ill. We walked miles each day, drank much cold, pure water, and ate much wholesome, well-cooked, delicious food—including blueberries three times a day; and our sleep was sound, sweet, and refreshing.

Dawson has about ten thousand inhabitants now; it once had twice as many, and it will have again. Mining in the Klondike is in the transition stage. It is passing from the individual owners to large companies and corporations which have ample capital to install expensive machinery and develop rich properties. It is the history of every mining district, and its coming to the Klondike was inevitable. Its first effect, however, is always "to ruin the camp."

"Dawson's a camp no longer," said one who "went in" in 1897, sadly. "It's all spoiled. The individual miner has let go and the monopolists are coming in to take his place. The good days are things of the past. Pretty soon they'll be giving you change when you throw down two-bits for a lead pencil!" he concluded, with a lofty scorn—as much as to say: "It will then be time to die."

Dawson is connected with the "outside" by telegraph. It has two daily newspapers,—which are metropolitan in style,—an electric-light plant, and a telephone system. Its streets are graded and sidewalked, and it is piped for water; but its lack of systematized sewerage—or what might be more appropriately called its systematized lack of sewerage—is an abomination. It is, however, not alone in its unsanitation in this respect, for Nome follows its example.

Both homes and public buildings are of exceeding plainness of style, owing to the excessive cost of building in a region bounded by the Arctic Circle. The interiors of both, however, are attractive and luxurious in finish and furnishings; and owing to the sway of the mounted police, the town has an air of cleanliness and orderliness that is admirable.

A creditable building holds the post-office and customs office, and there is a public school building which cost fifty thousand dollars. The handsome administration building, standing in a green, park-like place, cost as much. There is a large court-house, the barracks of the mounted police, and other public buildings. Only the ruins remain of the executive mansion on the bank of the river, which was destroyed by fire two years ago and has not been rebuilt. It was the pride of Dawson. It was a large residence of pleasing architecture, lighted by electricity and finished throughout in British Columbia fir in natural tones. It contained the governor's private office, palatial reception rooms and parlors, a library, a noble hall and stairway, a state dining room, a billiard room and smoking room, and spacious chambers.

The governor's office in the administration building is large and handsomely furnished. The commissioner of Yukon Territory is called by courtesy governor, and the present commissioner, Governor Henderson, is a gentleman of distinguished presence and courtly manners. He had just returned from an automobile tour of inspection among "the creeks."

Governors, elegant executive mansions and offices, and automobile tours—where eleven years ago was nothing but the creeks and the virgin gold which brought all that is there to-day! We did not rebel at anything but the automobile; somehow, it jarred like an insult. An automobile up among the storied creeks!

There is a railroad, also, on which daily trains are run for a distance of twenty miles through the mining district. Six and eight horse stages will make the trip in one day for a party of six for fifty dollars.

Thirty dollars is first asked. When that price is found to be satisfactory, it is immediately discovered that the small stage is engaged or out of repair; a larger one must be used, for which the price is forty dollars. When this price is agreed upon, some infirmity is discovered in the second stage; a third must be substituted, for whose all-day use the price is fifty dollars. If one cares to see the "cricks," with no assurance that he will stumble upon a clean-up, at this price, he meekly takes his seat and is jolted up into the hills, paying a few dollars extra for his meals.

He may, however, take an hour's walk up Bonanza Creek and see the great dredges at work and the steam-pipes thawing the frozen gravel; and if he should voyage on down to Nome, he may take an hour's run by railway out on the tundra and see thirty thousand dollars sluiced out any day. Almost anything is preferable to the "graft" that is worked by the stage companies upon the helpless cheechacos at Dawson.

The British Yukon is an organized territory, having a commissioner, three judges, and an executive legislature, of whose ten members five are elected and five appointed. The governor is also appointed. He presides over the sessions of the legislature, giving the appointed members a majority of one.

The Yukon has a delegate in parliament, a gold commissioner, a land agent, and a superintendent of roads. Three-fourths of the population of the territory are Americans, yet the town has a distinctly English, or Canadian, atmosphere. In incorporated towns there is a tax levy on property for municipal purposes.

Order is preserved by the well-known organization of Northwest Mounted Police, whose members might be recognized anywhere, even when not in uniform, by their stern eyes, set lips, and peculiar carriage.

The first station of mounted police in the Yukon was established at Forty-Mile, or Fort Cudahy, in 1895, when the discovery of gold was creating a mild excitement. Although so many boasts have been made by the British of their early settlement of the

Yukon, not only was Mr. Ogilvie compelled to cross in 1887 under protection of the American Commander Newell, but in 1895 the members of the first force of mounted police to come into the country were forced to ascend the Yukon, by special permission of the United States government, so difficult were all routes through Yukon Territory.

There are at the present time about sixty police stations in the territory, as well as garrisons at Dawson and White Horse. The smaller stations have only three men. They are scattered throughout the mining country, wherever a handful of men are gathered together. Between Dawson and White Horse, where travel is heavy, a weekly patrol is maintained, and a careful register is kept of all boats and passengers going up or down the river. On the winter trail passengers are registered at each road house, with date of arrival and departure, making it easy to locate any traveller in the territory at any time. In the larger towns the mounted police serve as police officers; they also assist the customs officers and fill the offices of police magistrate and coroner. A police launch to patrol the river in summer has been recommended.

Dawson is laid out in rectangular shape, with streets about seventy feet wide and appearing wider because the buildings are for the most part low. In 1897 town lots sold for five thousand dollars, when there was nothing but tents on the flat at the mouth of the Klondike. The half-dollar was the smallest piece of money in circulation, as the quarter is to-day. Saw-mills were in operation, and dressed lumber sold for two hundred and fifty dollars a thousand feet. Fifteen dollars a day, however, was the ordinary wage of men working in the mines; so that such prices as fifty cents for an orange, two dollars a dozen for eggs, and twenty-five cents a pound for potatoes did not seem exorbitant.

There are rival claimants for the honor of the first discovery of gold on the Klondike, but George Carmack is generally credited with being the fortunate man. In August, 1896, he and the Indians "Skookum Jim" and "Tagish Charlie,"—Mr. Carmack's brothers-in-law—were fishing one day at the mouth of the Klondike River. (This river was formerly called Thron-Dieuck, or Troan-Dike.) Not being successful, they concluded to go a little way up the river to prospect. On the sixteenth day of the month they detected signs of gold on what has since been named Bonanza Creek; and from the first pan they washed out twelve dollars. They staked a "discovery" claim, and one above and below it, as is the right of discoverers.

At that time the gold flurry was in the vicinity of Forty-Mile. The first building ever done on the site of Dawson was that of a raft, upon which they proceeded to Forty-Mile to file their claims. On the same day began the great stampede to the little river which was soon to become world-famous.

The days of the bucket and windlass have passed for the Klondike. Dredging and hydraulicking have taken their place, and the trains and steamers are loaded with powerful machinery to be operated by vast corporations. It is certain that there are extensive quartz deposits in the vicinity, and when they are located the good and stirring days of the nineties will be repeated. Ground that was panned and sluiced by the individual miner is now being again profitably worked by modern methods. Scarcity of water has been the chief obstacle to a rapid development of the mines among the creeks; but experiments are constantly being made in the way of carrying water from other sources.

It was perplexing to hear people talking about "Number One Above on Bonanza," "Number Nine Below on Hunker," "Number Twenty-six Above on Eldorado," and others, until it was explained that claims are numbered above and below the one originally discovered on a creek. Eldorado is one of the smallest of creeks; yet, notwithstanding its limited water supply, it has been one of the richest producers. One reach, of about four miles in length, has yielded already more than thirty millions of dollars in coarse gold.

The gold of the Klondike is beautiful. It is not a fine dust. It runs from grains like mustard seed up to large nuggets.

When one goes up among the creeks, sees and hears what has actually been done, one can but wonder that any young and strong man can stay away from this marvellous country. Gold is still there, undiscovered; it is seldom the old prospector, the experienced miner, the "sour-dough," that finds it; it is usually the ignorant, lucky "cheechaco." It is like the game of poker, to which sits down one who never saw the game played and holds a royal flush, or four aces, every other hand. How young men can clerk in stores, study pharmacy, or learn politics in provincial towns, while this glorious country waits to be found, is incomprehensible to one with the red blood of adventure in his veins and the quick pulse of chance. Better to dare, to risk all and lose all, if it must be, than never to live at all; than always to be a drone in a narrow, commonplace groove; than never to know the surge of this lonely river of mystery and never to feel the air of these vast spaces upon one's brow.

No one can even tread the deck of a Yukon steamer and be quite so small and narrow again as he was before. The loneliness, the mystery, the majesty of it, reveals his own soul to his shrinking eyes, and he grows—in a day, in an hour, in the flash of a thought—out of his old self. If only to be borne through this great country on this wide water-way to the sea can work this change in a man's heart, what miracle might not be wrought by a few years of life in its solitude?

The principle of "panning" out gold is simple, and any woman could perform the work successfully without instruction, success depending upon the delicacy of manipulation. From fifty cents to two hundred dollars a pan are obtained by this old-fashioned but fascinating method. Think of wandering through this splendid, gold-set country in the matchless summers when there is not an hour of darkness; with the health and the appetite to enjoy plain food and the spirit to welcome adventure; to pause on the banks of unknown creeks and try one's luck, not knowing what a pan may bring forth; to lie down one night a penniless wanderer, so far as gold is concerned, and, perhaps, to sleep the next night on banks that wash out a hundred dollars to the pan—could one choose a more fascinating life than this?

Rockers are wooden boxes which are so constructed that they gently shake down the gold and dispose of the gravel through an opening in the bottom. Sluicing is more interesting than any other method of extracting gold, but this will be described as we saw the process separate the glittering gold from the dull gravel at Nome.

CHAPTER XLIV

The two great commercial companies of the North to-day are the Northern Commercial Company and the North American Transportation and Trading Company. The Alaska Commercial Company and the North American Transportation and Trading Company were the first to be established on the Yukon, with headquarters at St. Michael, near the mouth of the river. In 1898 the Alaska Exploration Company established its station across the bay from St. Michael on the mainland; and during that year a number of other companies were located there, only two of which, however, proved to be of any permanency—the Empire Transportation Company and the Seattle-Yukon Transportation Company.

In 1901 the Alaska Commercial, Empire Transportation, and Alaska Exploration companies formed a combination which operated under the names of the Northern Commercial Company and the Northern Navigation Company, the former being a trading and the latter a steamship company. Owing to certain conditions, the Seattle-Yukon Transportation Company was unable to join the combination; and its properties, consisting principally of three steamers, together with four barges, were sold to the newly formed company. During the first year of the consolidation the North American Transportation and Trading Company worked in harmony with the Northern Navigation Company, Captain I. N. Hibberd, of San Francisco, having charge of the entire lower river fleet, with the exception of one or two small tramp boats.

By that time very fine combination passenger and freight boats were in operation, having been built at Unalaska and towed to St. Michael. In its trips up and down the river, each steamer towed one or two barges, the combined cargo of the steamer and tow being about eight hundred tons. It was impossible for a boat to make more than two round trips during the summer season, the average time required being fourteen days on the "up" trip and eight on the "down" for the better boats, and twenty and ten days respectively for inferior ones, without barges, which always added at least ten days to a trip.

After a year the North American Transportation and Trading Company withdrew from the combination and has since operated its own steamers.

Of all these companies the Alaska Commercial is the oldest, having been founded in 1868; it was the pioneer of American trading companies in Alaska, and was for twenty years the lessee of the Pribyloff seal rookeries. It had a small passenger and freight boat on the Yukon in 1869. The other companies owed their existence to the Klondike gold discoveries.

The two companies now operating on the Yukon have immense stores and warehouses at Dawson and St. Michael, and smaller ones at almost every post on the Yukon; while the N. C. Company, as it is commonly known, has establishments up many of the tributary rivers.

As picturesque as the Hudson Bay Company, and far more just and humane in their treatment of the Indians, the American companies have reason to be proud of their record in the far North. In 1886, when a large number of miners started for the Stewart River mines, the agent of the A. C. Company at St. Michael received advice from headquarters in San Francisco that an extra amount of provisions had been sent to him, to meet all possible demands that might be made upon him during the winter. He was further advised that the shipment was not made for the purpose of realizing profits beyond the regular schedule of prices already established, but for humane purposes entirely—to avoid any suffering that might occur, owing to the large increase in population. He was, therefore, directed to store the extra supplies as a reserve to meet the probable need, to dispose of the same to actual customers only and in such quantities as would enable him to relieve the necessities of each and every person that might apply. Excessive prices were prohibited, and instructions to supply all persons who might be in absolute poverty, free of charge, were plain and unmistakable.

Men of the highest character and address have been placed at the head of the various stations,—men with the business ability to successfully conduct the company's important interests and the social qualifications that would enable them to meet and entertain distinguished travellers through the wilderness in a manner creditable to the company. Tourists, by the way, who go to Alaska without providing themselves with clothes suitable for formal social functions are frequently embarrassed by the omission. Gentlemen may hasten to the company's store—which carries everything that men can use, from a toothpick to a steamboat—and array themselves in evening clothes, provided that they are not too fastidious concerning the fit and the style; but ladies might not be so fortunate. Nothing is too good for the people of Alaska, and when they offer hospitality to the stranger within their gates, they prefer to have him pay them the compliment of dressing appropriately to the occasion. If voyagers to Alaska will consider this advice they may spare themselves and their hosts in the Arctic Circle some unhappy moments.

Yukon summers are glorious. There is not an hour of darkness. A gentleman who came down from "the creeks" to call upon us did not reach our hotel until eleven o'clock. He remained until midnight, and the light in the parlor when he took his departure was as at eight o'clock of a June evening at home. The lights were not turned on while we were in Dawson; but it is another story in winter.

Clothes are not "blued" in Dawson. The first morning after our arrival I was summoned to a window to inspect a clothes-line.

"Will you look at those clothes! Did you ever see such whiteness in clothes before?"

I never had, and I promptly asked Miss Kinney what her laundress did to the clothes to make them look so white.

"I'm the laundress," said she, brusquely. "I come out here from Chicago to work, and I work. I was half dead, clerking in a store, when the Klondike craze come along and swept me off my feet. I struck Dawson broke. I went to work, and I've been at work ever since. I have cooks, and chambermaids, and laundresses; but it often happens that I have to be all three, besides landlady, at once. That's the way of the Klondike. Now, I must go and feed those malamute pups; that little yellow one is getting sassy."

She had almost escaped when I caught her sleeve and detained her.

"But the clothes—I asked you what makes them so white—"

"Don't you suppose," interrupted she, irascibly, "that I have too much work to do to fool around answering the questions of a cheechaco? I'm not travelling down the Yukon for fun!"

This was distinctly discouraging; but I had set out to learn what had made those clothes so white. Besides, I was beginning to perceive dimly that she was not so hard as she spoke herself to be; so I advised her that I should not release her sleeve until she had answered my question.

She burst into a kind of lawless laughter and threw her hand out at me.

"Oh, you! Well, there, then! I never saw your beat! There ain't a thing in them there clothes but soap-suds, renched out, and sunshine. We don't even have to rub clothes up here the way you have to in other places; and we never put in a *pinch* of blueing. Two-three hours of sunshine makes 'em like snow."

"But how is it in winter?"

She laughed again.

"Oh, that's another matter. We bleach 'em out enough in summer so's it'll do for all winter. Let go my sleeve or you won't get any blueberries for lunch."

This threat had the desired effect. Surely no woman ever worked harder than Miss Kinney worked. At four o'clock in the mornings we heard her ordering maids and malamute puppies about; and at midnight, or later, her springing step might be heard as she made the final rounds, to make sure that all was well with her family.

We were greatly amused and somewhat embarrassed on the day of our arrival. We saw at a glance that the only vacant room was too small to receive our baggage.

"I'll fix that," said she, snapping her fingers. "I just gave a big room on the first floor to two young men. I'll make them exchange with you."

It was in vain that we protested.

"Now, you let me be!" she exclaimed; "I'll fix this. You're in the Klondike now, and you'll learn how white men can be. Young men don't take the best room and let women take the worst up here. If they come up here with that notion, they soon get it taken out of 'em—and I'm just the one to do it. Now, you let me be! They'll be tickled to death."

Whatever their state of mind may have been, the exchange was made; but when we endeavored to thank her, she snapped us up with:—

"Anybody'd know you never lived in a white country, or you wouldn't make such a fuss over such a little thing. We're used to doing things for other people *up here*," she added, scornfully.

Miss Kinney gave us many surprises during our stay, but at the last moment she gave us the greatest surprise of all. Just as our steamer was on the point of leaving, she came running down the gangway and straight to us. Her hands and arms were filled with large paper bags, which she began forcing upon us.

"There!" she said. "I've come to say good-by and bring you some fruit. I'd given you one of those malamute puppies if I could have spared him. Well, good-by and good luck!"

We were both so touched by this unexpected kindness in one who had taken so much pains to conceal every touch of tenderness in her nature, that we could not look at one another for some time; nor did it lessen our appreciation to remember how ceaselessly and how drudgingly Miss Kinney worked and the price she must have paid for those great bags of oranges, apples, and peaches—for freight rates are a hundred and forty dollars a ton on "perishables." It set a mist in our eyes every time we thought about it. It was our first taste of Arctic kindness; and, somehow, its flavor was different from that of other latitudes.

Dawson is gay socially, as it has always been. In summer the people are devoted to outdoor sports, which are enjoyed during the long evenings. There is a good club-house for athletic sports in winter, and the theatres are well patronized, although, in summer, plays commence at ten or ten-thirty and are not concluded before one. As in all English and Canadian towns, business is resumed at a late hour in the morning, making the hours of rest correspond in length to ours.

Two young Yale men who were travelling in our party had been longing to see a dance-hall,—a "real Klondike dance-hall,"—but they came in one midnight, their faces eloquent with disgust.

"We found a dance-hall *at last*," said one. "They hide their light under such bushels now that it takes a week to find one; the mounted police don't stand any foolishness. Then—think of a dance-hall running in broad daylight! No mystery, no glitter, no soft, rosy glamour—say, it made me yearn for bread and butter. Do you know where Miss Kinney keeps her bread jar and blueberries? Honestly, I don't know anything or any place that could cultivate a taste in a young man for sane and decent things like one of these dance-halls here. I never was so disappointed in my life. I can go to church *at home*; I didn't come to the Klondike for *that*. Why, the very music itself sounded about as lively as 'Come, Ye Disconsolate!' Come on, Billy; let's go to bed."

No one should visit Dawson without climbing, on a clear day, to the summit of the hill behind the town, which is called "the Dome." The view of the surrounding country from this point is magnificent. The course of the winding, widening Yukon may be traced for countless miles; the little creeks pour their tawny floods down into the Klondike before the longing eyes of the beholder; and faraway on the horizon faintly shine the snow-peaks that beautify almost every portion of the northern land.

The wagon roads leading from Dawson to the mining districts up the various creeks are a distinct surprise. They were built by the Dominion government and are said to be the best roads to be found in any mining district in the world. A Dawson man will brag about the roads, while modestly silent about the gold to which the roads lead.

"You must go up into the creeks, if only to see the roads," every man to whom one talks will presently say. "You can't beat 'em anywheres."

Claim staking in the Klondike is a serious matter. The mining is practically all placer, as yet, and a creek claim comprises an area two hundred and fifty feet along the creek and two thousand feet wide. This information was a shock to me. I had always supposed, vaguely, that a mining claim was a kind of farm, of anywhere from twenty to sixty acres; and to find it but little larger than the half of a city block was a chill to my enthusiasm. They explained, however, that the gravel filling a pan was but small in quantity, that it could be washed out in ten minutes, and that if every pan turned out but ten dollars, the results of a long day's work would not be bad.

Claims lying behind and above the ones that front on the creeks are called "hill" claims. They have the same length of frontage, but are only a thousand feet in width. In staking a claim, a post must be placed at each corner on the creek, with the names of the claim and owner and a general description of any features by which it may be identified; the locator must take out a free miner's license, costing seven dollars and a half, and file his claim at the mining recorder's office within ten days after staking. No one can stake more than one claim on a single creek, but he may hold all that he cares to acquire by purchase, and he may locate on other creeks. Development work to the amount of two hundred dollars must be done yearly for three years, or that amount paid to the mining recorder; this amount is increased to four hundred dollars with the fourth year. The locator must secure a certificate to the effect that the necessary amount of yearly work has been done, else the claim will be cancelled.

Wreck of "Jessie," Nome Beach

CHAPTER XLV

When the *D. R. Campbell* drew away from the Dawson wharf at nine o'clock of an August morning, another of my dreams was "come true." I was on my way down the weird and mysterious river that calls as powerfully in its way as the North Pacific Ocean. For years the mere sound of the word "Yukon" had affected me like the clash of a wild and musical bell. The sweep of great waters was in it—the ring of breaking ice and its thunderous fall; the roar of forest fires, of undermined plunging cliffs, of falling trees, of pitiless winds; the sobs of dark women, deserted upon its shores, with white children on their breasts; the mournful howls of dogs and of their wild brothers, wolves; the slide of avalanches and the long rattle of thunder—for years the word "Yukon" had set these sounds ringing in my ears, and had swung before my eyes the shifting pictures of canyon, rampart, and plain; of waters rushing through rock walls and again loitering over vast lowlands to the sea; of forestated mountains, rose thickets, bare hills, pale cliffs of clay, and ranges of sublime snow-mountains. Yet, with all that I had read, and all that I had heard, and all that I had imagined, I was unprepared for the spell of the Yukon; for the spaces, the solitude, the silence. At last I was to learn how well the name fits the river and the country, and how feeble and how ineffectual are both description and imagination to picture this country so that it may be understood.

Six miles below Dawson the site of old Fort Reliance is passed, and forty-six miles farther Forty-Mile River pours its broad flood into the Yukon. About eight miles up this river, at the lower end of a canyon, a strong current has swept many small boats upon dangerous rocks and the occupants have been drowned. The head of the Forty-Mile is but a short distance from the great Tanana.

The settlement of Forty-Mile is the pioneer mining-camp of the Yukon. The Alaska Commercial Company established a station here soon after the gold excitement of 1887; and, as the international boundary line crosses Forty-Mile River twenty-three miles from its mouth and many of the most important mining interests depending upon the town for supplies are on the American side, a bonded warehouse is maintained, from which American goods can be drawn without the payment of duties. As late as 1895 quite a lively town was at the mouth of the river, boasting even an opera house; but the town was depopulated upon the discovery of gold on the Klondike. Six years ago the settlement was flooded by water banked up in Forty-Mile River by ice, and the residents were taken from upstairs windows in boats. The former name of this river was Che-ton-deg, or "Green Leaf," River.

Now there are a couple of dozen log cabins, a dozen or more red-roofed houses, and store buildings. The steamer pushed up sidewise to the rocky beach, a gang-plank was floated ashore, and a customs inspector came aboard. On the beach were a couple of ladies, some members of the mounted police in scarlet coats, and fifty malamute dogs, snapping, snarling, and fighting like wolves over the food flung from the steamer.

The dog is to Alaska what the horse is to more civilized countries—the intelligent, patient, faithful beast of burden. He is of the Eskimo or "malamute" breed, having been bred with the wolf for endurance; or he is a "husky" from the Mackenzie River.

Eskimo dogs are driven with harness, hitched to sleds, and teams of five or seven with a good leader can haul several hundred pounds, if blessed with a kind driver. In summer they have nothing to do but sleep, and find their food as best they may. Along the Yukon they haunt steamer-landings and are always fed by the stewards—who can thus muster a dog fight for the pleasure of heartless passengers at a moment's notice.

With the coming of winter a kind of electric strength seems to enter into these dogs. They long for the harness and the journeys over snow and ice; and for a time they leap and frisk like puppies and will not be restrained. They are about the size of a St.

Bernard dog, but of very different shape; the leader is always an intelligent and superior animal and his eyes frequently hold an almost human appeal. He is fairly dynamic in force, and when not in harness will fling himself upon food with a swiftness and a strength that suggest a missile hurled from a catapult. Nothing can check his course; and he has been known to strike his master to the earth in his headlong rush of greeting—although it has been cruelly said of him that he has no affection for any save the one that feeds him, and not for him after his hunger is satisfied.

The Eskimo dog seldom barks, but he has a mournful, wolflike howl. His coat is thick and somewhat like wool, and his feet are hard; he travels for great distances without becoming footsore, and at night he digs a deep hole in the snow, crawls into it, curls up in his own wool, and sleeps as sweetly as a pet Spitz on a cushion of down. His chief food is fish. If the Alaska dog is not affectionate, it is because for generations he has had no cause for affection. No dog with such eyes—so asking and so human-like in their expression—could fail to be affectionate and devoted to a master possessing the qualities which inspire affection and devotion.

In winter all the mails are carried by dogs, covering hundreds of miles.

Half a mile below Forty-Mile the town of Cudahy was founded in 1892 by the North American Trading and Transportation Company, as a rival settlement.

Fifty miles below Forty-Mile, at the confluence of Mission Creek with the Yukon, is Eagle, having a population of three or four hundred people. It has the most northerly customs office and military post, Fort Egbert, belonging to the United States, and is the terminus of the Valdez-Eagle mail route and telegraph line. It is also of importance as being but a few miles from the boundary.

Fort Egbert is a two-company post, and usually, as at the time of our visit, two companies are stationed there. The winter of 1904-1905 was the gayest in the social history of the fort. Several ladies, the wives and the sisters of officers, were there, and these, with the wife of the company's agent and other residents of the town, formed a brilliant and refined social club.

From November the 27th to January the 16th the sun does not appear above the hills to the south. The two "great" days at Eagle are the 16th of January,—"when the sun comes back,"—and the day "when the ice breaks in the river," usually the 12th of May. On the former occasion the people assemble, like a band of sun-worshippers, and celebrate its return.

The vegetable and flower gardens of Eagle were a revelation of what may be expected in the agricultural and floral line in the vicinity of the Arctic Circle. Potatoes, cabbages, cauliflower, lettuce, turnips, radishes, and other vegetables were in a state of spendthrift luxuriance that cannot be imagined by one who has not travelled in a country where vegetables grow day and night.

In winter Eagle is a lonely place. The only mail it receives is the monthly mail passing through from Dawson to Nome by dog sleds; and no magazines, papers, or parcels are carried.

It was from Eagle that the first news was sent out to the world concerning Captain Amundsen's wonderful discovery of the Northwest Passage; here he arrived in midwinter after a long, hard journey by dog team from the Arctic Ocean and sent out the news which so many brave navigators of early days would have given their lives to be able to announce.

Within five years a railroad will probably connect Eagle with the coast at Valdez; meantime, there is a good government trail, poled by a government telegraph line.

Eagle came into existence in 1898, and the fort was established in 1899.

"Woodings-up" are picturesque features of Yukon travel. When the steamer does not land at a wood yard, mail is tied around a stick and thrown ashore. Fancy standing, a forlorn and homesick creature, on the bank of this great river and watching a letter from home caught by the rushing current and borne away! Yet this frequently happens, for heart affairs are small matters in the Arctic Circle and receive but scant consideration.

On the Upper Yukon wood is five dollars a cord; on the Lower, seven dollars; and a cord an hour is thrust into the immense and roaring furnaces.

During "wooding-up" times passengers go ashore and enjoy the forest. There are red and black currants, crab-apples, two varieties of salmon-berries, five of huckleberries, and strawberries. The high-bush cranberries are very pretty, with their red berries and delicate foliage.

Nation is a settlement of a dozen log cabins roofed with dirt and flowers, the roofs projecting prettily over the front porches. The wife of the storekeeper has lived here twenty-five years, and has been "outside" only once in twelve years. Passengers usually go ashore especially to meet her, and are always cordially welcomed, but are never permitted to condole with her on her isolated life. The spell of the Yukon has her in thrall, and content shines upon her brow as a star. Those who go ashore to pity, return with the dull ache of envy in their worldly hearts; for there be things on the Yukon that no worldly heart can understand.

We left Eagle in the forenoon and at midnight landed at Circle City, which received this name because it was first supposed to be located within the Arctic Circle. We found natives building houses at that hour, and this is my most vivid remembrance of Circle. Gold was discovered on Birch Creek, within eight miles of the settlement, as early as 1892; and until the Klondike excitement this was the most populous camp on the Yukon, more than a thousand miners being quartered in the vicinity. Like other camps, it was then depopulated; but many miners have now returned and a brilliant discovery in this vicinity may yet startle the world. The output of gold for 1906 was two hundred and fifty thousand dollars. About three hundred miners are operating on tributaries up Birch Creek. The great commercial companies are established at all these settlements on the Yukon, where they have large stores and warehouses.

Early on the following morning we were on deck to cross the Arctic Circle. One has a feeling that a line with icicles dangling from it must be strung overhead, under which one passes into the enchanted realm of the real North.

"Feel that?" asked the man from Iowa of a big, unsmiling Englishman.

"Feel—er—what?" said the Englishman.

"That shock. It felt like stepping on the third rail of an electric railway."

But the Iowa humor was scorned, and the Englishman walked away.

We soon landed at Fort Yukon, the only landing in the Arctic Circle and the most northerly point on the Yukon. This post was established at the mouth of the Porcupine in 1847 by A. H. McMurray, of the Hudson Bay Company, and was moved in 1864 a mile lower on the Yukon, on account of the undermining of the bank by the wash of the river. During the early days of this post goods were brought from York Factory on Hudson Bay, four thousand miles distant, and were two years in transit. The whole Hudson Bay system, according to Dall, was one of exacting tyranny that almost equalled that of the Russian Company. The white men were urged to marry Indian, or native, women, to attach them to the country. The provisions sent in were few and these were consumed by the commanders of the trading posts or given to chiefs, to induce them to bring in furs. The white men received three pounds of tea and six of sugar annually, and no flour. This scanty supply was uncertain and often failed. Two suits of clothes were granted to the men, but nothing else until the furs were all purchased. If anything remained after the Indians were satisfied, the men were permitted to purchase; but Indians are rarely satisfied.

Fort Yukon has never been of importance as a mining centre, but has long been a great fur trading post for the Indians up the Porcupine. This trade has waned, however, and little remains but an Indian village and the old buildings of the post. We walked a mile into the woods to an old graveyard in a still, dim grove, probably the only one in the Arctic Circle.

CHAPTER XLVI

The Yukon is a mighty and a beautiful river, and its memory becomes more haunting and more compelling with the passage of time. From the slender blue stream of its source, it grows, in its twenty-three hundred miles of wandering to the sea, to a width of sixty miles at its mouth. In its great course it widens, narrows, and widens; cuts through the foot-hills of vast mountain systems, spreads over flats, makes many splendid sweeping curves, and slides into hundreds of narrow channels around spruce-covered islands.

It is divided into four great districts, each of which has its own characteristic features. The valley extending from White Horse to some distance below Dawson is called the "upper Yukon," or "upper Ramparts," the river having a width of half a mile and a current of four or five miles an hour, and the valley in this district being from one to three miles in width.

Following this are the great "Flats"—of which one hears from his first hour on the Yukon; then, the "Ramparts"; and last, the "lower Yukon" or "lower river."

The Flats are vast lowlands stretching for two hundred miles along the river, with a width in places of a hundred miles. Their very monotony is picturesque and fascinates by its immensity. Countless islands are constantly forming, appearing and disappearing in the whimsical changes of the currents. Indian, white, and half-breed pilots patrol these reaches, guiding one steamer down and another up, and by constant travel keeping themselves fairly familiar with the changing currents. Yet even these pilots frequently fail in their calculations.

At Eagle a couple of gentlemen joined our party down the river on the *Campbell*, expecting to meet the same day and return on the famous *Sarah*—as famous as a steamer as is the island of the same name on the inland passage; but they went on and on and the *Sarah* came not. One day, two days, three days, went by and they were still with us. One was in the customs service and his time was precious. Whenever we approached a bend in the river, they stood in the bow of the boat, eagerly staring ahead; but not until the fourth day did the cry of "*Sarah*" ring through our steamer. Hastening on deck, we beheld her, white and shining, on a sand-bar, where she had been lying for several days, notwithstanding the fact that she had an experienced pilot aboard.

Throughout the Flats lies a vast network of islands, estimated as high as ten thousand in number, threaded by countless channels, many of which have strong currents, while others are but still, sluggish sloughs. Mountains line the far horizon lines, but so far away that they frequently appear as clouds of bluish pearl piled along the sky; at other times snow-peaks are distinctly visible. Cottonwoods, birches, and spruce trees cover the islands so heavily that, from the lower deck of a steamer, one would believe that he was drifting down the single channel of a narrow river, instead of down one channel of a river twenty miles wide.

It is within the Arctic Circle that the Yukon makes its sweeping bend from its northwest course to the southwest, and here it is entered by the Porcupine; twenty miles farther, by the Chandelar; and just above the Ramparts, by the Dall. These are the three important rivers of this stretch of the Yukon.

Many complain of the monotony of the Flats; but for me, there was not one dull or uninteresting hour on the Yukon. In my quiet home on summer evenings I can still see the men taking soundings from the square bow of our steamer and hear their hoarse cries:—

"Six feet starboard! Five feet port! Seven feet starboard! Five feet port! Five feet starboard! Four feet port!" At the latter cry the silent watchers of the pilot-house came to attention, and we proceeded under slow bell until a greater depth was reached.

On the shores, as we swept past, we caught glimpses of dark figures and Indian villages, or, farther down the river, primitive Eskimo settlements; and the stillness, the pure and sparkling air, the untouched wilderness, the blue smoke of a wood-chopper's lonely fire, the wide spaces swimming over us and on all sides of us, charmed our senses as only the elemental forces of nature can charm. One longs to stay awake always on this river; to pace the wide decks and be one with the solitude and the stillness that are not of the earth, as we know it, but of God, as we have dreamed of him.

The blue hills of the Ramparts are seen long before entering them. The valley contracts into a kind of canyon, from which the rampart-like walls of solid stone rise abruptly from the water. The hills are not so high as those of the Upper Ramparts, which bear marked resemblance to the lower; and although many consider the latter more picturesque, I must confess that I found no beauty below Dawson so majestic as that above. Many of the hills here have a rose-colored tinge, like the hills of Lake Bennett.

In places the river does not reach a width of half a mile and is deep and swift. The shadows between the high rock-bluffs and pinnacled cliffs take on the mysterious purple tones of twilight; many of the hills are covered with spruce, whose dark green blends agreeably with the gray and rose color. The bends here are sharp and many; at the Rapids the current is exceedingly rapid, and Dall reported a fall of twelve feet to the half mile, with the water running in sheets of foam over a granite island in the middle of the stream. This was on June 1, 1866. In August, 1883, Schwatka, after many hours of anxiety and dread of the reputed rapids, inquired of Indians and learned that he had already passed them. They were not formidable at the time of our voyage,—August,—and it is only during high stages of water that they present a bar to navigation.

We reached Rampart at six o'clock in the morning. After Tanana, this is the loveliest place on the Yukon. Its sparkling, emerald beauty shone under a silvery blue sky. There was a long street of artistic log houses and stores on a commanding bluff, up which paths wound from the water. Roofs covered with earth and flowers, carried out in brilliant bloom over the porches, added the characteristic Yukon touch. Every door-yard and window blazed with color. Narrow paths ran through tall fireweed and grasses over and around the hill—each path terminating, like a winding lane, in a pretty log-cabin home. There was an atmosphere of cleanliness, tidiness, and thrift not found in other settlements along the Yukon.

Captain Mayo, who, with McQuesten, founded Rampart in 1873, still lives here. The two commercial companies have large stores and warehouses; and residences were comfortably, and even luxuriously, furnished.

Rampart is two hundred and thirty miles below Fort Yukon, and is about halfway between Dawson and the sea. It has a population of four or five hundred people—when they are in from the mines!—and almost as many fighting, hungry dogs. Its street winds, and the buildings follow its windings; sometimes it stops altogether, and the buildings stop with it—then both go on again; and in front of all the public buildings are clean rustic benches, where one may sit and "look to the rose about him." The river here is half a mile wide, and on its opposite shore the green fields of the government experimental station slope up from the water.

Gold was discovered on Minook Creek, half a mile from town, in 1895, and the camp is regarded as one of the most even producers in Alaska. In 1906, despite an unusually dry season, the output of the district was three hundred and fifty thousand dollars.

In the afternoon of the same day we reached Tanana, which is, as I have said, the most beautiful place on the Yukon. It has a splendid site on a level plateau; and all the springlike greenness, the cleanliness and order, the luxuriant vegetation, of Dawson, are outdone here. One walks in a maze of delight along streets of tropic, instead of arctic, bloom. The log houses are set far back from the streets, and the deep dooryards are seas of tremulous color, through which neat paths lead to flower-roofed homes. Cleanliness, color, and perfume are everywhere delights, but on the lonely Yukon their unexpectedness is enchanting.

In 1900 Fort Gibbon was established here, and this post has the most attractive surroundings of any in Alaska. Tanana is situated at the mouth of the Tanana River, seventy-five miles below Rampart, and passengers for Fairbanks connect here with luxurious steamers for a voyage of three hundred miles up the Tanana. It is a beautiful voyage and it ends at the most progressive and metropolitan town of the North.

CHAPTER XLVII

In the autumn of 1902 Felix Pedro, an experienced miner and prospector, crossed the divide between Birch and McManus creeks and entered the Tanana Valley.

Previous to that year many people had travelled through the valley, on their way to the Klondike, by the Valdez route; and a few miners from the Birch Creek and Forty-Mile diggings had wandered into the Tanana country, without being able to do any important prospecting because of the distance from supplies; but Pedro was the first man to discover that gold existed in economic quantities in this region, and his coming was an event of historical importance.

One of the best tests of the importance and value of geological survey work lies in the significant report of Mr. Alfred H. Brooks for the year of 1898—four years before the discoveries of Mr. Pedro:—

"We have seen that the little prospecting which has been done up to the present time has been too hurried and too superficial to be regarded as a fair test of the region. Our best information leads us to believe that the same horizons which carry gold in the Forty-Mile and Birch Creek districts are represented in the Tanana and White River basins.... I should advise prospectors to carefully investigate the small tributary streams of the lower White and of the Tanana from Mirror Creek to the mouth."

Pedro's discovery was on the creek which bears his name, and before another year gold was discovered on several other creeks. In 1901 a trading post was established by Captain E. T. Barnette, on the present site of Fairbanks, and the development of the country progressed rapidly. The Fairbanks Mining District was organized and named for the present Vice-President of the United States. In the autumn of 1903 eight hundred people were in the district, and about thirty thousand dollars had been produced, the more important creeks at that time being Pedro, Goldstream, Twin Creek, Cleary, Wolf, Chatham, and Fairbanks. In the fall of 1904 nearly four thousand miners had come in, and the year's output was three hundred and fifty thousand dollars. Fairbanks and Chena had grown to thriving camps, and a brilliant prosperity reigned in the entire district. Roads were built to the creeks, sloughs were bridged, and Fairbanks' "boom" was in full swing. It was the old story of a camp growing from tents to shacks in a night, from shacks to three-story buildings in a month. The glory of the Klondike trembled and paled in the brilliance of that of Fairbanks. Every steamer for Valdez was crowded with men and women bound for the new camp by way of the Valdez trail; while thousands went by steamer, either to St. Michael and up the Yukon, or to Skaguay and down the Yukon, to the mouth of the Tanana.

Fairbanks is now a camp only in name. It has all the comforts and luxuries of a city, and is more prosperous and progressive than any other town in Alaska or the Yukon. It started with such a rush that it does not seem to be able to stop. It is the headquarters of the Third Judicial District of Alaska, which was formerly at Rampart; it has electric light and water systems, a fire department, excellent and modern hotels, schools, churches, hospitals, daily newspapers, a telegraph line to the outside world which is operated by the government, and a telephone system which serves not only the city, but all the creeks as well.

The Tanana Mines Railway, or Tanana Valley Railway, as it is now called, was built in 1905 to connect Fairbanks with Chena and the richest mining claims of the district; and two great railroads are in course of construction from Prince William Sound.

In 1906 the output of gold was more than nine millions of dollars, and had it not been for the labor troubles in 1907, this output would have been doubled. In the earlier days of the camp the crudest methods of mining were employed; but with the improved transportation facilities, modern machinery was brought in and the difficulties of the development were greatly lessened.

Upon a first trip to Fairbanks, the visitor is amazed at the size and the metropolitan style and tone of this six-year-old camp in the wilderness.

It is situated on the banks of the Chena River, about nine miles from its confluence with the Tanana. It has a level town site, which looks as though it might extend to the Arctic Circle. The main portion of the town is on the right bank of the river, the railway terminal yards, saw-mills, manufacturing plants, and industries of a similar nature being located on the opposite shore, on what is known as Garden Island, the two being connected by substantial bridges. The city is incorporated and, like other incorporated towns of Alaska, is governed by a council of seven members, who elect a presiding officer who is, by courtesy, known as mayor. The executive officers of the municipal government consist of a clerk, treasurer, police magistrate, chief of police, chief of the fire department, street commissioner, and physician.

The municipal finances are derived from a share in federal licenses, from the income derived from the local court, from poll taxes, and from local taxation of real and personal property. From all these sources the municipal treasury was enriched during the year of 1906 by about ninety-five thousand dollars.

Each of the three banks operates an assay office under the supervision of an expert. The population of the district is from fifteen to twenty thousand, of which five thousand belong permanently to the town. The climate is dry and sparkling; the summers are delightful, the winters still and not colder than those of Minnesota, Montana, and the Dakotas, but without the blizzards of those states. In 1906 the coldest month was January, the daily mean temperature being thirty-six degrees below zero, but dry and still. Travel over the trail by dog team is continued throughout the winter, skating and other outdoor sports being as common as in Canada.

Five saw-mills are in operation, with an aggregate daily capacity of a hundred and ten thousand feet, the entire product being used locally. There is an abundance of poplar, spruce, hemlock, and birch; an unlimited water supply; a municipal steam-heating plant; two good hospitals; two daily newspapers; graded schools,—the four-year course of the high school admitting the student to the Washington State University and to high educational institutions of other states; a Chamber of Commerce and a Business Men's Association; twelve hotels, five of which are first class; while every industry is represented several times over.

This is Fairbanks, the six-year-old mining-camp of the Tanana Valley.

Sunrise on Behring Sea

CHAPTER XLVIII

At Tanana our party was enlarged by a party of four gentlemen, headed by Governor Wilford B. Hoggatt, of Juneau, who was on a tour of inspection of the country he serves.

Our steamer, too, underwent a change while we were ashore. We now learned why its bow was square and wide. It was that it might push barges up and down the Yukon; and it now proceeded, under our astonished eyes, to push four, each of which was nearly as large as itself. All the days of my life, as Mr. Pepys would say, I have never beheld such an object floating upon the water. The barges were fastened in front of us and on both sides of us; two were flat and uncovered, one was covered, but open on the sides, while the fourth was a kind of boat and was crowned with a real pilot-house, in which was a real wheel.

We viewed them in open and hostile dismay, not yet recognizing them as blessings in disguise; we then laughed till we wept, over our amazing appearance as we went sweeping, bebarged, down to the sea. Four barges to one steamboat! One barge would have seemed like an insult, but four were perfectly ridiculous. The governor was told that they constituted his escort of honor, but he would not smile. He was in haste to get to Nome; and barges meant delay.

We swept down the Yukon like a huge bird with wide wings outspread; and those of us who did not care whether we went upon a sand-bar or not soon became infatuated with barges. Straight in front of our steamer we had, on one barge, a low, clean promenade a hundred feet long by fifty wide; on the others were shady, secluded nooks, where one might lie on rugs and cushions, reading or dreaming, ever and anon catching glimpses of native settlements—tents and cabins; thousands of coral-red salmon drying on frames; groups of howling dogs; dozens of silent dark people sitting or standing motionless, staring at their whiter and more fortunate brothers sweeping past them on the rushing river.

Poor, lonely, dark people! As lonely and as mysterious, as little known and as little understood, as the mighty river on whose shores their few and hard days are spent. Little we know of them, and less we care for them. The hopeless tragedy of their race is in their long, yearning gaze; but we read it not. We look at them in idle curiosity as we flash past them; and each year, as we return, we find them fewer, lonelier,—more like dark sphinxes on the river's banks. As the years pass and their numbers diminish, the mournfulness deepens in their gaze; it becomes more questioning, more haunting. The day will come when they will all be gone, when no longer dark figures will people those lonely shores; and then we will look at one another in useless remorse and cry:—

"Why did they not complain? Why did they not ask us to help them? Why did they sit and starve for everything, staring at us and making no sign?"

Alas! when that day comes, we will learn—too late!—that there is no appeal so poignant and so haunting as that which lies in the silence and in the asking eyes of these dark and vanishing people.

Below Rampart the hills withdraw gradually until they become but blue blurs on the horizon line during the last miles of the river's course. It is now the lower river and becomes beautifully channelled and islanded. Across these low, wooded, and watered plains the sunset burns like a maze of thistle-down touched with ruby fire—burns down, at last, into the rose of dawn; and the rose into emerald, beryl, and pearl.

Not far above Nulato the Koyukuk pours its tawny flood into the Yukon. For many years the Koyukuk has given evidences of great richness in gold, but high prices of freight and labor have retarded its progress. During the past winter, however, discoveries have been made which promise one of the greatest stampedes ever known. Louis Olson, after several seasons in the district, experienced a gambler's "hunch" that there "was pay on Nolan Creek." He and his associates started to sink, and the first bucket they got off bedrock netted seven dollars; the bedrock, a slate, pitched to one side of the hole, and when they had followed it down and struck a level bedrock, they got two hundred and sixty dollars.

"Our biggest pan," said Mr. Olson, telling the story when he came out, one of the richest men in Alaska, "was eighteen hundred dollars. You can see the gold lying in sight."

Captain E. W. Johnson, of Nome, who had grub-staked two men in the Koyukuk, "fell into it," as miners say. They struck great richness on bedrock, and Captain Johnson promptly celebrated the strike by opening fifteen hundred dollars' worth of champagne to the camp.

Within ten days three pans of a thousand dollars each were washed out. Coldfoot, Bettles, Bergman, and Koyukuk are the leading settlements of this region, the first two lying within the Arctic Circle. Interest has revived in the Chandelar country which adjoins on the east.

Really, Seward's "land of icebergs, polar bears, and walrus," his "worthless, God-forsaken region," is doing fairly well, as countries go.

Nulato, nearly three hundred miles below Tanana, is one of the most historic places on the Yukon, and has the most sanguinary history. It was founded in 1838 by a Russian half-breed named Malakoff, who built a trading post. During the following winter, owing to scarcity of provisions, he was compelled to return to St. Michael, and the buildings were burned by natives who were jealous of the advance of white people up the river. The following year the post was reëstablished and was again destroyed. In 1841 Derabin erected a fort at this point, and for ten years the settlement flourished. In 1851, however, Lieutenant Bernard, of the British ship *Enterprise*, arrived in search of information as to the fate of Sir John Franklin. Unfortunately, he remarked that he intended to "send for" the principal chief of the Koyukuks. This was considered an insult by the haughty chief, and it led to an assault upon the fort, which was destroyed. Derabin, Bernard and his companions, and all other white people at the fort were brutally murdered, as well as many resident Indians. The atrocity was never avenged.

Nulato is now one of the largest and most prosperous Indian settlements on the river. A large herd of reindeer is quartered there. There was, as every one interested in Alaska knows, a grave scandal connected with the reindeer industry a few years ago. Many of the animals imported by the government from Siberia at great expense, for the benefit of needy natives and miners, were appropriated by missionaries without authority; but after an investigation by a special agent of the government there was an entire reorganization of the system. In all, Congress appropriated more than two hundred and twenty thousand dollars, with which twelve hundred reindeer have, at various times, been imported. There are now about twelve thousand head in Alaska, of which the government owns not more than twenty-five hundred. There are also stations at Bethel, Beetles, Iliamna, Kotzebue, St. Lawrence Island, Golovnin, Teller, Cape Prince of Wales, Point Barrow, and at several other points. They are used for sledding purposes and for their meat and hides, really beautiful parkas and mukluks—the latter a kind of skin boot—being made of the hides.

Surf at Nome

A native woman named Mary Andrewuk has a large herd, is quite wealthy, and is known as the "Reindeer Queen."

We reached Anvik at seven in the evening. Anvik is like Uyak on Kadiak Island, and I longed for the frank Swedish sailor who had so luminously described Uyak. If there be anything worth seeing at Anvik—and they say there is a graveyard!—they must first kill the mosquitoes; else, so far as I am concerned, it will forever remain unseen. Under a rocky bluff two dozen Eskimo, men and women, sat fighting mosquitoes and trying to sell wares so poorly made that no one desired them. Eskimo dolls and toy parkas were the only things that tempted us; and hastily paying for them, we fled on board to our big, comfortable stateroom, whose window was securely netted from the pests which made the very air black.

We left Anvik at midnight. We were to arrive at Holy Cross Mission at four o'clock the same morning. Expecting the *Campbell* to arrive later in the day, the priest and sisters had arranged a reception for the governor, in which the children of the mission were to take part. Thinking of the disappointment of the children, the governor decided to go ashore, even at that unearthly hour, and we were invited to accompany him. We were awakened at three o'clock.

The dawn was bleak and cheerless; it was raining slightly, and the mosquitoes were as thick and as hungry as they had been at the Grand Canyon. Of all the passengers that had planned to go ashore, there appeared upon the sloppy deck only four—the governor, a gentleman who was travelling with him, my friend, and myself. We looked at one another silently through rain and mosquitoes, and before we could muster up smiles and exchange greetings, an officer of the boat called out:—

"Governor, if it wasn't for those damn disappointed children, I'd advise you not to go ashore."

We all smiled then, for the man had put the thought of each of us into most forcible English.

We were landed upon the wet sand and we waded through the tall wet grasses of the beach to the mission. At every step fresh swarms of mosquitoes rose from the grass and assailed us. A gentleman had sent us his mosquito hats. These were simply broad-brimmed felt hats, with the netting gathered about the crowns and a kind of harness fastening around the waist.

The governor had no protection; and never, I am sure, did any governor go forth to a reception and a "programme" in his honor in such a frame of mind and with such an expression of torture as went that morning the governor of "the great country." It was a silent and dismal procession that moved up the flower-bordered walk to the mission—a procession of waving arms and flapping handkerchiefs. At a distance it must have resembled a procession of windmills in operation, rather than of human beings on their way to a reception in the vicinity of the Arctic Circle.

So ceaseless and so ferocious were the attacks of the mosquitoes that before the sleeping children were aroused and ready for their programme, my friend and I, notwithstanding the protection of the hats, yielded in sheer exhaustion, and, without apology or farewell, left the unfortunate governor to pay the penalty of greatness; left him to his reception and his programme; to the earnest priests, the smiling, sweet-faced sisters, and the little solemn-eyed Eskimo children.

This mission is cared for by the order of Jesuits. Two priests and several brothers and sisters reside there. Fifty or more children are cared for yearly,—educated and guided in ways of thrift, cleanliness, industry, and morality. They are instructed in all kinds of useful work. About forty acres of land are in cultivation; the flowers and vegetables which we saw would attract admiration and wonder in any climate. The buildings were of logs, but were substantially built and attractive, each in its setting of brilliant bloom. How these sisters, these gentle and refined women, whose faces and manner unconsciously reveal superior breeding and position, can endure the daily and nightly tortures of the mosquitoes is inconceivable.

"They are not worth notice now," one said, with her sweet and patient smile. "Oh, no! You should come earlier if you would see mosquitoes."

"Our religion, you know," another said gently, "helps us to bear all things that are not pleasant. In time one does not mind."

In time one does not mind! It is another of the lessons of the Yukon; and reading, one stands ashamed. There those saintly beings spend their lives in God's service. Nothing save a divine faith could sustain a delicate woman to endure such ceaseless torment for three months in every year; and yet, like the lone woman at Nation, their faces tell us that we, rather than they, are for pity. The stars upon their brows are the white and blessed stars of peace.

The steamer lands at neither Russian Mission nor Andreaofsky; but at both may be seen, on grassy slopes, beautiful Greek churches, with green, pale blue, and yellow roofs, domes and bell-towers, chimes and glittering crosses.

Down where the mouth of the Yukon attains a width of sixty

miles we ran upon a sand-bar early in the afternoon, and there we remained until nearly midnight. It was a weird experience. Dozens of natives in bidarkas surrounded our steamer, boarded our barges, and offered their inferior work for sale. The brown lads in reindeer parkas were bright-eyed and amiable. Cookies and gum sweetened the way to their little wild hearts, and they would hold our hands, cling to our skirts, and beg for "more."

A splendid, stormy sunset burned over those miles of water-threaded lowlands at evening. Rose and lavender mists rolled in from the sea, parted, and drifted away into the distances stretching on all sides; they huddled upon islands, covering them for a few moments, and then, withdrawing, leaving them drenched in sparkling emerald beauty in the vivid light; they coiled along the horizon, like peaks of rosy pearl; and they went sailing, like elfin shallops, down poppy-tinted water-ways. Everywhere overhead geese drew dark lines through the brilliant atmosphere, their mournful cries filling the upper air with the weird and lonely music of the great spaces. Up and down the water-ways slid the bidarkas noiselessly; and along the shores the brown women moved among the willows and sedges, or stood motionless, staring out at their white sisters on the stranded boat. There were times when every one of the millions of sedges on island and shore seemed to flash out alone and apart, like a dazzling emerald lance quivering to strike.

They are dull of soul and dull of imagination who complain of monotony on the Yukon Flats. There is beauty for all that have eyes wherewith to see. It is the beauty of the desert; the beauty and the lure of wonderful distances, of marvellous lights and low skies, of dawns that are like blown roses, and as perfumed, and sunsets whose mists are as burning dust. When there is no color anywhere, there is still the haunting, compelling beauty that lies in distance alone. Vast spaces are majestic and awesome; the eye goes into them as the thought goes into the realm of eternity—only to return, wearied out with the beauty and the immensity that forever end in the fathomless mist that lies on the far horizon's rim. It is a mist that nothing can pierce; vision and thought return from it upon themselves, only to go out again upon that mute and trembling quest which ceases not until life itself ceases.

The northernmost mouth of the Yukon has been called the Aphoon or Uphoon, ever since the advent of the Russians, and is the channel usually selected by steamers, the Kwikhpak lying next to it on the south. By sea-coast measurement the most northerly mouth is nearly a hundred miles from the most southerly, and five others between them assist in carrying the Yukon's gray, dull yellow, or rose-colored floods out into Behring Sea, whose shallow waters they make fresh for a long distance. It is not without hazard that the flat-bottomed river boats make the run to St. Michael; and the pilots of steamers crossing out anxiously scan the sea and relax not in vigilance until the port is entered.

CHAPTER XLIX

We were released from the sand-bar near midnight, and at eight o'clock on the following morning we steamed around a green and lovely point and entered Norton Sound, in whose curving blue arm lies storied St. Michael.

St. Michael is situated on the island of the same name, about sixty miles north of the mouth of the Yukon. It was founded in 1833 by Michael Tebenkoff, and was originally named Michaelovski Redoubt. The Russian buildings were of spruce logs brought by sea from the Yukon and Kuskoquim rivers, as no timber grows in the vicinity of St. Michael or Nome. Some of the original Russian buildings yet remain,—notably, the storehouse and the redoubt. The latter is an hexagonal building of heavy hewn logs, with sloping roof, flagstaff, door, and port-holes. It stands upon the shore, within a dozen steps of the famous "Cottage,"—the residence of the managers of the Northern Commercial Company, under whose hospitable roof every traveller of note has been entertained for many years,—and in front of it the shore slopes green to the water. Inside lie half a dozen rusty Russian cannons, mutely testifying to the sanguinary past of the North.

The redoubt was attacked in 1836 by the hostile Unaligmuts of the vicinity, but it was successfully defended by Kurupanoff. The Russians had a temporary landing-place built out to deep water to accommodate boats drawing five feet; this was removed when ice formed in the bay. The tundra is rolling, with numerous pools that flame like brass at sunset; only low willows and alders grow on the island and adjacent shores. The island is seven miles wide and twenty-five long, and is separated from the mainland by a tortuous channel, as narrow as fifty feet in places. The land gradually rises to low hills of volcanic origin near the centre of the island. These hills are called the Shaman Mountains. The meadow upon which the main part of the town and the buildings of the post are situated is as level as a vast parade-ground; but the land rises gently to a slender point that plunges out into Behring Sea, whose blue waves beat themselves to foam and music upon its tundra-covered cliffs.

On the day that I stood upon this headland the sunlight lay like gold upon the island; the winds were low, murmurous, and soothing; flowers spent their color riotously about me; the tundra was as soft as deep-napped velvet; and the blue waves, set with flashes of gold, went pushing languorously away to the shores of another continent. Scarcely a stone's throw from me was a small mountain-island, only large enough for a few graves, but with no graves upon it. In all the world there cannot be another spot so noble in which to lie down and rest when "life's fevers and life's passions—all are past." There, alone,—but never again to be lonely!—facing that sublime sweep of sapphire summer sea, set here and there with islands, and those miles upon miles of glittering winter ice; with white sails drifting by in summer, and in winter the wild and roaring march of icebergs; with summer nights of lavender dusk, and winter nights set with the great stars and the magnificent brilliance of Northern Lights; with the perfume of flowers, the songs of birds, the music of lone winds and waves, out on the edge of the world—could any clipped and cared-for plot be so noble a place in which to lie down for the last time? Could any be so close to God?

The entire island is a military reservation, and it is only by concessions from the government that commercial and transportation companies may establish themselves there. Fort St. Michael is a two-company post, under the command of Captain Stokes, at whose residence a reception was tendered to Governor Hoggatt. The filmy white gowns of beautiful women, the uniforms of the officers, the music, flowers, and delicate ices in a handsomely furnished home made it difficult for one to realize that the function was on the shores of Behring Sea instead of in the capital of our country.

There is an excellent hotel at St. Michael, and the large stores of the companies are well supplied with furs and Indian and Eskimo wares. Beautiful ivory carvings, bidarkas, parkas, kamelinkas, baskets, and many other curios may be obtained here at more reasonable prices than at Nome. There are public bath-houses where one may float and splash in red-brown water that is never any other color, no matter how long it may run, but which is always pure and clean.

No description of St. Michael is complete that does not include "Lottie." No liquors are sold upon the military reservation, and Lottie conducts a floating groggery upon a scow. It has been her custom each fall to have her barge towed up the canal just beyond the line of the military reservation, ten miles from the flagstaff at the barracks, thus placing herself beyond the control of the authorities, greatly to their chagrin. In summer she anchors her barge in one of the numerous bights along the shore, and they are again powerless to interfere with her brilliantly managed traffic, since it has been decided that their sway extends over the land only.

Moonlight on Behring Sea

It is Lottie's practice to have the barge made fast in such a way that a boat can be run to it from the shore on an endless line. One desiring a bottle of whiskey approaches the boat and drops his money and order into the bottom of it. The boat is then drawn out to the barge, whiskey is substituted for the money, and the purchaser pulls the boat ashore, where it is left for the next customer.

There is no witness to the transaction and it has been impossible to prove, the authorities claim, who put the money and the whiskey into the boat, or took either therefrom.

Lottie's barge has operated for many years. Its illicit transactions could easily have been stopped had the civil authorities on shore taken a firm stand and worked in conjunction with the military; but there was the usual jealousy as to the rights of the different officials—and Lottie has profited by these conditions. Furthermore, many people of the vicinity entertained a friendly feeling for Lottie—not only those who were wont to draw the little boat back and forth, but others in sheer admiration of the ingenuity and skill with which she carried on her business. She was careful in preserving order in her vicinity, was very charitable, and frequently provided for natives who would have otherwise suffered. Thus, by her diplomacy, self-control, good business sense, and many really worthy traits of character, Lottie has been able to outwit the officials for years. Her barge still floats upon the blue waves of Norton Sound. However, it seems, even to a woman, that Lottie must be blessed with "a friend at court."

We had been invited to voyage from St. Michael to Nome—a distance of a hundred and eleven miles—on the *Meteor*, a very small tug; being warned, however, that, should the weather prove to be unfavorable, our hardships would be almost unendurable, as there was only an open after-deck and no cabin in which to take refuge. We boldy took our chances, remaining three days at St. Michael.

Never had Behring Sea, or Norton Sound, been known to be so beautiful as it was on that fourteenth day of August. We started at nine in the morning, and until evening the whole sea, as far as the eye could reach in all directions, was as smooth as satin, of the palest silvery blue. Never have I seen its like, nor do I hope ever to see it again. To think that such seductive beauty could bloom upon a sea whereon, in winter, one may travel for hundreds of miles on solid ice! At evening it was still smooth, but its color burned to a silvery rose.

The waters we sailed now were almost sacred to some of us. Over them the brave and gallant Captain Cook had sailed in 1778, naming Capes Darby and Denbigh, on either side of Norton Bay; he also named the bay and the sound and Besborough, Stuart, and Sledge islands; and it was in this vicinity that he met the family of cripples.

But of most poignant interest was St. Lawrence Island, lying far to our westward, discovered and named by Vitus Behring on his voyage of 1728. If he had then sailed to the eastward for but one day!

Every one has read of the terrors of landing through the pounding surf of the open roadstead at Nome. Large ships cannot approach within two miles of the shore. Passengers and freight are taken off in lighters and launches when the weather is "fair"; but fair weather at Nome is rough weather elsewhere. When they call it rough at Nome, passengers remain on the ships for days, waiting to land. Frequently it is necessary to transfer passengers from the ships to dories, from the dories to tugs, from the tugs to flat barges. The barges are floated in as far as possible; then an open platform—miscalled a cage—is dropped from a great arm, which looks as though it might break at any moment; the platform is crowded with passengers and hoisted up over the boiling surf, swinging and creaking in a hair-crinkling fashion, and at last depositing its large-eyed burden upon the wharf at Nome. I had pitied *cattle* when I had seen them unloaded in this manner at Valdez and other coast towns!

We anchored at eleven o'clock that night in the Nome

roadstead. In two minutes a launch was alongside and a dozen gentlemen came aboard to greet the governor. We were hastily transferred in the purple dusk to the launch. The town, brilliantly illuminated, glittered like a string of jewels along the low beach; bells were ringing, whistles were blowing, bands were playing, and all Nome was on the beach shouting itself hoarse in welcome.

There was no surf, there was not a wave, there was scarce a ripple on the sea. The launch ran smoothly upon the beach and a gangway was put out. It did not quite reach to dry land and men ran out in the water, picked us up unceremoniously, and carried us ashore.

The most beautiful landing ever made at Nome was the one made that night; and the people said it was all arranged for the governor.

There was an enthusiastic reception at the Golden Gate Hotel, followed by a week's brilliant functions in his honor.

Three days later the *Meteor* came over from St. Michael, with a distinguished Congressman aboard. The weather was rough, even for Nome, and for three blessed days the *Meteor* rolled in the roadstead, and with every roll it went clear out of sight.

There were those at the hotel who differed politically from the Congressman aboard the little tug; and, like the people of Nome when the senatorial committee was landed under such distressful circumstances a few years ago, their faces did not put on mourning as they watched the *Meteor* roll.

CHAPTER L

Nome! Never in all the world has been, and never again will be, a town so wonderfully and so picturesquely built. Imagine a couple of miles of two and three story frame buildings set upon a low, ocean-drenched beach and, for the most part, painted white, with the back doors of one side of the main business street jutting out over the water; the town widening for a considerable distance back over the tundra; all things jumbled together—saloons, banks, dance-halls, millinery-shops, residences, churches, hotels, life-saving stations, government buildings, Eskimo camps, sacked coal piled a hundred feet high, steamship offices, hospitals, schools—presenting the appearance of having been flung up into the air and left wherever they chanced to fall; with streets zigzagging in every conceivable and inconceivable, way—following the beach, drifting away from it, and returning to it; one building stepping out proudly two feet ahead of its neighbor, another modestly retiring, another slipping in at right angles and leaving a V-shaped space; board sidewalks, narrow for a few steps, then wide, then narrow again, running straight, curving, jutting out sharply; in places, steps leading up from the street, in others the streets rising higher than the sidewalks; boards, laid upon the bare sand in the middle of the streets for planking, wearing out and wobbling noisily under travel; every second floor a residence or an apartment-house; crude signs everywhere, and tipsy telephone poles; the streets crowded with men at all hours of the day and night; and a blare of music bursting from every saloon. This is Nome at first sight; and it was with a sore and disappointed heart that I laid my head upon my pillow that night.

But Nome grows upon one; and by the end of a week it had drawn my heartstrings around it as no orderly, conventional town could do. From the very centre of the business section it is but twenty steps to the sea; and there, day and night, its surf pounds upon the beach, its musical thunder and fine mist drifting across the town.

Ten years ago there was nothing here save the golden sands, the sea that broke upon them, and the gray-green tundra slopes; there is not a tree for fifty miles or more. To-day there is a town of seven thousand people in summer, and of three or four thousand in winter—a town having most of the comforts and many of the luxuries to be obtained in cities of older civilization. Nome sprang into existence in the summer of 1899, and grew like Fairbanks and Dawson; but it is more wonderfully situated than, probably, any town in the world. For eight months of the year it is cut off from steamship service, and its front door-yard is a sea of solid ice stretching to the shores of Siberia, while its back yard is a gold-mine. There are many weeks when the sun rises but a little way, glimmers faintly for three or four hours, and fades behind the palisades of ice, leaving the people to darkness and unspeakable loneliness until it returns to its full brilliance in spring and opens the way for the return of the ships.

Nome is picturesque by day or by night and at any season. Its streets are constantly crowded with traffic and thronged by a cosmopolitan population. The Eskimo encampment is on the "sand-spit" at the northern end of the main street, where Snake River flows into the sea; and the men, women, and children may be seen at all hours loitering about the streets in reindeer parkas and mukluks. Especially in the evenings do they haunt the streets and the hotels, offering their beautifully carved ivories for sale.

Both the Eskimos and the Indians are lovers of music, and the former readily yield to emotion when they hear melodious strains. When a "Buluga," or white whale, is killed, a feast is held and the natives sing their songs and dance. The music of stringed instruments invariably moves them to tears. At a recent Thanksgiving service in Fairbanks, some visiting Indians were invited to sing "Oh, Come, All Ye Faithful." With evident pleasure, they sang it as follows:—

"Oni, tsenuan whuduguduwhuta yilh;Oni, yuwhun dutlish, oni nokhlhan,Oni, dodutalokhlho,Oni, dodutalokhlho,Oni, dodutalokhlho,Lud."

At Point Barrow, three hundred miles northeast of Behring Strait, an old Eskimo who could not speak one word of English was heard to whistle "The Holy City," and it filled the hearer's heart with home-loneliness. A trader had sold the old native music-lover a phonograph, receiving in pay two white polar bear-skins, worth several hundred dollars.

Some one gave an ordinary French harp to a little Eskimo lad on our steamer; and from early morning until late at night he sat on a companionway, alone, indifferent to all passers-by, blowing out softly and sweetly with dark lips the prisoned beauty of his soul.

All the islands of Behring Sea, as well as the coast of the Arctic Ocean, are inhabited by Eskimos. From the largest island, St. Lawrence, to the small Diomede on the American side, they have settlements and schools. St. Lawrence is eighty miles long by fifteen in width; while the Diomede is only two miles by one. The natives beg pitifully for education—"to be smart, like the white man." We shrink from their filth and their immorality, but we teach them nothing better; yet we might see through their asking eyes down into their starved souls if we would but look.

In many ways Nome is the most interesting place in Alaska. It is at once so pagan and so civilized; so crude and so refined. It is the golden gateway through which thousands of people pass each summer to and from the interior of Alaska. Treeless and harborless it began and has continued, surmounting all obstacles that lay in its way of becoming a city. It has a water system that supplies its household needs, with steam pipes laid parallel to the water pipes, to thaw them in winter—and then it has not a yard of sewerage. It has a wireless telegraph station, a telephone service, and electric-light plant; and it is seeking municipal steam-heating. Electric lighting is excessively high, owing to the price of coal, and many use lamps and candles. There are three good newspapers, which play important parts in the politics of Alaska—the *Nugget*, the *Gold-Digger*, and the *News*; three banks, with capital stocks ranging from one to two hundred thousand dollars, each of which has an assay-office; two good public schools; three churches; hospitals; and a telephone system connecting all the creeks and camps within a radius of fifty miles with Nome. The orders of Masons, Odd Fellows, Knights of Pythias, Eagles, and Arctic Brotherhood have clubs at Nome. The Arctic Brotherhood is the most popular order of the North, and the more important entertainments are usually given under its auspices and are held in its club-rooms; the wives of its members form the most exclusive society of the North.

The spirit of Nome is restless; it is the spirit of the gold-seeker, the seafarer, the victim of wanderlust; and it soon gets into even the visitor's blood. Millions of dollars have been taken out of the sands whereon Nome is now built, and millions more may be waiting beneath it. It seemed as though every man in Nome should be digging—on the beach, in the streets, in cellars.

"Why are not all these men digging?" I asked, and they laughed at me.

"Because every inch of tundra for miles back is located."

"Then why do not the locators dig, dig, day and night?"

"Oh, for one reason or another."

If I owned a claim on the tundra back of Nome, nothing save sudden death could prevent my digging.

New strikes are constantly being made, to keep the people of Nome in a state of feverish excitement and dynamic energy. When we landed, we found the town wild over a thirty-thousand-dollar clean-up on a claim named "Number Eight, Cooper Gulch." Four days later an excursion was arranged to go out on the railroad—for they have a railroad—to see another clean-up at this mine.

We started at nine o'clock, and we did not return until five; and it rained steadily and with exceeding coldness all day. There was a comfortable passenger-car, but despite the wind and the rain we preferred the box-cars, roofed, but open at the sides. The country which we traversed for six miles possessed the indescribable fascination of desolation. Behind us rolled the sea; but on all other sides stretched wide gray tundra levels, varied by low hills. Hills they call them here, but they are only slopes, or mounds, with here and there a treeless creek winding through them. The mist of the rain drove across them like smoke.

We were received at the mine by Captain and Mrs. Johnson and Mr. Corson, the owners. The ladies were entertained in the Johnsons' cabin home and the gentlemen at a near-by cabin, there being twelve ladies and twenty gentlemen in the party. An immense bowl of champagne punch—the word "punch" being used for courtesy—stood outside the ladies' cabin and was not allowed to grow empty. Late in the afternoon the heap of empty champagne bottles outside the gentlemen's cabin resembled in size one of the numerous gravel dumps scattered over the tundra; yet not a person showed signs of intoxication. They told us that one may drink champagne as though it were water in that latitude; and this is one northern "story" which I am quite willing to believe.

At noon a bountiful and delicious luncheon was served at the mess-house. It was this same fortunate Captain Johnson, by the way, who opened fifteen hundred dollars' worth of champagne when bedrock was reached in his Koyukuk claim.

Sluicing is fascinating. A good supply of water with sufficient fall is necessary. Some of the claims are on creeks, but the owners of others are compelled to buy water from companies who supply it by pumping-plants and ditches. Boxes, or flat-bottomed troughs, are formed of planks with slats, or "riffles," fastened at intervals across the bottom. Several boxes are arranged on a gentle slope and fitted into one another. The boxes at "Number Eight" were twenty feet in length and slanted from the ground to a height of twelve feet on scaffolding. A narrow planking ran along each side of the telescoped boxes, and upon these frail foundations we stood to view the sluicing. The gravel is usually shovelled into the boxes, but "Number Eight" has an improved method. The gravel is elevated into an immense hopper-like receptacle, from which it sifts down into the sluice-boxes on each side, and a stream of water is kept running steadily upon it from a large hose at the upper end. Men with whisk brooms sweep up the gold into glistening heaps, working out the gravel and passing it on, as a housewife works the whey out of the yellowing butter. The gold, being heavy, is caught and held by the riffles; if it is very fine, the bottoms of the boxes are covered with blankets, or mercury is placed at the slats to detain it.

The clean-up that day was twenty-nine thousand dollars, and each lady of the party was presented with a gold nugget by Mrs. Johnson. We were taken down into the mine, where we went about like a company of fireflies, each carrying his own candle. The ceiling was so low that we were compelled to walk in a stooping position. On the following morning we went to a bank and saw this clean-up melted and run into great bricks.

The lure and the fascination of virgin gold is undeniable. It catches one and all in its glistening, mysterious web. A man may sell his potato patch in town lots and become a millionnaire, without attracting attention; but let him "strike pay on bedrock"—and instantly he walks in a golden mist of glory and romance before his fellow-men. It may be because the farmer deposits his money in the bank, while the miner "sets up" the champagne to his less fortunate friends. Be that as it may, it is a sluggish pulse that does not quicken when one sees cones of beautiful coarse gold and nuggets washed and swept out of the gravel in which it has been lying hundreds of years, waiting. If Behring had but landed upon this golden beach, Alaska—despite all the eloquence and the earnestness of Seward and Sumner-might not now be ours.

To the Nome district have been gradually added those of Topkuk, Solomon, and Golovin Bay, forty-five miles to eastward on the shores of Norton Sound, Cripple Creek, Bluff, Penny, and a chain of diggings extending up the coast and into the Kotzebue country, including the rich Kougarok and Blue Stone districts, Candle Creek, and Kowak River.

When gold was discovered at Nome, prospectors scattered over the Seward Peninsula in all directions. Some drifted west into the York district, near Cape Prince of Wales, the extreme western point of the North American continent. In this region they found gold in the streams, but sluicing was so difficult, owing to a heavy gravel which they encountered, that they abandoned their claims, not knowing that the impediment was stream-tin. Wiser prospectors later recognized the metal and located claims. The tin is irregularly distributed over an area of four hundred and fifty square miles, embracing the western end of the peninsula. The United States uses annually twenty million dollars' worth of tin, which is obtained largely from the Straits Settlement, although much comes from Ecuador, Bolivia, Australia, and Cornwall. Tin cannot at present be treated successfully in this country, owing to the lack of smelter facilities; but now that it has been discovered in so vast quantities and of so pure quality in the Seward Peninsula, smelters in this country will doubtless be equipped for reducing tin ores.

The centre of the tin-mining industry is at Tin City, a small settlement three miles west of Teller, Cape Prince of Wales, and is reached by small steamers which ply from Nome. Several corporations are developing promising properties with large stamp-mills. Both stream-tin and tin ore in ledges are found throughout the district.

The Council district is the oldest of Seward Peninsula, the first discovery of gold having been made there in 1898, by a party headed by Daniel P. Libby, who had been through the country with the Western Union's Expedition in 1866. Hearing of the Klondike's richness, he returned to Seward Peninsula and soon found gold on Fish River. He and his party established the town of Council and built the first residence; it now has a population of eight hundred. This district is forestated with spruce of fair size and quality.

The Ophir Creek Mines are of great value, having produced more than five millions of dollars by the crudest of mining methods. The Kougarok is the famous district of the interior of the peninsula. Mary's Igloo—deriving its name from an Eskimo woman of some importance in early days—is the seat of the recorder's office for this district. It has a post-office and is an important station. May it never change its striking and picturesque name!

The entire peninsula, having an area of nearly twenty-three thousand miles, is liable to prove to be one vast gold-mine, the extreme richness of strikes in various localities indicating that time and money to install modern machinery and develop the country are all that are required to make this one of the richest producing districts of the world.

The leading towns of the peninsula are Council, Solomon, Teller, Candle, Mary's Igloo, and Deering, on Kotzebue Sound. Solomon is on Norton Sound, at the mouth of Solomon River; a railroad runs from this point to Council.

The early name of Seward Peninsula was Kaviak—the name of the Innuit people inhabiting it.

Gold was discovered on Anvil Creek in the hills behind Nome in September, 1898, by Jafet Lindeberg, Erik Lindblom, and John Brynteson, the "three lucky Swedes." In the following summer gold was discovered on the beach, and in 1900 occurred the memorable stampede to Nome, when fifteen thousand people struggled through the surf during one fortnight. Then began the amazing building of the mining-camp on the northwesternmost point of the continent. Anvil Creek, Dexter, Dry and Glacier creeks, Snow and Cooper gulches, have yielded millions of dollars. The tundra reaching back to the hills five or six miles from the sea is made up of a series of beach lines, all containing deposits of gold. Five millions of dollars in dust were taken from the famous "third" beach line in one season; and its length is estimated at thirty or forty miles. The hills are low and round-topped, and beyond them—thirty miles distant—are the Kigluaik Mountains, known to prospectors by the name of Sawtooth. Among their sharp and austere peaks is the highest of the peninsula, rising to an altitude of four thousand seven hundred feet by geological survey.

There are several railroads on the peninsula. Some are but a few miles in length, the rails are narrow and "wavy," the trains run by starts and plunges and stop fearsomely; but they are railroads. One can climb into the box-cars or the one warm passenger-coach and go from Nome out among the creeks,—to Nome River, to Anvil Creek, to Kougarok and Hot Springs, from Solomon to the Council Country,—and Nome is only ten years old.

Nome has a woman's club. It is federated and it owns its club-house, a small but pretty building. Its name is Kegoayah Kosga, or Northern Lights. It held an open meeting while we were in Nome. Bishop Rowe described a journey by dog sled and canoe, Congressman Sulzer gave an informal talk, and the ladies of the club presented an interesting programme. The afternoon was the most profitable I have spent at a woman's club.

For two or three months in summer it is all work at Nome; but when the snow begins to drive in across the town; when the last steamer drifts down the roadstead and disappears before the longing eyes that follow it; when the ice piles up, mile on mile, where the surf dashed in summer, and the wind in the chimneys plays a weird and lonely tune; then the people turn to cards and dance and song to while away the long and dreary months of darkness. The social life is gay; and poker parties, whereat gambling runs high, are frequent.

"I'd like to give a poker party for you," said a handsome young woman, laughing, "but I suppose it would shock you to death."

We confessed that we would not be shocked, but that, not

knowing how to play the game, we declined to be "bluffed" out of all our money.

"Oh, we are easy on cheechacos," said she, lightly. "Do come. We'll play till two o'clock, and then have a little supper; curlew, plovers, and champagne—the 'big cold bottle and the small hot bird.'"

When we still declined, she looked bored as she said politely:—

"Oh, very well; let us call it a five-hundred party. Surely, that is childlike enough for you. But the men!"

I laughed at the thought of the men I had met in Nome playing the insipid game of five-hundred.

"Then," said she, dolefully, "there's nothing left but bridge—and we just gamble our pockets inside out on bridge; it's worse than poker, and we play like fiends."

We suggested that, as General Greeley had come down the river with us and would be over from St. Michael the next day, they should wait for him; when the first player has led the first card, General Greeley knows in whose hand every deuce lies, and I wickedly longed to see the inside of Nome's composite pocket by the time General Greeley had sailed away.

There was no party for us that night; but there is a wide, public porch behind a big store by the life-saving station. It projects over the sea and about ten feet above it, and upon this porch are benches whereon one may sit alone and undisturbed until midnight, or until dawn, for that matter, but alone—with the glitter of Nome and the golden tundra behind one, and in front, the far, faint lights of the ships anchored in the roadstead and the tumultuous passion of waves that have lapped the shores of other lands.

Sitting here, what thoughts come, unbidden, of the brave and shadowy navigators of the past who have sailed these waters through hardships and sufferings that would cause the stoutest hearts of to-day to hesitate. Read the descriptions of the ships upon which Arctic explorers embark at the present time—of their stores and comforts; and then turn back and imagine how

Simeon Deshneff, a Cossack chief, set sail in June, two hundred and sixty years ago, from the mouth of the Kolyma River in Siberia in search of fabled ivory. In company with two other "kotches," which were lost, he sailed dauntlessly along the Arctic sea-coast and through Behring Strait from the Frozen Ocean. His "kotch" was a small-decked craft, rudely and frailly fashioned of wood; in September of that year, 1648, he landed upon the shores of the Chukchi Peninsula and saw the two Diomede Islands, between which the boundary line now runs. He must have seen the low hills of Cape Prince of Wales, for it plunges boldly out into the sea, within twenty miles of the Diomedes, but probably mistook them for islands. Half a century later Popoff, another Cossack, was sent to East Cape to persuade the rebellious Chukchis—as the Siberian natives of that region are called—to pay tribute; he was not successful, but he brought back a description of the Diomede Islands and rumors of a continent said to lie to the east. The next passage of importance through the strait was that of Behring, who, in 1728, sailed along the Siberian coast from Okhotsk, rounded East Cape, passed through the strait, and, after sailing to the northeast for a day, returned to Okhotsk, marvellously missing the American continent. Geographers refused to accept Behring's statement that Asia and North America were not connected until it was verified in 1778 by Cook, who generously named the strait for the illustrious Dane.

Less than a day's voyage from Nome is the westernmost point of our country—Cape Prince of Wales, the "Kingegan" of the natives. It is fifty-four miles from this cape to the East Cape of Siberia, and like stepping-stones between lie Fairway Rock and the Diomedes. Beyond is the Frozen Ocean. These islands are of almost solid stone. They are snow-swept, ice-bound, and ice-bounded for eight months of every year. But ah, the auroral magnificence that at times must stream through the gates of frozen pearl which swing open and shut to the Arctic Sea! What moonlights must glitter there like millions of diamonds; what sunrises and sunsets must burn like opaline mist! How large the stars must be—and how bright and low! And in the spring—how this whole northern world must tremble and thrill at the mighty march of icebergs sweeping splendidly down through the gates of pearl into Behring Sea!

APPENDIX

In the preparation of this volume the following works have been consulted, which treat wholly, or in part, of Alaska. After the narratives of the early voyages and discoveries, the more important works of the list are Bancroft's "History," Dall's "Alaska and Its Resources," Brooks' "Geography and Geology," Davidson's "Alaska Boundary," Elliott's "Arctic Province," Mason's "Aboriginal Basketry," Miss Scidmore's "Guide-book," and "Proceedings of the Alaska Boundary Tribunal."

Abercrombie, Captain. Government Reports.

Alaska Club's Almanac. 1907, 1908.

Bales, L. L. Habits and Haunts of the Sea-otter. Seattle Post-Intelligencer. April 7, 1907.

Bancroft, Hubert H. History of the Pacific States. Volumes on Oregon, Washington, Idaho, Montana, Alaska, and Northwest Coast. The volume on Alaska is a conscientious and valuable study of that country, the material for which was gathered largely by Ivan Petroff.

Beattie, W. G. Alaska-Yukon Magazine. October, 1907.

Blaine, J. G. Twenty Years of Congress. Two volumes. 1884.

Brady, J. G. Governor's Reports. 1902, 1904, 1905.

Brooks, Alfred H. The Geography and Geology of Alaska. 1906. Also, Coal Resources of Alaska.

Butler, Sir William. Wild Northland. 1873.

Clark, Reed P. Mirror and American.

Cook, James. Voyage to the Pacific Ocean. 1784.

Coxe, William. Russian Discoveries. Containing diaries of Steller, the naturalist, who accompanied Behring and Shelikoff, who made the first permanent Russian settlement inAmerica; also, an account of Deshneff's passage through Behring Strait in 1648. Fourth Edition. Enlarged. 1803.

Cunningham, J. T. Encyclopædia Britannica.

Dall, William Healy. Alaska and Its Resources. An accurate and important work. This volume and Bancroft's Alaska are the standard historical works on Alaska.

Davidson, George. The Alaska Boundary. 1903. Also, Glaciers of Alaska. 1904. Mr. Davidson's work for Alaska covers many years and is of great value.

Dixon, George. Voyage Around the World. 1789.

Dorsey, John. Alaska-Yukon Magazine. October, 1907.

Dunn, Robert. Outing. February, 1908.

Elliott, Henry W. Our Arctic Province. 1886. This book covers the greater part of Alaska in an entertaining style and contains a comprehensive study of the Seal Islands.

Georgeson, C. C. Report of Alaska Agricultural Experimental Work. 1903, 1904, 1905, 1906.

Harriman. Alaska Expedition. 1904.

Harrison, E. S. Nome and Seward Peninsula.

Holmes, W. H. Report of the Bureau of American Ethnology. 1907.

Irving, Washington. Astoria.

Jewitt, John. Adventures. Edited by Robert Brown. 1896. John Jewitt was captured and held as a slave by the Nootka Indians from 1803 until 1805.

Jones, R. D. Alaska-Yukon Magazine. October, 1907.

Kinzie, R. A. Treadwell Group of Mines. 1903.

Kostrometinoff, George. Letters and Papers.

La Pérouse, Jean François. Voyage Around the World. 1798.

Mackenzie, Alexander. Voyages to the Arctic in 1789 and 1793. Two volumes.

McLain, J. S. Alaska and the Klondike. 1905.

Mason, Otis T. Aboriginal American Basketry. An exquisite and poetic work.

Moser, Commander. Alaska Salmon Investigations.

Muir, John. The Alaska Trip. Century Magazine. August, 1897.

Müller, Gerhard T. Voyages from Asia to America. 1761 and

1764.

Nord, Captain J. G. Letters and papers.

Portlock, Nathaniel. Voyage Around the World. 1789.

Proceedings of the Alaska Boundary Tribunal. Seven volumes. 1904.

Schwatka, Frederick. Along Alaska's Great River. 1886. Lieutenant Schwatka voyaged down the Yukon on rafts in 1883 and wrote an interesting book. His namings were unfortunate, but his voyage was of value, and many of his surmises have proven to be almost startlingly correct.

Scidmore, Eliza Ruhamah. Guide-book to Alaska. 1893. Miss Scidmore's style is superior to that of any other writer on Alaska.

Seattle Mail and Herald. March 7, 1903.

Seattle Post-Intelligencer. 1906, 1907, 1908.

Seattle Times. 1908.

Seward, Frederick W. Inside History of Alaska Purchase. Seward Gateway. March 17, 1906.

Shaw, W. T. Alaska-Yukon Magazine. October, 1907.

Simpson, Sir George. Journey Around the World. 1847.

Sumner, Charles. Oration on the Cession of Russian America to the United States. 1867.

Tuttle, C. R. The Golden North. 1897.

Vancouver, George. Voyage of Discovery to the North Pacific Ocean. Three volumes. 1798.

Printed in the USA
CPSIA information can be obtained
at www.ICGtesting.com
LVHW040330210923
758795LV00002B/46

9 781528 704731